AF616639

6-6-78
19.95

Man-Made RADIO NOISE

Edward N. Skomal

VAN NOSTRAND REINHOLD COMPANY
NEW YORK CINCINNATI ATLANTA DALLAS SAN FRANCISCO
LONDON TORONTO MELBOURNE

Van Nostrand Reinhold Company Regional Offices:
New York Cincinnati Atlanta Dallas San Francisco

Van Nostrand Reinhold Company International Offices:
London Toronto Melbourne

Library of Congress Catalog Card Number: 77-19109
ISBN: 0-442-27648-6

Manufactured in the United States of America

Published by Van Nostrand Reinhold Company
135 West 50th Street, New York, N.Y. 10020

Published simultaneously in Canada by Van Nostrand Reinhold Ltd.

15 14 13 12 11 10 9 8 7 6 5 4 3 2 1

Library of Congress Cataloging in Publication Data

Skomal, Edward N.
Man-made radio noise.

Includes index.
1. Radio noise. I. Title.
TK6553.S288 621.3841'1 77-19109
ISBN 0-442-27648-6

To
Beth
and
Mr. Dewey H. Miner

Preface

During the past decade and a half, the field of radio-frequency–interference investigation that specializes in man-made incidental noise has undergone rapid expansion. Activity, both experimental and theoretical, has been intense. From an elementary understanding at the beginning of this growth period, the technology has evolved to a sophisticated appreciation of the underlying physical processes. As the profundity of these noise processes has become more evident, wider dissemination of the critical information and tested concepts has become necessary. Thus, the present work aims to provide an organized presentation of the major extant information and primary concepts of the incidental radio-noise field.

By initially introducing the reader to the known relationships that exist between man-made incidental noise and naturally occurring interference, and subsequently to the dominant sources of unintentionally generated interference, a foundation is prepared herein for a systematic treatment of the field. Upon this foundation, the detailed features of interfering emissions–their amplitude and their temporal, spatial, and spectral properties–are assembled.

The deterministic (to the extent possible) and the stochastic aspects of incidental noise are organized within a framework that first introduces the reader to the features of the various noise sources, to their variability, and, ultimately, to an appreciation of their interactions in large sets, which creates the composite noise environment emanating from and enshrouding all urban industrial areas.

For the reader requiring information in order to conduct a wireless system design or performance analyses applicable within or near a metropolitan area, the discussion of consolidated man-made incidental interference in Chapters 6 and 7 will be of major interest. Those who desire data on incidental radiation levels to accomplish the design of either a radio-system assembly or a component that must function in a specifiable man-made noise environment will find Chapters 2 through 4 of value. Therein, the characteristics of the expected ambient noise levels of the major sources are presented.

In Chapters 2, 5, and 6, both equipment designers and wireless-systems analysts will find theories describing the statistical properties of radiated man-made noise, which will aid in quantitatively assessing the influence that this form of radio interference is likely to have upon specific wireless apparatus.

Government regulation of the radio-noise environment, which may be considered a form of urban pollution, also depends upon an exact understanding of the cause, intensity, distribution, and growth of man-made incidental interference. Several sections in this volume provide an understanding and insight that will be valuable in intelligently planning a metropolitan-area noise-abatement-and-control program.

Finally, by assembling the present experimental and theoretical information on the man-made incidental-noise phenomenon into a single volume, those areas of this field that are substantially incomplete or imperfectly understood will stand in clear contrast and serve as a guide for further exploration.

EDWARD N. SKOMAL
Redlands, California

Contents

1 Definition, Distribution, and Sources of Radio Noise

DEFINITION OF MAN-MADE RADIO NOISE

The man-made radio-interference environment of a metropolitan area is the composite emission from four classes of human-produced radio-noise sources. The noise class of greatest intensity consists of coherent transmitters used in the broadcast services; in the aerospace, land, and maritime mobile-radio services; in fixed-point communication services; for radio navigation and position determination; for the transmission of standard time, standard frequencies, radio telemetry, and control signals; and in meterological monitoring and observation. Following in subsidiary positions are restricted radiation devices providing localized control and communication functions; out-of-band emissions of industrial, scientific, and medical equipment; and incidental, electrical or electromechanical, radio-noise sources.

Radio interference, arising from out-of-band emissions of coherent transmitters because of its great potential for reception disruption, exists as the major wireless-communications electromagnetic-compatibility problem. A large amount of technology dedicated to minimizing the adverse effects of coherent emissions upon sensitive receiving equipments is available. The equipment design and systems-design techniques that may be used to reduce adverse equipment interactions primarily depend upon frequency stabilization, filtering, radiated-power programming, antenna pattern shaping, antenna pointing control, terrain shielding, signal cancellation, and signal blanking.

In all instances, the level and variability of coherent interference are specifically related to the number and types of transmitters and receivers and the nature of the intervening terrain. Consequently, the man-made noise environment that arises from complexes of coherent transmitters is locally unique and not amenable to a generalized solution, rather, it requires specific treatment.

The subsidiary sources of man-made interference, although usually of lesser intensity than coherent carrier emissions, are nonetheless of major concern in wireless-system and equipment design. Restricted-radiation devices (for examples, proximity-radio signposts employed in automatic vehicle-location systems and short-range radio-control transmitters used in door-opening installations) are permitted to function without Federal Communications Commission (FCC) licensing and to produce moderate radiation fields within a limited area. Specifically, at a maximum source range of $\lambda/2\pi$, a field strength of 15 μV/m is allowable,[1] where λ is the signal wavelength. Furthermore, restricted-radiation equipment often transmits modulated carrier frequencies and thus generates a broad interference spectrum. Many applications of wireless transmitters conforming to the definition of a restricted-radiation device are introduced annually; however, the emission spectra of very few been investigated and published. Because the locations of restricted-radiation devices are uncontrolled, their radio emissions are geographically undifferentiable from the noise produced by the third and fourth noted subsidiary noise classes (i.e., out-of-band emissions of industrial, scientific, and medical equipment, and incidental radio-noise sources, respectively). Thus they coalesce to produce the composite, incoherent radiated-noise fields found in all metropolitan areas.

Industrial, scientific, and medical (ISM) equipments that are functionally dependent upon the radiation of power have been provided with seven allocations in the radio spectrum within which each may operate without restriction, contingent upon the exercise of good design practices that minimize the radiated fields.[2] Of these, a limited group consisting of industrial heating equipment, medical diathermy, and radio-frequency (RF) stabilized arc welders are licensable outside the seven bands, provided each conforms to rigidly prescribed radiated-field limitations. ISM equipment with an operating frequency confined to one of the seven authorized bands is prohibited from adversely affecting the operation of authorized radio equipment in other portions of the spectrum. Because the out-of-band emissions of ISM equipment are diverse in modulation characteristics and signal level, much of the radio interference produced passes unnoticed into the composite noise environment of a metropolitan area. As licensing permits substantial latitude in the selection of the central operating frequency, ISM emissions extend over many decades of the radio spectrum.

Incidental-radiation devices, the most commonly encountered of which are automotive ignition systems and electric power lines, comprise the fourth class of human-produced radio-noise sources. For this class, the radiation of electric energy is unnecessary to the successful performance of the equipment. Typically, radiation occurs because it is less expensive for the manufacturer to accept its presence than to suppress its emission. Prevailing incidental-noise sources may produce broad band radiation arising (1) from impulsive current surges present in automotive ignition circuits or (2) from gas discharge and insulating film breakdown between high-potential points on power-line supporting elements, collectively called *gap breakdown noise.* Line spectra may also be encountered, arising as the harmonics of an oscillating electric potential. Most often, line-spectra noise is harmonics of either 50 or 60 Hz generated by electric motors, generators, rectifiers, consumer appliances, and gaseous-discharge light sources.

In the chapters that follow, attention is restricted to the characteristics of the radio interference produced by ISM equipment and by restricted-radiation and incidental-radiation devices. Although existing literature employs various terms to refer to these noise-source classes (the most common of which are *man-made noise*, *incidental noise*, and *unintentionally generated radio noise*), in this text *unintentionally generated radio noise* is used.

PREVALENCE AND DISTRIBUTION OF UNINTENTIONALLY GENERATED MAN-MADE NOISE

Geographical Distribution

Unintentionally generated man-made noise been observed beneath, on, and above the surface of industrialized and urbanized areas. At subsurface points, the noise signal produced by surface-located sources is normally heavily attenuated by the soil and thus very low. Where appreciable underground noise levels occur such as in some tunnel and shaft mines, the sources have been traced to power-distribution facilities and to mining equipment located within the shafts and tunnels.

Surface distributions of unintentionally generated noise always coincide with the regional penetration of industrialization and urbanization, varying in proportion to the density of the major sources. Typically, the metropolitan-area noise maxima exist at the center of an urban area coincident with the greatest concentration of either vehicular traffic or industrial facilities. With increasing distance from an urban center, unintentionally generated noise decreases, although not necessarily in a uniform manner. Localized concentrations of noise sources prevailing at suburban business centers, industrial facilities, and along roadways perturb the inverse dependence of noise level upon urban-center separation distance. In remote rural areas, large unintentionally generated noise levels occur near major roadways and electric-power generation-and-distribution facilities. Farm equipment, mainly gasoline-engine-driven, raises the noise level of an otherwise uncontaminated region.

Rural areas have been noted to be affected by unintentionally generated noise propagated from remote metropolitan areas. However, the resulting levels of propagated noise are typically low and may be insignificant to all but radio-astronomy installations. The ground-wave attenuation loss, which provides appreciable shielding for remote rural locations from metropolitan area noise is appreciably reduced if a body of water intervenes. Offshore locations along seaports and lakeports may therefore be exposed to substantial noise levels caused by coastal metropolitan complexes.

Remote installations provided with transportation, electrification, and construction equipment create localized man-made noise environments. Thus, the existence of polar and desert, military and scientific research stations has extended the unintentionally generated man-made noise environment to all continents.

The dominant population of man-made noise sources, which produce the unintentionally generated radio noise of a metropolitan area, are located either at or within 300 feet of the surface. The source-radiation fields possess little directivity and, in conjunction with structure and surface reflections, cause the noise to envelop the space above an urban area. The resulting noise levels over metropolitan centers are large even at modern aircraft cruising altitudes. Airborne receiving equipment employing low-directivity earth-oriented antennas is particularly vulnerable to the unintentionally generated man-made noise of a major city and may be adversely affected when the plane is within line-of-sight range of a metropolitan center. In the higher radio bands, above the ionospheric cutoff frequency, the unintentionally generated man-made noise produced by the largest metropolitan complexes is detectable at subsynchronous satellite altitudes.

Spectral Distribution. Unintentionally generated noise of a metropolitan area may arise within any portion of the radio spectrum between 30 Hz and 7 GHz. The character of the noise waveform shows distinctive variations with spectral interval as the relative intensities of the sources that create the composite interference environment shift. Impulsive emission patterns (that is, aperiodically occurring transients) are present throughout most of the spectrum with impulse widths and magnitudes noticeably greater at the lower frequencies. Average impulse widths and amplitudes typically decrease with increasing frequency, while the average occurrence rate of the largest pulses may increase.

Line spectra are most often encountered in the unintentional noise spectrum at frequencies below 500 MHz. These arise as harmonics of a fundamental signal and, in the cases of power-distribution lines and certain industrial equipment, are often of a high-integer order.

With increasing radio frequency, composite unintentionally generated radio noise displays a decrease in peak electric field strength and average power. The spectral level variation is not uniform, and it arises, as do signal-pattern changes, from dissimilar alterations in the emissions generated by the several noise-source types coexisting in an area. In the lower portion of the radio spectrum, the dominant noise sources are electric transmission facilities and ISM equipment. Above the high-frequency (HF)-band, automotive ignition noise dominates in urban complexes but is occasionally superseded in rural areas by power-distribution lines. Ignition systems and gas-discharge noise sources found on power lines emit impulsive patterns accompanied by broad spectrum radiation. Corona discharges occurring on electric power facilities likewise yield broad spectrum radiation, although the associated amplitude distribution is more nearly Gaussian than impulsive.

Temporal Distribution. Temporal variations in unintentionally generated man-made in metropolitan areas are observed to occur in synchronism with two cyclical processes: business activity and meteorological changes.

Business-activity variations in automotive traffic density and numbers of operating industrial electric equipment give rise to a periodic change in the urban noise level. As seen in Fig. 1.1, during workdays, the level peaks in the morning and late afternoon, thus revealing a midday plateau significantly greater than the midnight minimum.[3] The average maximum power variation occurring during this cycle decreases with increasing radio frequency from more than 12 dB in the low VHF band to 5 dB at frequencies above 400 MHz. Weekend and holiday VHF-band variations manifest during a 24-hour period are diminished relative to a business day by 5 dB—that is, to approximately 7 dB. It is recognized that most of the business-cycle

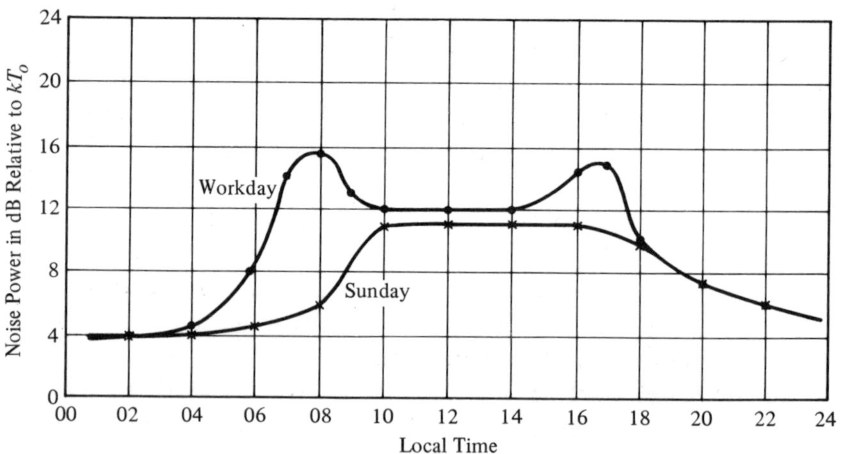

Fig. 1-1. Diurnal variation of 73 MHz incidental radio noise power for a workday and a Sunday. Noise power in db relative to kT_o. (After Buehler, King, and Lunden)

variation in unintentionally generated noise level is associated with changes in vehicle traffic density.

Meterologically produced changes in unintentionally generated man-made noise arise from the effects of rain, moisture, and sunlight upon power-distribution and transmission lines. Gap-breakdown and corona-discharge noise-generation mechanisms are affected differently by the natural processes. Gap-discharge breakdown, requiring for its initiation a high potential between two electrically insulated points, is suppressed during rain and high relative humidity because of the short-circuiting action caused by the appearance of water films on insulating surfaces and across microgaps in line-supporting hardware. Condensed droplets produce the converse effect upon corona discharges associated with high-voltage transmission lines. The presence of water droplets resting on high-potential conductors or falling in their immediate vicinity increases the electric field in the neighboring airspace as a consequence of the dielectric-field displacement occurring in the water. An increase in the point-electric field of a conductor elevates both the corona-discharge rate and radiated-pulse rate. Upon cessation of a rain shower, high-voltage lines normally display reduced interference levels caused by the cleansing action that has removed contaminant particles from the conductors and towers.

RELATIVE INTENSITIES OF RADIO-NOISE SOURCES

Man-Made Noise Sources

An appreciation of the identities and relative emission levels existing for the dominant unintentional noise sources with respect to one another and to the major naturally occurring noise sources is a necessary introduction to the topic of man-made noise. Three unintentional noise sources of primary importance, and for which ample data are available, are automotive ignition systems, power lines, and RF-stabilized arc welders.

The causes of radiated interference arising from vehicular ignition systems are known to be gap discharges produced by the distributor and breaker points and radiation from various leads, each enhanced by the presence of circuit and engine-cavity resonances. Other sources of automotive noise exist, such as horns and alternators. Typically, these auxiliary items generate interference that is indistinguishable from ignition emissions primarily because the latter dominate in intensity and frequency of occurrence.

Power-line-radiated interference is produced by one or more of the following mechanisms: (1) gap-discharge or insulating-film breakdown, (2) high-voltage corona discharges from conductor points with large potential gradients, and (3) line reradiation of interference injected into the transmission system by electrical equipment loads. Electrical equipment loads normally produce the lowest levels of radiated interference, in part because they are conventionally isolated by band rejection filtering from the distribution network. Gas-discharge and corona-interference sources vary in relative importance because each depends in a unique way upon line voltage, line construction, and line maintenance procedures, as well as upon observation frequency.

RF-energized metal-welding equipment superimposes 1- to 3-A arc stabilizing HF-band oscillating currents upon either direct-frequency or very-low-frequency, high-intensity welding currents of from 100 to 300 A. A spark-gap circuit, normally used to produce the stabilizing oscillations, generates a spectrum commonly rich in harmonics that radiate into the environment via unshielded leads or the welding electrodes.

For the three dominant unintentional man-made noise sources, the levels of radiated electric-field strength measured as a function of frequency using linearly polarized antennas and peak envelope detectors are shown in Fig. 1-2. The notation provided on each entry designates the source of the data (e.g., Pakala et al.[4] for power-line noise under fair weather conditions). Figure 1-2 presents peak detected electric-field strength per megahertz of impulse bandwidth. Translation of these results to another observation bandwidth is subject to uncertainties existing in the proportion of the overlapping impulses in the measured waveform. Were the impulses nonoverlapping, the peak electric-field strength would vary proportionally with the receiver's bandwidth. Pulse overlapping, however, does occur in the emissions from the sources shown; as a consequence, it is not possible to accurately translate the data of Fig. 1-2 from one bandwidth to another without additional information. The required additional information is available for the noise data observed from RF-stabilized arc

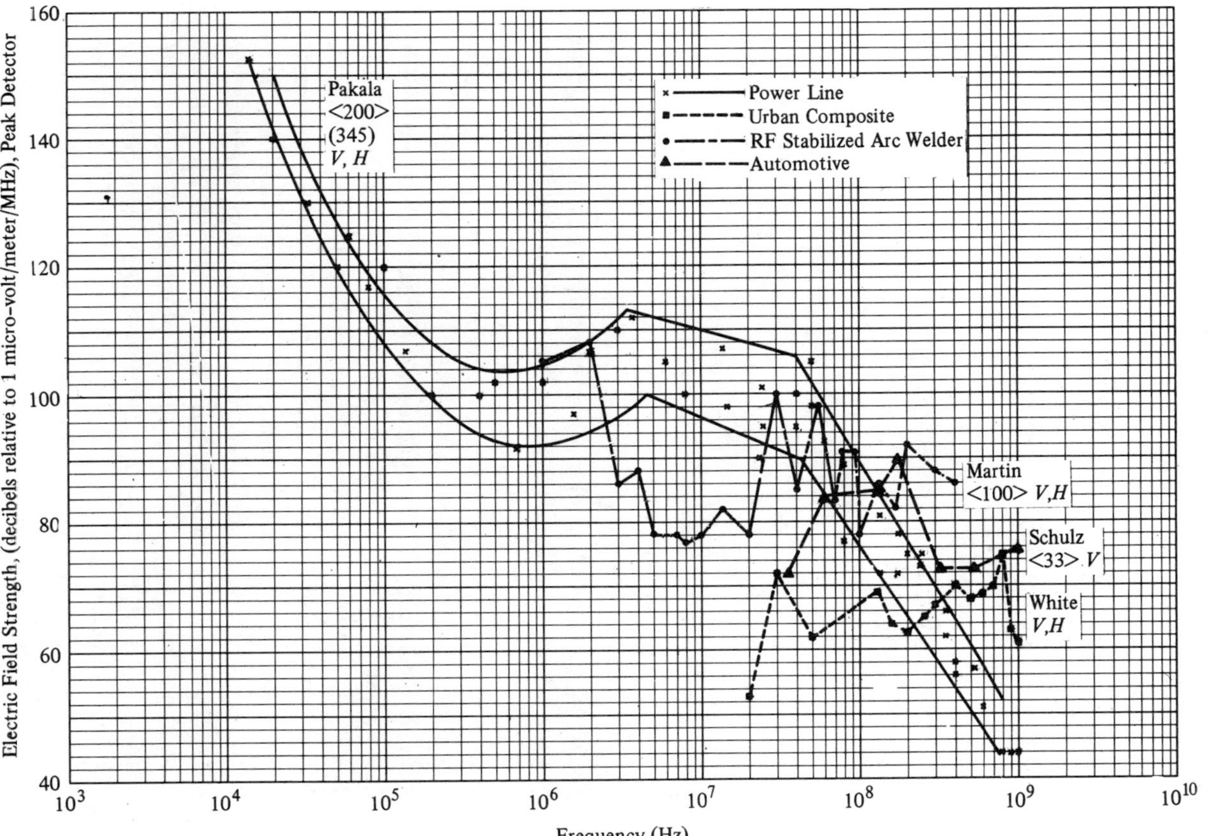

Fig. 1-2. Electric-field strength (peak detected) man-made radio-noise sources. (*345*) indicates power-line voltage in kilovolts; ⟨*200*⟩ indicates horizontal separation of noise source and observation point feet; observing antenna polzarization: *V* (vertical) or *H* (horizontal).

welders.[5] During this study, the bandwidth dependence of welder emissions was investigated, which established that the electric-field strength varied as the square root of the noise bandwidth for bandwidths of 5 kHz and less. Further, it was determined that for bandwidths ranging from 5 kHz to at least 300 kHz, the electric-field intensity displayed a direct dependence upon bandwidth.

Only in rare instances have reported investigations of unintentionally generated man-made radio noise included subsidiary studies establishing the detection-bandwidth dependence of the emission levels. Currently lacking are reliable and generally applicable experimental or theoretical bases for detector-bandwidth translations applicable to unintentionally generated radio-noise data observed with peak, quasi-peak, or average voltage detectors. Only for average power detectors may bandwidth translation be accurately performed, provided the detection bandwidth is small compared to the spectral range of the noise.

In the lower decades of the radio spectrum (below 1 MHz), power-line interference arising from gap discharges exceeds all other unintentional sources for both fair and foul weather conditions. Notice automotive ignition noise becomes a major contributor to the urban noise environment above 30 MHz, achieving a position of dominance at and above 100 MHz.[6] This dominance of automotive ignition noise is attributable both to the intensity of the individual vehicle radiation sources and to their numbers.

The radiated electric field strength of an RF-stabilized arc welder attains its greatest peak detected value below 50 MHz. Within this interval, the peak radiated-field components of both power lines and automotive ignition systems may be eclipsed by RF welder emissions.

The peak, detected, composite, urban-noise electric-field strengths shown in Fig. 1-2 for frequencies above 20 MHz were measured in the Detroit urban area, approximately 4 miles from the business center.[7] The data represent the average of observations made with both vertically and horizontally polarized antennas. The composite curve above 500 MHz bears a general resemblance to the automotive data by which it is dominated. In addition to providing a measure of the total field strength from several concurrently active types of unintentional noise sources common to an urban area, the inclusion of composite noise plots in Fig. 1-2 and later in Fig. 1-5 offers a vehicle for comparing the relative emission intensities of individual man-made and naturally occurring radio-noise sources. For both figures the composite urban-noise data measurements were performed within 5 miles of the business centers of several major urban areas.

Natural Radio-Noise Sources. The naturally occurring noise sources that dominate the radio-noise spectrum are presented in Fig. 1-3. Radiated power relative to 1 W/Hz of detection bandwidth, per unit area of receiving antenna aperture, is plotted in decibels versus frequency for atmospheric, solar, and cosmic noise sources. Below 20 MHz, atmospheric noise predominates over other natural sources in the temperate latitudes, a consequence of both the temporal variation in both thunderstorm intensity and the radio-path loss. Thunderstorm intensity throughout the world reaches peak level during the mid-afternoon, local time. The consequence of this fact is threefold. At frequencies for which the propagation-path loss is low, specifically in the vicinity of 10 kHz, the noise signals are propagated over intercontinental ranges, arriving at a local time corresponding to the time offset of the equatorial midafternoon storm period.[8] In the low-frequency (LF)-and mid-frequency (MF)-bands (30 kHz to 3 MHz), as well as in the lower portion of the HF band up to approximately 20 MHz, D-region attenuation during the local daytime suppresses intercontinental propagation of thunderstorm noise, resulting in a relative enhancement of locally generated atmospheric noise from storm fronts in the vicinity of the observer. Long-distance transmission of thunderstorm noise arising in African, South American, and Southeast Asian equatorial centers is never totally suppressed by high D-region loss. Some portion of the transcontinental great-circle path between an equatorial storm center and a temperately positioned observer will always lie in an evening shadow, and thus will experience low attenuation. The interaction of these processes is shown in Fig. 1-4 where atmospheric average radio-noise power per Hz, in decibels relative to kT_o (T_o = 290°K) is plotted versus local time for a temperate, north-latitude location at several frequencies between 51 kHz and 20 MHz.[9] Local day-night time transitions are noted by the vertical dashed lines. In the radio bands above the ionospheric cutoff frequency, observed atmospheric noise is of local origin, arising from lightning discharges associated with

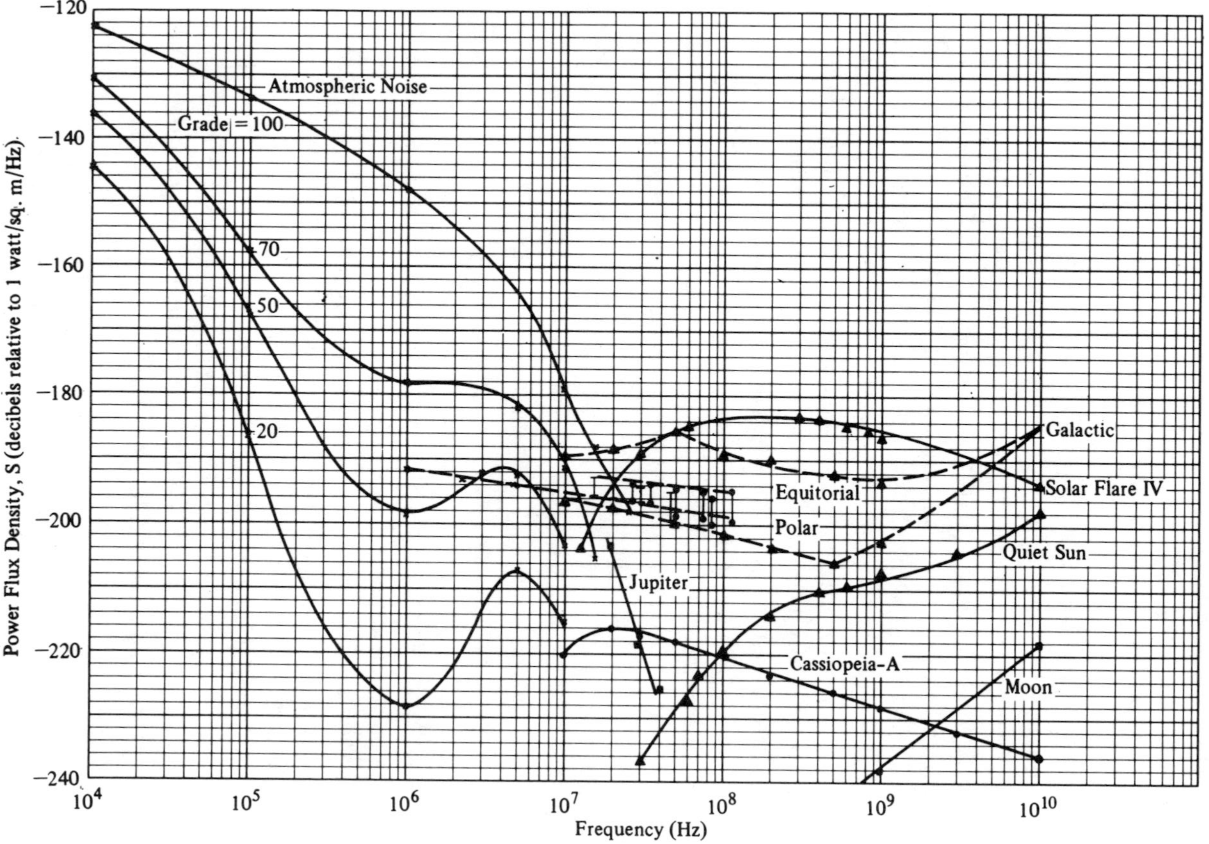

Fig. 1-3. Power flux density of natural radio-noise sources.

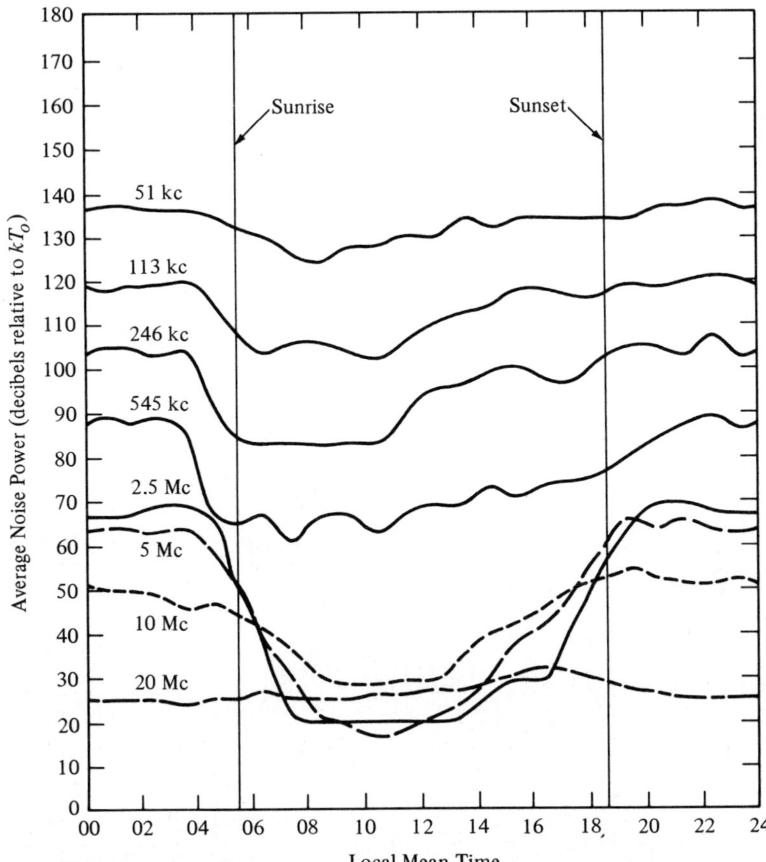

Fig. 1-4. Average noise power of atmospheric radio noise as a function of local time and frequency for a north latitude site. (From URSI special report No. 7, *The Measurement of Characteristics of Terrestrial Radio Noise*. Amsterdam: Elsevier Publishing Co., 1962).

thunderstorm activity occurring within line-of-sight range of the observation point.

During the prenoon period, (0800 to 1200 hours), the noise power below 10 MHz observed in the plane of the galactic equator usually exceeds atmospheric noise in temperate and polar regions.[10] Note in Fig. 1-5 that the maximum noise grade of 100 represents a limiting, rarely attained level, whereas grades of between 30 and 80 bound the range of atmospheric noise commonly observed.[11]

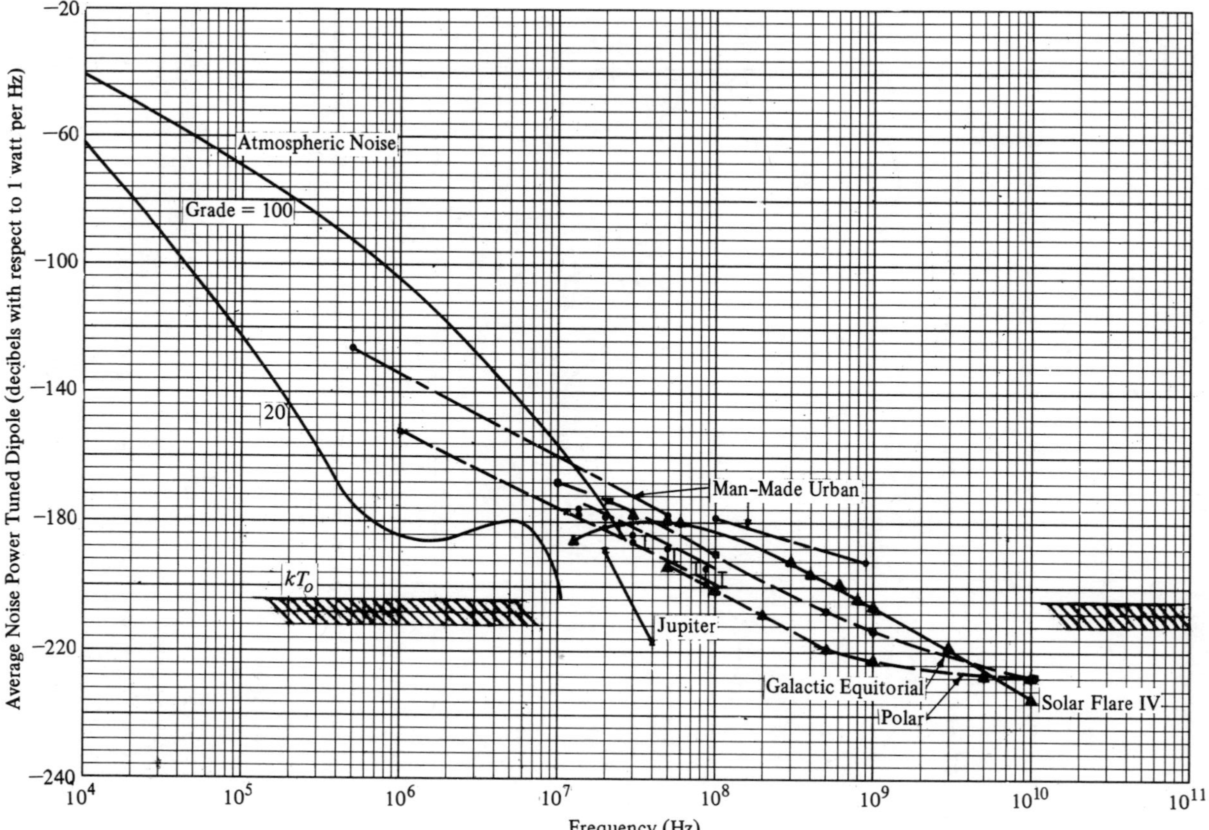

Fig. 1-5. Average power of natural and man-made radio noise sources observed with a tuned dipole antenna.

During periods of high sunspot activity, solar disturbances are accompanied by sharp rises in radio-noise emission. Solar-flare radio-noise radiations vary in spectral content, duration, and polarization; they can be characterized by feature to one of the five flare types. The radio spectrum for a Type IV flare is shown in Fig. 1-3 and may be compared to the emission of a quiet sun, which is also plotted. The noise flux density of a Type IV flare is not the largest of the five flare types. Types I, II, and III have been observed to emit greater power densities; however, Type IV emissions are stable and broadband, capable of persisting for periods of several days.[12,13]

Radio interference, arising from synchrotron oscillations of charged particles composing the Jovian atmosphere, and reflected solar emissions from the moon are the two additional significant sources of natural noise in the solar system.[12] The spectra of each is distinct. Jovian emissions are of nonthermal origin; they are produced by an accelerated electron plasma in the planet's magnetic field. Ionospheric opacity of the earth's atmosphere produces the apparent extinction at the earth's surface of the Jovian flux at frequencies below 20 MHz.

The lunar emission spectra presented in Fig. 1-3 and manifesting a distinct positive slope is that of a black body whose temperature varies between 100°K and 300°K during the lunar cycle.

External to the solar system there exist two radio sources of significance: sky background or galactic noise, and the stellar object, Cassiopeia A. Galactic noise, shown in Fig. 1-3 for frequencies of from 1 MHz to 1 GHz, is generated within our galaxy by numerous, unresolved, discrete sources plus a continuum emission concentrated in the plane of the galactic equator. Within the VHF and UHF bands, collective asymmetries result in an approximate 10-dB increase of galactic equatorial power flux density relative to galactic-pole emission levels. The continuum emission arises jointly from the presence of ionized hydrogen in interstellar space, the spectral distribution of which conforms to black-body radiation, and linearly polarized, electron synchrotron radiation displaying a nonthermal spectrum.

Continuum emission level attains a value of 3°K and is independent of galactic direction in the frequency range of 10 GHz and above.[14]

The supernova remnant Cassiopeia A included in Fig. 1-3 is the most intense extrasolar radio point source. The negative slope of the spectral distribution establishes its radiation arises from a nonthermal mechanism.[15]

Comparison of Man-Made and Naturally Occurring Radio-Noise Sources

By converting the power-flux density observed for naturally occurring radio-noise sources into the available power at the terminals of a tuned dipole antenna, it becomes possible to directly compare these results with those obtained for unintentional man-made noise sources typically reported for tuned dipole antennas.[16] In Fig. 1-5, atmospheric, galactic, disturbed solar, and Jovian noise power per Hz bandwidth, in decibels relative to 1 W, is presented for a matched, tuned dipole antenna within the frequency range of 10 kHz to 10 GHz. Added for comparison are average, composite, unintentionally generated noise plots assembled from measurements in several urban areas in North America, Europe, Asia, and Australia.[16] At frequencies higher than the ionospheric cutoff, approximately 30 MHz, composite man-made noise observed in the central zone of a typical urban center exceeds the natural noise source intensity; and unlike the spatially localized sources—the sun and Jupiter—it usually envelops the observer from all azimuthal directions. Labeling in Fig. 1-5 for galactic noise distinguishes between the intensity in the direction of the galactic poles and the galactic equator, as in Fig. 1-3. Also shown for reference is the noise level equal to kT_o, with $T_o = 290°K$.

UNINTENTIONAL SOURCES OF MAN-MADE NOISE

As noted in Fig. 1-2, noise-emission levels of unintentional man-made radio-noise sources vary with frequency for the most important source types. The emission-level–frequency-variation is unique to each category of sources and sufficiently disparate to produce a reordering of the dominant sources within different portions of the radio spectrum. An example of a shift in relative strength of emission sources may be noted in Fig. 1-2 for power-line and automotive ignition noise within the frequency interval 10 MHz to 1 GHz. Athough similar variations occur for all other incidental-noise sources, an approximate ranking of noise-emission intensity still exists as a useful guide for assessing the potential importance, as radio-system performance threats, of the many known unintentional sources of man-made radio noise. Such a ranking is presented in Table 1-1.

Table 1-1. Ranking of unintentional man-made radio-noise sources.

I. Automotive Sources
- Ignition circuitry
- Alternators, generators, and electric motors
- Buzzers, switches, regulators, and horns

II. Power Transport and Generating Facilities
- Distribution lines
- Transmission lines
- AC transformer substations
- DC rectifier stations
- Generator stations

III. Industrial Equipment
- RF stabilized arc welders
- Electric discharge machines
- Resistance welding machines
- Induction heating equipment
- RF soldering machines
- Dielectric welder and cutting machines
- Dielectric and plastic preheaters
- Wood gluing equipment
- Silicon control rectifiers (SCR)
- Electric motors and inverters
- Circuit breakers
- Circuit switches
- Microwave heaters
- Electric calculators and office machines
- Cargo-loading cranes

IV. Consumer Products
- Appliance motors
- Fluorescent, sodium vapor, and mercury vapor lights
- Vibrators
- Citizen-band AM transmitters (spurious emissions)
- Electric door openers
- Television local oscillator radiation

V. Lighting Systems
- Neon, mercury, argon, and sodium vapor lights
- Fluorescent light fixtures

VI. Medical Equipment
- Diathermy

VII. Electric Trains and Buses
- DC-drive motors
- Pantograph and third-rail contacts

ORGANIZATION OF THE SUBJECT

For many of the known incidental noise sources, measured emission spectra are available. Pertaining to categories I, II, and III of Table 1-1, experimental spectral data and envelope statistics are especially plentiful. Moreover, because the magnitude of the noise radiated by automotive ignition systems and electric power facilities may be very intense, substantial theoretical attention has been directed toward both categories I and II. By this combined effort, two sets of analytical models have been developed, which creditably represent the dependence of mean emission level upon source characteristics and observer-source geometry. Concurrently, for automotive ignition interference, modeling of the radiated envelope statistics has achieved significant success in recent investigations. It is the purpose of Chapters 2 and 3 to present the available experimental data on automotive ignition and electric power facilities that are of the most utility and the greatest reliability, with the expectation that these measurements, drawn from numerous studies, will provide a detailed assessment of the features of the radiated fields of categories I and II. Included in the respective chapters are the development and presentation of the best analytical representation of either the mean radiated signal power or field intensity, depending upon availability and utility, for categories I and II incidental-noise sources. Chapter 2 includes the development of a competent model for the envelope statistics of automotive ignition noise applicable to both stationary vehicles and automotive traffic. Deferred, however, until Chapter 5 is a general and comprehensive theoretical treatment of the noise-envelope statistics for a broad range of unintentional-radio-noise types, which encompasses but is not limited to categories I and II.

Chapter 4 provides the available experimental data, which are precise enough to yield a useful characterization of the remaining categories of unintentional-noise sources. The quantity and quality of these data vary dramatically among the types of known sources. For instance, much information of acceptable utility and accuracy exists for industrial equipment (category III), yet little exists on electric trains and buses.

Chapter 4 concludes the presentation of radiated-noise data for individual types of incidental sources. Chapter 5 contains a comprehensive analysis of the noise-envelope statistics based upon and summarized from the works of Middleton and, to a lesser extent, the works of Furutsu and Ishida. Developed in this chapter are the amplitude probability distribution and the lower-order moments of the noise envelope for two classes of noise emitters defined in relation to the bandwidth of an observing or affected radio receiver. To permit an assessment of the validity of the models, their flexibility, and constraints, selected experimental comparisons have been included.

Chapter 6 and 7 treat man-made incidental noise from the viewpoint of a composite interference that is most commonly encountered by wireless systems operating in or above urban areas. In composite incidental interference, individual sources and source types are indiscernible, and their multiplicity and variety generate a noise-emission field that pervades and enshrouds large geographical regions of an industrialized area. Although arising from source heterogeneity, the composite metropolitan-area noise fields are definable quantitatively, either by activity designations (for example, business and residential) or by a surface distance computed from the urban center. Either characterization permits presenting composite surface-noise data as a function of frequency; yielding the parametric representations provided in Chapter 6. The mean-noise-power variations observed for composite incidental radiations are shown to be describable in terms of the propagation processes that govern radio transmission from low-height antennas deployed in irregular terrain and a distribution of random independent impulsive sources representable by a Poisson distribution of occurrences.

The analytical representation of either composite surface-noise power or electric-field strength as a function of metropolitan-center displacement distance may be employed, as presented in Chapter 7, to compute the observed noise fields at points above an urban complex. It is sufficient to know the surface dependence of composite noise power or radiated field intensity and the radiation pattern of the observing antenna to complete the determination of the ambient incidental-noise level at a specified frequency above an arbitrarily selected metropolitan area. Several examples of such computations are provided in Chapter 7 and compared with airborne measurements. Various polarizations, altitudes, frequencies, and center offset distances are provided, and the dependence of incidental-noise level upon each parameter is determined using the surface representations of composite noise developed in Chapter 6. Sufficient cases are included to establish confidence in the reliability of the modeling concept and to provide a tool for other applications.

References

1. Federal Communication Commission. *Rules and Regulations*. Volume II, Part 15.
2. Ibid. Volume II, Part 18.
3. Buehler, W. K., King, C. H., and Lunden, C. D., "VHF city noise," 1968 *IEEE Electromagnetic Compatibility Symposium Record*, pp. 113–118.
4. Pakala, W. E., Taylor, E. R., Jr., and Harrold, R. T. Radio noise measurements on high voltage lines. *IEEE Electromagnetic Compatibility Symposium Record 1968*, 96–107.
5. Martin, H. and Tabor, F. Radio Frequency Emission Characteristics and Measurement Procedures of Incidental Radiation Devices and Industrial, Scientific, and Medical Equipment. Report FAA-RD-72-80-1. Electromagnetic Compatibility Analysis Center, Annapolis, Maryland, 1972.
6. Schulz, R., Southwick, R., and Smith, P. C. Measurement of Electromagnetic Emissions Generated by Vehicle Ignition Systems. Southwest Research Institute, 1973.
7. White Electromagnetics, Inc. Ambient Electromagnetic Survey Detroit, Michigan. Report 69-15. Prepared for Automobile Manufacturers Association, Detroit, Michigan, 1969.
8. McKerrow, C. A., Some measurements of atmospheric noise levels at low and very low frequencies in Canada. *J. Research* **65** (7): 1911–1926 (1960).
9. International Union of Radio Scientists, *The Measurement of Characteristics of Terrestrial Radio Noise*. URSI Special Report No. 7. Amsterdam: Elsevier Publishing Co., 1962.
10. Horner, F. Radio noise from thunderstorms. *Radio Science, Advances in Research*, 2: 121–204 (1964).
11. CCIR. World Distribution and Characteristics of Atmospheric Radio Noise. CCIR Report 322. Geneva: International Telecommunications Union, 1964.
12. Kraus, J. D. *Radio Astronomy*. New York: McGraw-Hill Book Co., 1966.
13. Boischot, A. Bruits cosmiques in microondea. *L'Onde Electrique* 252–257 (1967).
14. Boischot, A. Les bruits cosmiques. *NATO/AGARD Conference Proceedings Electromagnetic Noise, Interference and Compatibility* 4.1 to 4.12 (November 1975).
15. Andrew, B. H. The flux densities of twenty radio sources at 13.1 MC/S. *Astrophysical Journal*, No. 2, pp. 423–432 (1967).
16. Skomal, E. N. An analysis of metropolitan incidental radio noise data. *IEEE Trans. on Electromagnetic Compatibility* **EMC-15** (2): 45–57 (1973).

2 Automotive Noise

Impulsive noise from automobiles is generated by the ignition system, the battery-charging circuitry, accessory motors, electric warning devices, and starter motors. Certain auxiliary devices, depending upon their design, may emit a noticeable level of unintentionally generated noise. However, because the operational-duty-cycle of warning devices and accessory and starter motors is extremely low, their contribution to the total automotive noise emission is negligible. It is primarily the ignition system of gasoline-engine vehicles and the battery-charging components that are the major contributors to the emitted noise. Furthermore, since a battery circuit operates at very low voltage, radiated noise intensity from generators and charging components is low, as long as the elements have not seriously deteriorated (as might occur to unmaintained brushes of a direct-current generator). Thus, the ignition system of a gasoline engine remains as the only source of high potentials and currents functioning continuously during vehicle operation. Radiation produced by the presence of high-pulse currents and voltages in cabling and at points of ignition-circuit discontinuity are the primary source of automotive radio noise.

IGNITION-SYSTEM NOISE SOURCES

The sequence of events that results in the emission of an ignition-noise-pulse pattern is indicated in Fig. 2-1. The chain of events begins in Fig. 2-1(a), immediately prior to opening of the breaker points. The breaker bypass capacitor and stray circuit capacitances

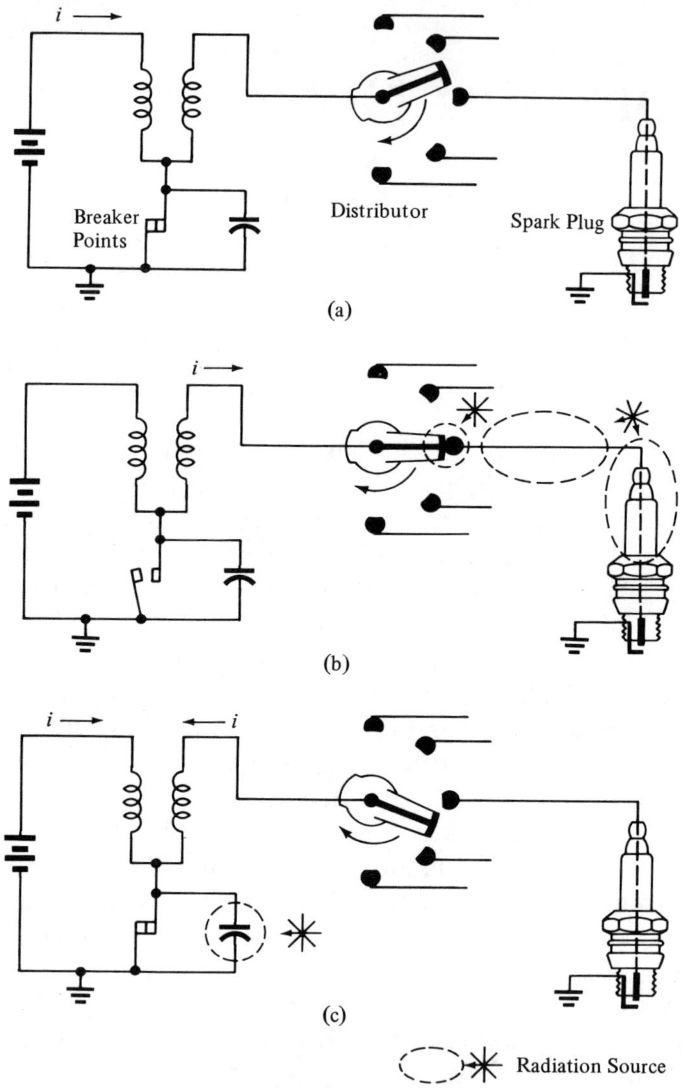

Fig. 2-1. Event sequence in an automotive ignition circuit producing radio-frequency interference. (*a*) Breaker points closed, secondary circuit at low voltage. No radiation. (*b*) Breaker points open. Distributor gaps are excited by inductive surge in coil secondary. Spark-plug gaps are ignited. Radiation initiated from distributor and unshielded sections of the secondary circuit. (*c*) Breaker points closed. Distributor and Spark-plug arcs extinguished. Breaker point by-pass capacitor discharges through breaker points. Radiation produced by capacitor-connecting circuitry.

have discharged, and the magnetic field of the ignition coil is approaching its steady-state value; no voltage exists on the spark plug, and the potential across the distributor points is low, far below gas-discharge threshold.

The breaker points open in Fig. 2-1(b), as the distributor rotor passes a cap contact attached by a length of secondary wiring to a spark plug. The abrupt decrease in battery current flow through the ignition coil primary induces in the high-turns secondary a peak pulse in excess of 15 KV, which simultaneously charges the breaker-point bypass capacitor. The rotor gap typically designed with a spacing of 0.03 in. (special gap widths to 0.125 in. may be used) experiences a gap-discharge breakdown analogous to the discharge process that occurs on power-line gap faults. The air-gap arc in the distributor presents for approximately 1 msec, depending upon engine speed, a low resistance path between the coil secondary and the spark plug. Furthermore, during this period of approximately 1 msec, a strong variable radiation field is created in the distributor gap, the variability of which is a direct result of the secondary circuit resonant response to an impulsive discharge. Within less than 10 msec, the impulsive current surge has traveled from the coil secondary, through the distributor, to the spark plug with little diminution, and at the spark plug has initiated a discharge in the air-fuel mixture in the cylinder. The low resistance path at the spark-plug electrodes discharges the coaxial capacitor formed by the plug-center conductor and metal base that began to charge when the distributor-point arc was ignited. The plug capacitance discharge spike extinguishes within 5 nsec. The plug-gap arc persists for a fraction of 1 μsec, sustaining in the secondary circuit a time-variable current pattern, modified, as noted earlier, by the existing circuit resonances. The unshielded secondary-wiring-plug tower radiates a pattern of impulsive interference in synchronism with the circuit current. The total radiated field is modified by the presence of the engine compartment equipment, the compartment geometry, the road surface, and vehicle road clearance before the noise signal reaches a remotely located observer.

As the distributor rotor moves past the cap contact in Fig. 2-1(c), the breaker points close, thus initiating a discharge of the distributor by pass capacitor and various stray capacitances in the breaker-to-distributor leads. There follows from the postfiring capacitive discharge a weak radiated field emanating from the wiring, as secondary-circuit voltages are now insufficient to sustain an arc discharge.

Ignition Waveforms

Figure 2-2 shows the conducted current pattern produced by an unsuppressed secondary-ignition circuit at, and subsequent to, initiation of the spark-plug discharge.[1] The time-base increments represent 10 nsecs. The sharp spike lasting 5 nsec or less arises from the discharge of the coaxial capacitor of the spark plug and is known to be readily suppressable by inserting an in-line center conductor resistor near the plug tip, at the gap end, without detracting from engine performance.[2] One may observe in Fig. 2-2(a) the oscillations impressed upon the radiated field by the secondary-circuit resonances that follow the capacitive spike. In Fig. 2-2(b), a larger fraction of the

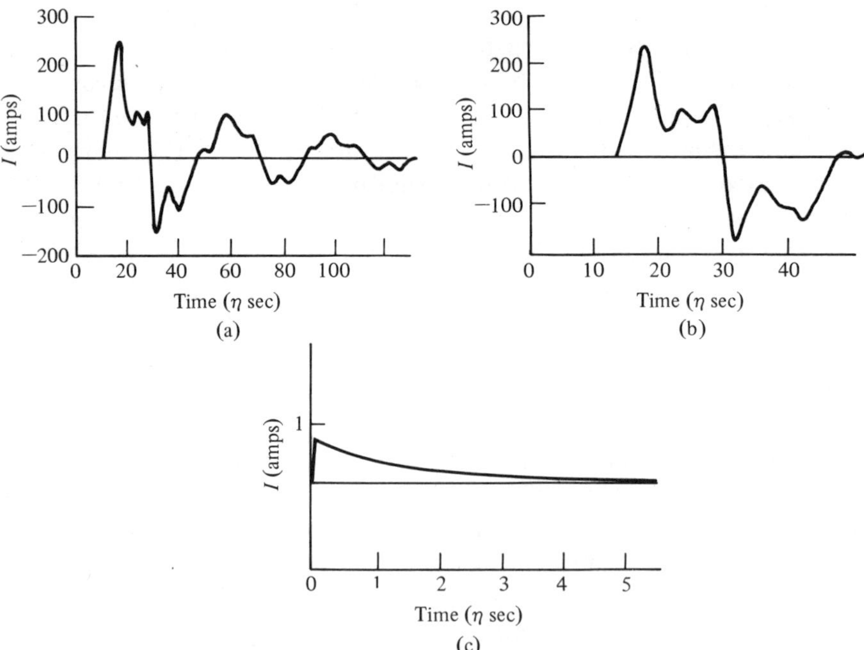

Fig. 2-2. Secondary ignition circuit current waveforms, 12-Kv applied peak potential. (*a*) and (*b*) Unsuppressed resistor located at the spark plug. (From Ball and Nethercot, *Proc. Inst. of Mech. Eng.*, London, 1952)

initial portion of the radiated pattern is shown, and, in this sample, much longer resonances are visible. Notice that the peak current levels exceed 200 A, whereas during the subsequent oscillations, the pulse current exceeds 50 A. Figure 2-2(c) demonstrates the reduction of peak secondary-circuit current produced by the inclusion of a 15 kΩ resistor at the spark plug. The maximum current is less than 1 A, and the exponential decay is seen to persist for 6 microsec.

Automotive-Ignition-System Noise Measurements

Studies of automotive-ignition-noise radio emissions have been experimentally conducted by investigators using primarily four types of detecting circuits: peak-envelope, quasi-peak, rms-voltage, and envelope-voltage detectors.

Observations performed over varying sampling periods commonly yield the limiting values of the variables plus a measure of either the mean or median. Peak- and envelope-voltage detectors in a very limited number of studies have been supplemented with data-processing circuitry, which permitted the measurement of certain statistics of the detector output. The most common statistics accumulated in this manner for ignition noise have been the cumulative distribution of detected-signal-envelope voltage or of signal-envelope peak amplitude. In isolated cases, signal-envelope, voltage-threshold crossing rate and envelope-width distributions above specified signal levels have also been reported.

By application of these detector functions and processing circuitry, radiated-noise-signature data have been assembled versus frequency from single and multiple, stationary and moving vehicles operating at various engine speeds. Several manufacturers, years of manufacture, ignition-system-interference suppression designs, antenna positions, observation distances, and road conditions are represented in the results. The large body of data separates into two principal categories: stationary vehicle studies and automotive traffic distributions The procedure used to present the most important results of this appreciable body of data segregates the measurements into three categories by detector type. Within each category, the data have been assembled into one of the two groups: stationary vehicles or automotive traffic.

In presenting peak and quasi-peak detector measurements, some published results provide data in terms of electric-field strength at the antenna aperture without specifying a detection bandwidth, whereas others report field strength per unit of bandwidth (which has been adopted in the present work). As noted in Chapter 1, no detector function provides a measure of impulsive, unintentionally generated noise that may be accurately translated from one bandwidth to another for all bandwidths of interest. Therefore, it is important to know the bandwidth of the detector used in obtaining each assembly of data and, with this information, to assess the degree of comparability of the detection bandwidth of interest and that of the previously used noise-measuring equipment. When it has been necessary to convert observed peak or quasi-peak noise field strength into field strength per unit of bandwidth, the procedure has been to divide the former by the impulse bandwidth. Impulse bandwidths are provided in all cases, making it possible to perform the inverse transformation and return to the original meter reading of electric-field strength observed by the receiving antenna, if so desired.

Peak Envelope Detector Measurements. Several extensive studies of automotive ignition noise have been performed using peak-envelope detectors. To avoid possible intermingling of the ignition-noise emissions with broadband noise from other sources, most of the data accumulations have been confined to frequencies above 30 MHz. The undesirable commingling that might occur at lower frequencies is primarily between atmospheric and power-line noise (see, for example, Figs. 1-2 and 1-5). Care is also required in selecting test locations at frequencies of 30 MHz and higher to prevent the intrusion of interference from other industrial sources and nearby power lines.

If the radiated-noise pattern primarily consists of nonoverlapping pulses as observed in a selected detection bandwidth, the resulting detected peak-envelope voltage will not alter when the number of radiating sources is reduced. Such a situation may be encountered during the field-intensity measurements on one or a few stationary vehicles with a peak-envelope detector having a bandwidth of several hundred kilohertz. When the bandwidth of the peak-envelope detector is small compared to the reciprocal of the average pulse width, or if the number of independently radiating vehicle sources is large, the

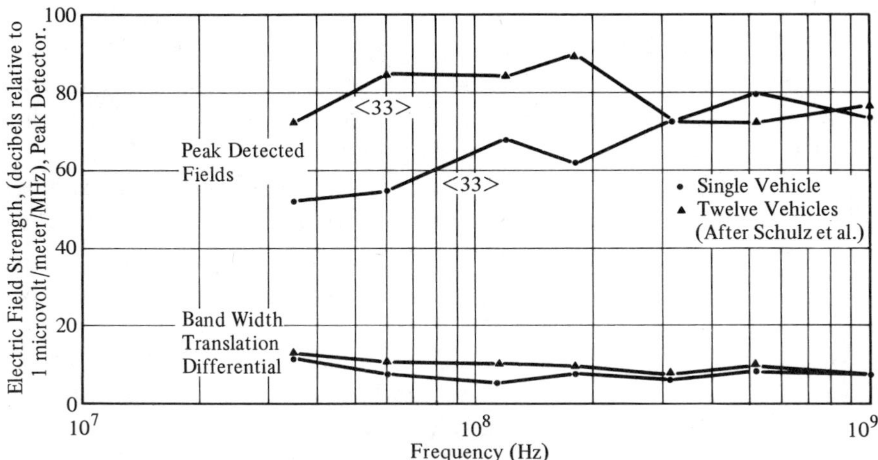

Fig. 2-3. Peak detected radiated electric-field strength from single and multiple vehicles. Variation of peak electric field strength with detector bandwidth, bandwidth translation differential, Equation 2-1.

detected voltage and the derived radiated electric-field strength will increase with the number of operating vehicles. Figure 2-3 presents vertically polarized, peak detected noise field strength as a function of frequency for a single stationary vehicle located 10 m (33 ft) from the observing antenna and also for a group of 12 stationary vehicles with minimum observer separation distance of 10 m. During these measurements, vehicle engines were running at 1500 rpm, and the hoods were closed. The test site, a remote rural field, was free of other noise signals.[3] The detector used in the measurements had an impulse bandwidth of 500 kHz. One observes that the radiated field strength does display an increase for 12 vehicles relative to a single vehicle in the lower part of the spectrum. However, at and above 300 MHz, the differences are comparable to the measurement accuracy.

The same vehicle groupings under identical conditions were measured using a peak-envelope detector possessing an impulse bandwidth of 50 KHz. A bandwidth translation differential ΔBW, may be defined for a pair of electric field intensity values $E(\ )_{BW1}$ and $E(\ )_{BW2}$ as:

$$\Delta BW = E(\mathrm{dB}\mu\mathrm{v/m/MHz})_{BW1} - E(\mathrm{dB}\mu\mathrm{v/m/MHz})_{BW2}, \qquad (2\text{-}1)$$

where BW1 = 500 kHz, and BW2 = 50 kHz, is plotted in the lower part of Fig. 2-3, for both one and 12 stationary vehicles. The quantity ΔBW measures the deviation from strict proportionality of impulsive electric-field strength given in units of dB relative to 1 μv/m per MHz bandwidth and the receiver-impulse bandwidth. Notice in Fig. 2-3 that values of ΔBW are less for the single-vehicle data. This occurrence is consistent with the behavior of peak-envelope detectors whose response is linearly dependent upon bandwidth, provided the detected incident impulses do not overlap appreciably within the detection filter.[4]

Radiated-electric fields produced by automotive ignition sources are affected by the vehicle structure and the presence of the earth.

The effect of these and possibly other factors upon observed-signal polarization is shown in Fig. 2-4 for stationary single vehicles where the ratio of the vertically to horizontally polarized electric-field strengths is presented for an observation distance of 10 m.[5]

There is evidence in the data that vertically polarized radiated noise is approximately 3dB greater than the horizontal component for the band of frequencies 30 MHz to 1 GHz. These results are representative of the polarization dependence that has been reported for both traffic and stationary-vehicle ignition-system emissions by other observers.[6, 7] The waveform characteristics of automotive traffic noise are related to stationary-vehicle emission patterns; however, their pulse amplitudes display greater variability, and pulse

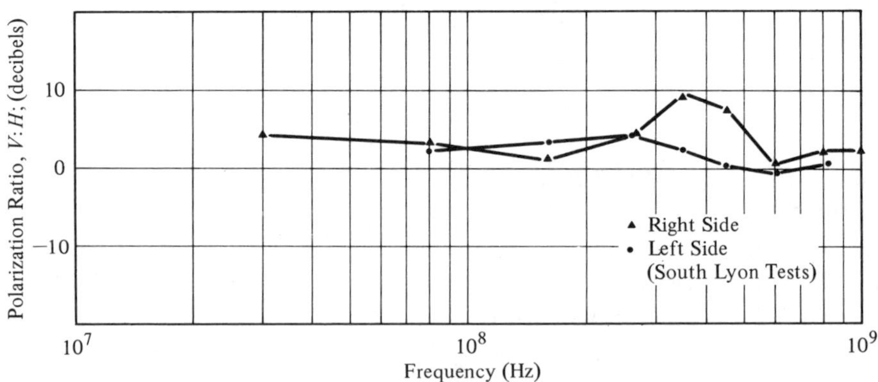

Fig. 2-4. Ratio (in dB) of the vertically to the horizontally polarized ignition-system radiation from single vehicles at a distance of 33 ft. Peak detector.

overlapping is more in evidence. Unchanged is the fact that impulsiveness of the radiation remains a central feature.

Two substantial bodies of vehicular-traffic, ignition-noise, median peak detected data are summarized in Fig. 2-5 for both vertically and horizontally polarized electric-field strength. Notice separation distances between the observer and traffic-flow overlap, with Shepherd et al.'s results including a larger range interval directly ahead of the detector.[8] Both series employed a combination of antenna types—low-gain broadbeam dipoles at the lower end of the frequency range, and higher-gain narrowbeam parabolic or horn antennas at and above 1.4 GHz. The vehicles contributing to the noise emissions were comparable with respect to their ignition-circuit radiation-suppression designs, although the vehicle models differed in age by 12 years. Approximately 1000 passing vehicles were observed by the Frederick Research Corp.[9] at an unrecorded passing rate, while the traffic density during the 15-min. observation period used by Shepherd et al. varied between 50 and 60 vehicle/min. The respective impulse bandwidths of the observing equipment were 199 kHz for the Frederick Research Corp. data and 300 kHz for Shepherd et al.'s results. In each series of tests, the equipment used was calibrated with an impulse generator, and the signal response was referenced to an impulse bandwidth of 1 MHz.

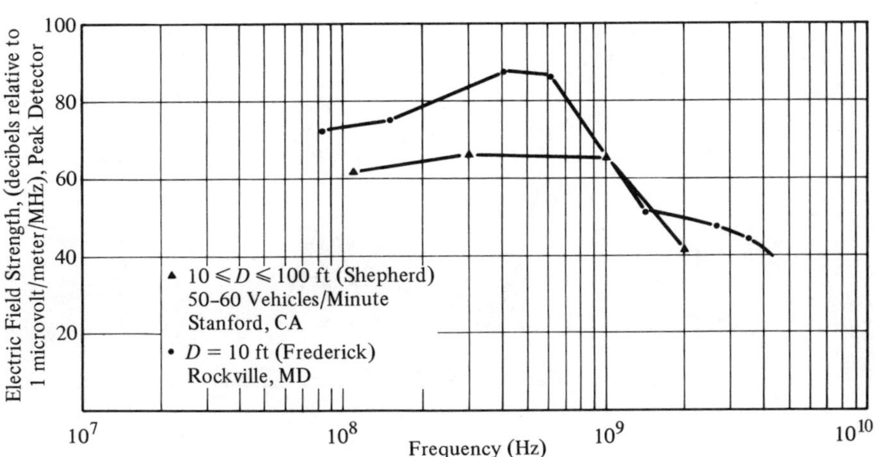

Fig. 2-5. Median values of peak detected electric-field strength for automotive traffic. Stanford, California,[8] and Rockville, Maryland[9].

One observes in Fig. 2-5 the distinct decrease in peak field intensity beginning at or somewhat below 1 GHz. The radiated-noise spectrum from automotive ignition systems is affected by several circuit and enclosure factors as noted earlier. Collectively, these mechanisms produce an electric-field spectrum distinguished by the presence of three maxima, two of which are shown in Fig. 2-6 where the multiple curves represent the noise spectra from several vehicles.[10] Notice that one resonance appears in the MF and one in the HF band. The third resonance rests in the UHF band (Fig. 2-5), where a broad maximum is found in the interval of 300 to 700 MHz. The existence of the UHF-band ignition-noise maximum is considered to be the major cause of the spectrum alteration manifest in composite, urban-area, UHF-band, average-power-versus-frequency data, which is discussed at greater length in Chapter 6. An example of the alteration may be seen in Fig. 1-5 by inspecting the two curves labeled *man-made urban noise.* The slope of the lower-frequency segment is -23 dB per decade frequency change, while the slope of the segment in the UHF band containing the third automotive-ignition-noise resonance peak is -13 dB per decade change in frequency.

Samples of vehicular-traffic-radiated noise data obtained at Stanford, California, have been reduced by Shepherd et al. to yield the time statistics of the peak envelope as shown in Fig. 2-7 for a detec-

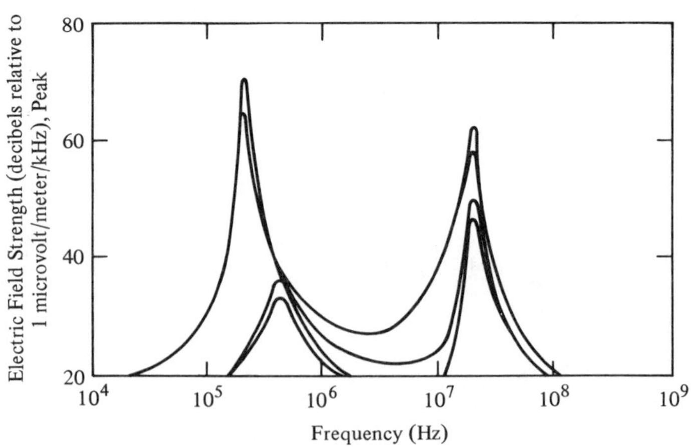

Fig. 2-6. Radiated electric-field strength from automobiles showing the lower-frequency ignition-radiation maxima. Separate vehicle data.

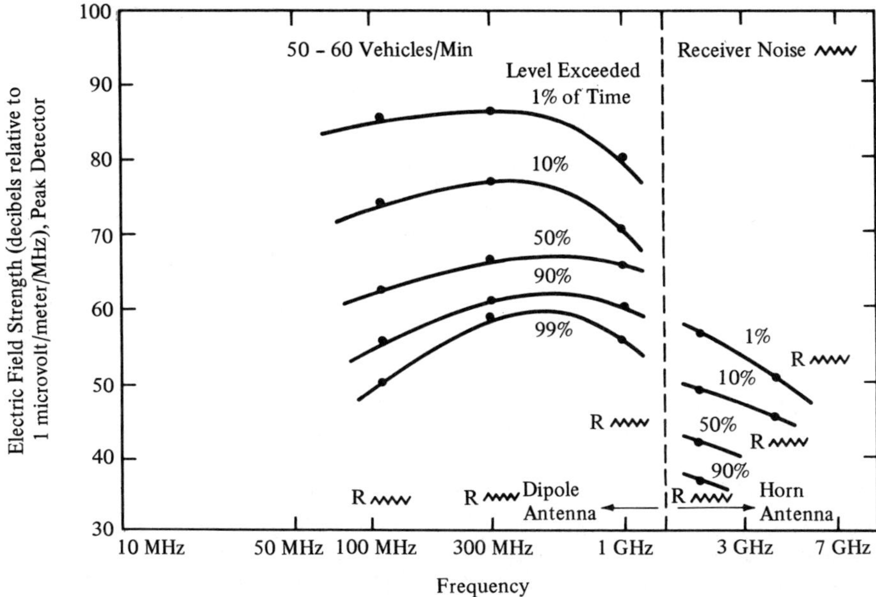

Fig. 2-7. Peak envelope distributions for a 15-min. sample of automotive traffic radio noise as a function of frequency. Observing antenna types: dipole 1 GHz and below, horn above 1 GHz. Noise bandwidth, 300 kHz. (After Shepherd et al., 1975)

tion bandwidth of 300 kHz. The peak-envelope values exceeded for varying percentages of the 15-min. observation interval are given. A vertical dashed line has been inserted to partition the data on the basis of the type of observing antenna used. It is possible that a portion of the abrupt decrease in each of the percentile curves occurring above 1 GHz is attributable to a reduction in the antenna field of coverage when a dipole, either vertically or horizontally polarized, was replaced by a horn.

When comparing the peak radiated electric-field strength of automotive ignition noise, either observed in or presented for differing observation bandwidths, BW (e.g., Figs. 2-6 and 2-7), the bandwidth transformation law to be used is electric-field strength, $E_{peak} \propto BW$.

The time statistics of traffic-radiated noise for the frequencies 110 MHz, 300 MHz, and 1 GHz deviate from a log distribution, as is evident when the data of Fig. 2-7 are replotted as shown in Fig. 2-8.

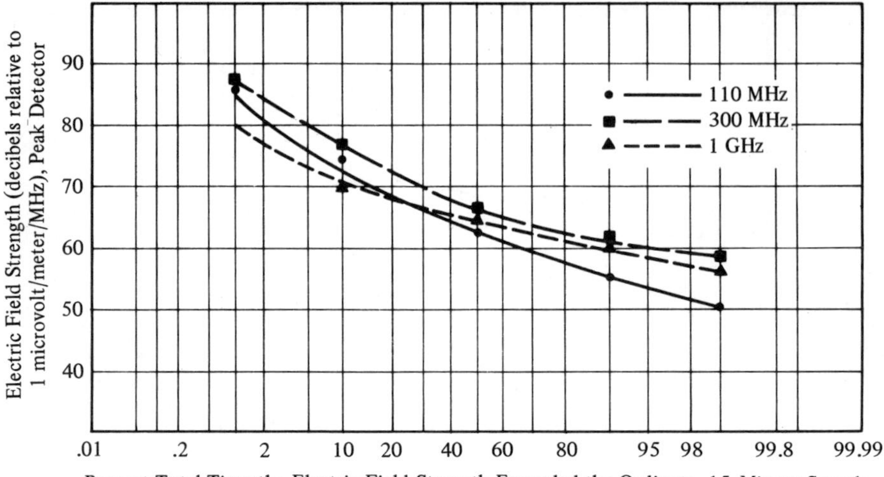

Fig. 2-8. Cumulative distributions for automotive traffic noise, electric-field-strength peak-envelope detector. Fifteen-minute sample. Traffic density: 50–60 vehicles/min. Vertical and Horizontal Polarization. (After Shepherd et al., 1975)

Notice the population of high-pulse envelope values is greater than would have occurred were the data log normally distributed.

Quasi-Peak Detector Measurements. The large charge and discharge time constants, T_c and T_d, present in the detection circuit of quasi-peak field-measuring instrumentation produce a partial integration or approximate peak value storage of the applied signal-envelope voltage. The operation of storage or integration causes a quasi-peak detector to respond in a nonlinear fashion to the magnitude, duration, and repetition rate of impulsive noise.[4]

Variations in noise-impulse magnitude, pulse-duration and pulse-repetition rates usually prevent accurate bandwidth translations. Such is the commonly encountered problem when employing quasi-peak detectors for the measurement of man-made impulsive noise emissions. The added complication arising from the existence of quasi-peak detectors possessing differing ratios of T_c/T_d does not materially affect the interpretation of automotive-ignition-noise data, as most of the reported results and all data presented here have been obtained with instrumentation designed with T_c/T_d = 1 msec/600 msec, and not with instruments designed to the CISPR standards of either T_c/T_d = 1 msec/160 msec employed below 30 MHz, or T_c/T_d = 1 msec/550 msec for the range above 25 MHz.[11, 12]

Plotted in Fig. 2-9 are quasi-peak detected, ignition-noise data reported by Schultz et al.[3] for single and multiple stationary vehicles with engines running at 1500 rpm. A separation distance of 10 m (33 ft) existed between the nearest vehicle and a vertically polarized observing antenna. These measurements were obtained using a detector with a 500-kHz impulse bandwidth and were converted to a 1-MHz bandwidth by calibrating the receiver and detector with an impulse generator using the procedure documented in SAE Standard J551a.[13]

The bandwidth-translation differentials (Equation 2-1), shown in Fig. 2-9 by the lower pair of lines, correspond to the cases of one and 12 vehicles. Each represents a measure of the deviation of de-

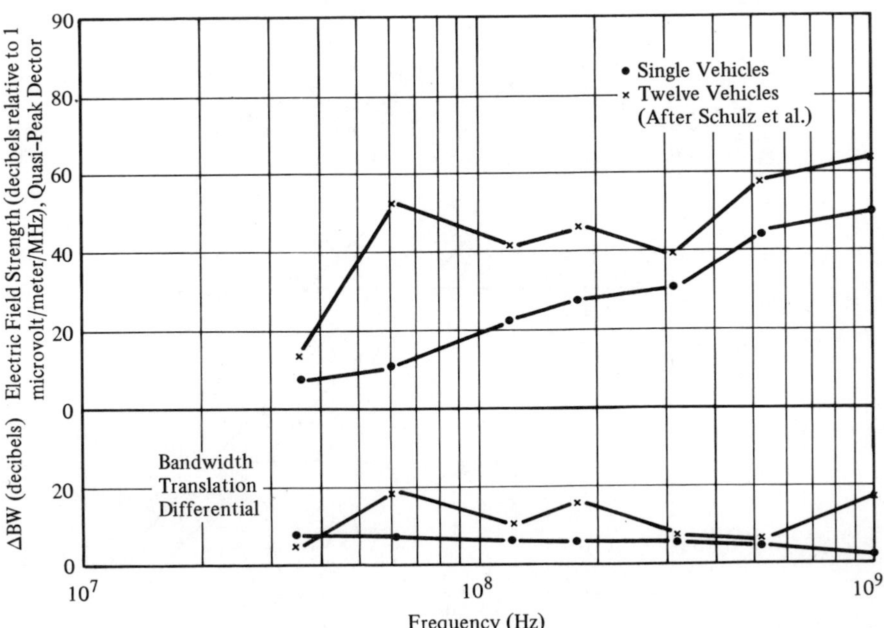

Fig. 2-9. Quasi-peak detected radiated electric-field strength from single and multiple vehicles, detector time-constant ratio T_c/T_d = 1 msec/600 msec. Variation of quasi-peak electric-field strength with detector bandwidth. Bandwidth translation differential, Equation 2-1. (After Schulz, et al., 1973)

tected electric-field strength from bandwidth proportionality. Note that at 60 MHz for 12 vehicles the value of ΔdB is 19 dB. This large deviation from bandwidth proportionality appearing in the detected-radiation field, exemplifies the effect of pulse-width extension and pulse-overlapping occuring in the narrow 50-kHz bandwidth receiver circuitry.

Horizontally polarized, quasi-peak detected radiation emitted by vehicles moving in urban traffic patterns are depcited in Fig. 2-10 for an observation frequency of 700 MHz and a traffic-lane separation

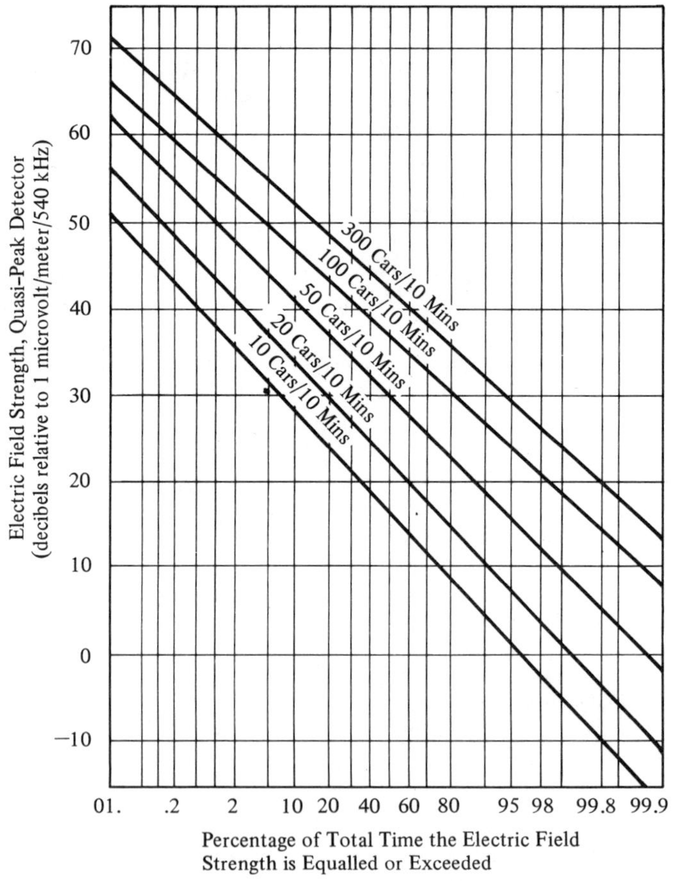

Fig. 2-10. Cumulative distribution of electric field strength as a function of automotive traffic density for an observation frequency of 700-MHz, 540 kHz impulse bandwidth and T_c/T_d = 1 msec/600 msec. (After Suzuki, 1963)

distance of 10 m (33 ft).[14] The equipment used to obtain the results possessed an impulse bandwidth of 540 kHz and a time constant ratio of T_c/T_d = 1:600. This observation frequency is of interest as it falls near the peak of the third automotive-ignition-noise resonance. These data are of particular importance because a sufficient number of measurements were performed in three Japanese metropolitan areas, with comparable traffic and urban structural features, to yield a dependence of radio-noise electric-field intensity upon traffic volume. From the results, it may be concluded that quasi-peak noise-field intensity is log-normally distributed for the range of traffic volume spanning 10 to 300 vehicles per minute. The mean value of the quasi-peak field intensity is observed to increase with traffic volume, a consequence of both detector integration and a large receiver bandwidth. Notice that, concurrently with a traffic-volume increase, a decrease in the variance of the noise-field intensity is evident. The log-normal distribution of quasi-peak traffic-noise electric-field strength shown in Fig. 2-10 may be contrasted with the cumulative distribution of peak-detected traffic noise presented in Fig. 2-8 for comparable values of frequency and traffic density.

RMS Detector Measurements. In contrast to the abundance of reported data on motor-vehicle noise emissions measured with peak and quasi-peak detectors, only a small amount of information obtained using rms detectors is available. Studies that have been published on either stationary vehicles with engines running or automotive traffic, principally have been confined to observation frequencies below 100 MHz. Unavailability of adequate UHF-band instrumentation, however, has not been a deterrent to such studies; rather, a lack of applicational need is implied.

Measurements of radio-interference fields performed using rms detectors are commonly converted to average noise power, p_n, and referenced to the terminals of an ideal lossless antenna. A noise factor may be defined for an ideal lossless antenna, F_a, as the ratio, expressed in decibels, of p_n to the product $kT_o b$ in which k = Boltzman's constant equaling 1.38×10^{-23} J/°K, T_o = the reference temperature (290°K), and b = receiver noise bandwidth. Thus,

$$f_a = \frac{p_n}{kT_o b} \qquad (2\text{-}2)$$

or

$$F_a = 10 \log \left(\frac{p_n}{kT_o b} \right), \text{ in dB.}$$

Occasionally, incidental noise power, referenced to the receiver input or equivalently at the output of the antenna-coupling circuit, is reported as p_{no} rather than p_n. When p_{no} is provided, there results a modified value of the noise factor, which, when expressed in decibels, is denoted by $\overline{F}_a$. The relationships between F_a, f_a, $\overline{F}_a$, p_n, and p_{no} are obtained as follows by referring to Fig. 2-11. Figure 2-11(a) represents the antenna and receiver coupling circuit. Ohmic losses in the antenna impedance matching circuitry and connecting transmission line attenuate the signal power available to the receiver. When both sources of circuit loss are consolidated into a single quantity, ℓ_c, defined as the ratio $\ell_c = p_n/p_{no}$, Fig. 2-11(b) is applicable. In Fig. 2-11(b), the equivalent antenna circuit is represented as existing at temperature t_c in °K.

The observed noise power p_{no} appearing at the input to the receiver is related to the available antenna noise power p_n and the noise generated by the antenna circuit losses p_{no} by

$$p_{no} = p_n/\ell_c + p_{nc}. \tag{2-3}$$

The expression p_{nc}, equal to the thermal noise power radiated into the transmission line by the dissipative loss ℓ_c, is numerically equal to the power absorbed by the equivalent circuit; that is,

$$p_{nc} = kt_c b - kt_c b/\ell_c = kt_c b(1 - 1/\ell_c) \tag{2-4}$$

with the equivalent circuit approximated by a black body radiator. Combining Equations 2-3 and 2-4 yields

$$p_{no} = f_a kT_o b/\ell_c + kt_c b(1 - 1/\ell_c). \tag{2-5}$$

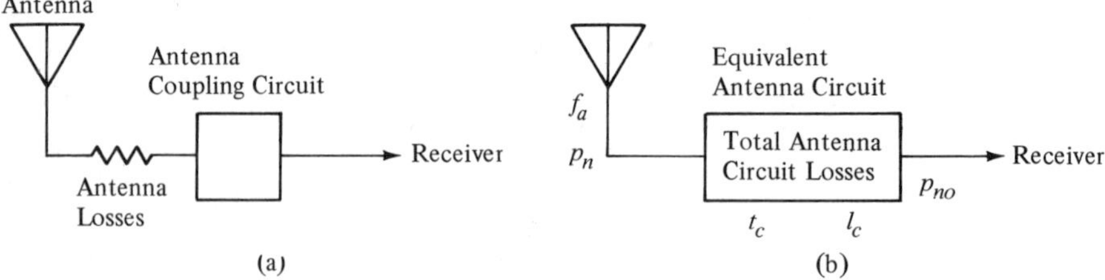

Fig. 2-11. Antenna and antenna coupling circuit for determination of F_a in terms of $\overline{F}_a$ and circuit quantities.

Setting $\overline{f_a}$ equal to

$$\overline{f_a} = p_{no}/kT_o b, \tag{2-6}$$

and using Equation 2-5 after rearranging, one obtains

$$f_a = \overline{f_a}\,\ell_c - t_c(\ell_c - 1)/T_o. \tag{2-7}$$

Individual stationary vehicle studies using rms detectors have been performed at several frequencies in the upper HF band.[15] Figure 2-12 displays the average power observed at the terminals of a vertically polarized dipole antenna positioned 10 m (33 ft) in front of one vehicle. The antenna efficiency was not determined preventing a conversion of the results to the quantity F_a. Two engine speeds of 600 and 2000 rpm were measured with instrumentation possessing a noise bandwidth of 16 kHz.

Although the frequency interval presented in Fig. 2-12 is limited, it is possible to note the onset at 29 MHz of the second automotive ignition noise resonance, which lies between 20 and 40 MHz for the majority of vehicles.

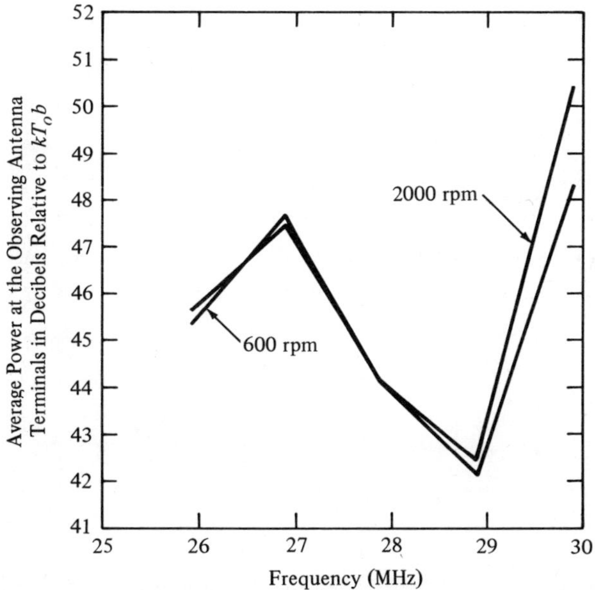

Fig. 2-12. RMS detected average radiated interference power from a single vehicle observed at engine speeds of 600 rpm and 2000 rpm. (After Shepherd et al., 1975)

Extension of rms-detector measurements of average radiated ignition noise to automotive traffic patterns yields, for the frequency range 2.5 to 48 MHz, the results of Fig. 2-13. In Fig. 2-13, the mean values of F_a as a function of traffic density and frequency are shown; the values were obtained from a series of measurements performed on 26 thoroughfares in San Antonio, Texas.[16] The recording equipment employed a vertically polarized antenna and a receiver with 10-kHz noise bandwidth. The linear dependences were derived by performing a least-squares linear regression analysis on the multiple-location data sets. Shown for each frequency is the standard error of estimate, the root variance of the data about the linear regression line.

Examination of the average power data of Fig. 2-13 does not immediately reveal evidence of the second automotive-ignition-noise resonance, which, when using peak and quasi-peak detectors, appears in the range 20 to 40 MHz. Instead of a distinctive maxima in the interval 20 to 40 MHz, average noise-power measurements for traffic interference manifest a change of slope of the function F_a versus frequency, which becomes apparent when the preceding results are cross-plotted using vehicle density as a family parameter (Fig. 2-14).

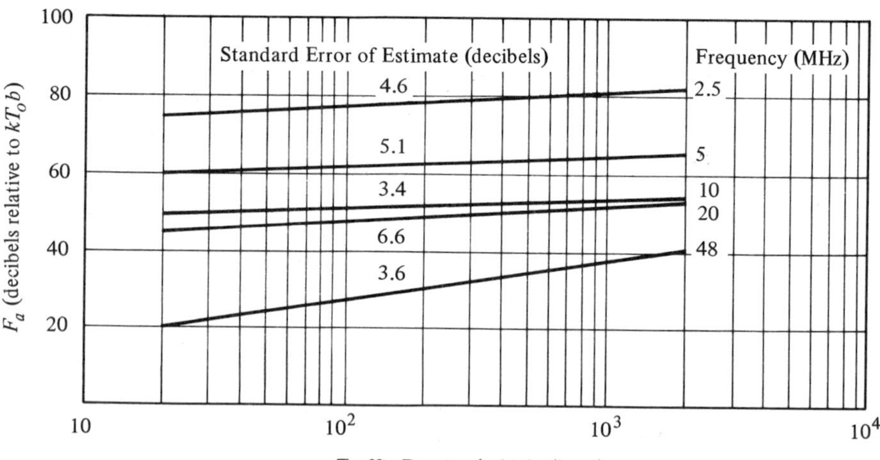

Fig. 2-13. Average automotive-traffic-noise power expressed as F_a for various traffic densities and frequencies. Detector-noise bandwidth, 10 kHz. (After Spaulding, 1972)

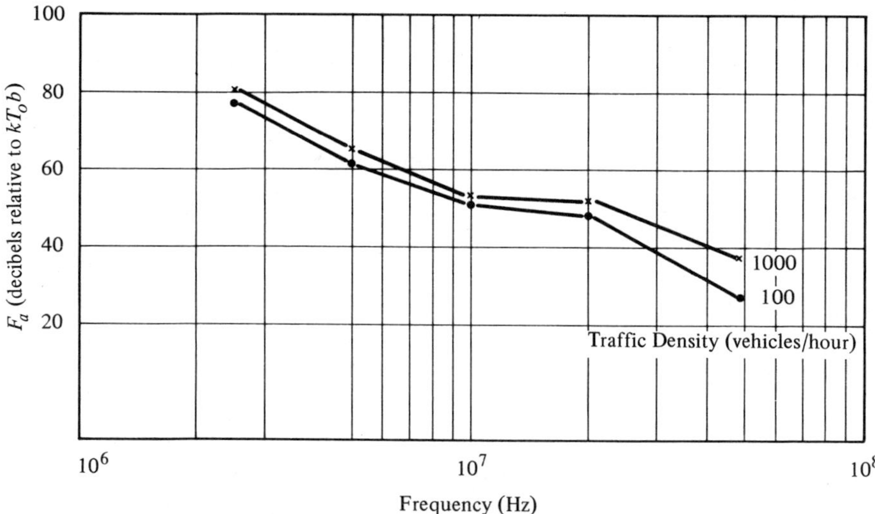

Fig. 2-14. Average automotive-traffic-noise power expressed as F_a for two traffic densities, as a function of frequency. Detector-noise bandwidth, 10 kHz.

Notice the nearly zero slope occurring in the range of 10 to 20 MHz (Fig. 2-14), followed by the reappearance of a high negative slope.

Amplitude Probability Distribution Measurements. Unintentionally generated man-made radio noise is not completely characterized by the first and second envelope-amplitude moments or by a limiting value such as the peak-envelope voltage or by a single mixed parameter (for example, the quasi-peak field strength). The incompleteness of description residing in the attempted representation of man-made unintentionally generated noise by the preceding quantities derives directly from the complicated statistical nature of this noise; namely, the distribution function of incidental noise is not simply represented analytically. Inadequacy of representation arising from the use of these quantities alone lies in their failure to provide sufficient information for determining either the behavior of radio equipment exposed to unintentionally generated noise or a means for modifying the noise sources, other than their complete elimination, and thus yielding emission fields less harmful to radio-equipment performance.

In considering which signal-distribution function or functions would be most valuable for enlarging the present representation of unintentionally generated radio noise, the first-order distributions offer the most appeal for two reasons. First, the instrumentation, the data-acquisition time and the data-reduction investments are minimized if the higher order and mixed distributions are avoided. Second, interpretation of radio-equipment interactions with impulsive noise and the diagnosis of appropriate noise-source modifications to mitigate equipment interference are initially facilitated by addressing the simplest statistical distribution, i.e., the envelope-amplitude-probability distribution. However, distribution functions of envelope-crossing rate and pulse duration at specific amplitude values, referenced to the rms voltage, present additional information, which may ultimately be recognized as essential in both equipment-susceptibility analysis and noise-source diagnostics.

The radio-noise signal $X(t)$ observed at the output terminals of a receiving-antenna circuit tuned to carrier frequency $\omega_o/2\pi$ has a temporal representation:

$$X(t) = V(t) \cos [\omega_o t + \phi(t)] \tag{2-8}$$

where both the voltage-envelope amplitude, $V(t)$, and the carrier phase, $\phi(t)$ are time-dependent random variables. The cumulative probability that the instantaneous value of the envelope voltage $V(t)$ exceeds a specified level V during a measurement period T is designated the *amplitude probability distribution (APD).* The APD for the noise-envelope voltage is written as the integral of the envelope-density function, $p(V)$:

$$APD_T = \int_{V(t)}^{\infty} p(V)\, dV. \tag{2-9}$$

The time T specified for determining the APD_T must not exceed the period of time the noise source may be considered stationary, nor too brief to assure that a sufficiently large number of high envelope values have been recorded to correctly characterize the noise distribution in the low-events portion of the distribution. When these conditions are met, the resulting APD is representative of the noise source, independent of the observation interval, and the subscript T may be conveniently omitted:

$$APD_{T \geqslant T_{\min}} \equiv APD.$$

Experimental determination of the APD for man-made unintentionally generated noise normally proceeds by quantizing the output of a band-limited detector, e.g., rms, linear or logarithmic into a series of levels. An associated timing-circuit, event triggered by the occurrence of a quantization-level passage of the envelope voltage, supplies a time-duration indication to an accumulation processor, which totals the time the envelope is superior to each quantization threshold during a measurement interval T. The predetection filter bandwidth, the minimum quantization increment, and the dynamic range of the receiver-detector-quantizer directly affect the resulting APD and the rms voltage to which it is usually referenced. The envelope excursions of unintentionally generated man-made noise often exceed 60 dB; thus to achieve a measurement accuracy of 0.5 dB, a 70-dB dynamic range is needed in the predetection circuitry. To preserve a comparable accuracy through the detector circuit requires for a linear detector a 70-dB dynamic range and a somewhat lesser range for rms and logarithmic detectors.[17]

The standard deviation of the quantization error, ϵ_q, introduced by a selected minimum threshold separation of $\hat{\ell}$ dB among any adjacent pair of N levels, is:

$$\epsilon_q = \hat{\ell}/2\sqrt{3} \text{ (in dB)},$$

which derives from the presumption the increments δV_i

$$\delta V_i = V(t) - V_i,$$

where $i = 1$ to N, for $V_i \leqslant V_{(t)} \leqslant V_{i+1}$, are uniformly distributed.

The choice of predetection-filter bandwidth is crucial in influencing the ultimate character of the APD. Because no generally applicable way presently exists for transforming an unintentionally generated man-made noise APD, determined in one bandwidth, into an APD that would be measured in another bandwidth and for concurrently estimating a measure of the distortion of form introduced by the process, the interpretation and use of an APD is restricted to the bandwidth of measurement. Some progress in determining the functional form for special cases of man-made unintentionally generated noise has recently been made and offers promise of ultimately yielding the needed bandwidth transformation relationship. This matter is discussed in Chapter 5.

The envelope-voltage level V (Equation 2-8) may be arbitrarily chosen and left unreferenced to any absolute measure without affecting the form of the APD. However, to achieve maximum utility, detected voltage is commonly referenced to kT_ob by determining the rms voltage of the signal $X(t)$, thus yielding $\bar{f}_a$ or F_a in decibels. By means of Equation 2-7 and knowledge of the equivalent antenna-circuit quantities of Fig. 2-11, F_a may be computed.

Figure 2-15 presents two examples of the APD for a stationary motor vehicle whose engine emissions were observed at 29.91 MHz.[18] The important experimental variable values were 3-kHz detector-noise bandwidth, a vertically polarized antenna placed 10 m in front of the vehicle, and engine speeds of both 600 rpm (idle) and 1500 rpm (cruise). The data samples used to construct the APDs of Fig. 2-15 were 9.5 min in duration for the 600-rpm samples and 11.6 min for the 1500 rpm results. The signal-envelope magnitudes have been referenced to kT_ob by means of separately measured values of the average signal power, which were found to be $\overline{F}_a$ = 45.4 dB (idle) and 48.3 dB (cruise). Either APD may be referenced to RMS envelope voltage by subtracting 45.4 or 48.3 dB from the respective ordinates. For example, the plotted point of greatest amplitude on the engine-idle APD represents a signal-envelope voltage approximately 24 dB above V_{rms} of the signal $X(t)$.

Several features of these APD are of general as well as specific importance and justify explanation. The sharp increase in negative slope beginning at abscissa values of between 5 and 50 percent is a dramatic manifestation of the impulsive character of ignition-system noise. The area between the APD and a leftward extension of the low amplitude, low-slope portion of each APD is an indication of the many more high-amplitude pulses present in the waveform $X(t)$ of ignition noise, as contrasted with thermal noise. A thermal-noise envelope-amplitude distribution is plotted in the lower left corner of each part of Fig. 2-15. The slope of the thermal-noise APD is approximately equal to the slope of the low-amplitude, low-slope portion of the ignition-noise APD. The inference then is that automotive ignition noise is a mixed statistical process consisting of at least two distributions: (1) a thermal-noise distribution, which contributes a large population of low- and medium-amplitude envelope peaks and (2) an impulsive distribution, which produces the relative numerous high-amplitude pulses.

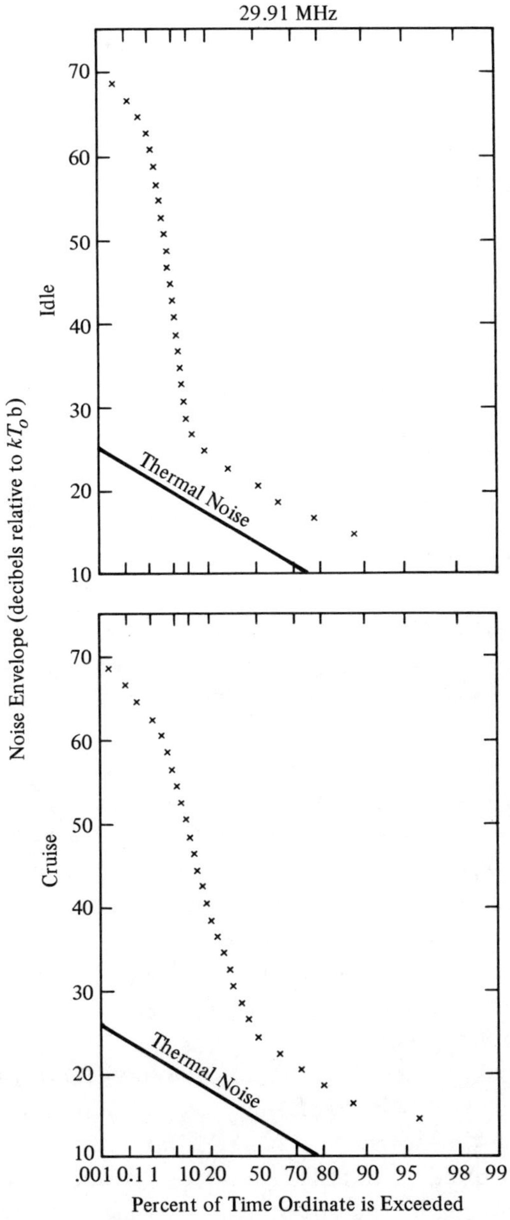

Fig. 2-15. APDs for a single stationary vehicle, 19.91 MHz; engine speeds of 600 rpm and 1500 rpm. $\overline{F_a}$ = 45.4 dB (Idle) and 48.3 dB (cruise). (After Shepherd et al., 1975)

Pulse amplitudes are bounded, as must be the case in any real process. This point is evidenced in Fig. 2-15 by the decrease in APD slope shown on the left.

The coordinate grid used to present Fig. 2-15 and subsequent figures employs a highly folded abscissa. The equation for a straight line in the coordinate system used is

$$y = a - \frac{k}{2} x \qquad (2\text{-}10)$$

in which

$$y = \log_{10} V$$
$$x = -\tfrac{1}{2} \log_{10} |\log P(V > V_o)|$$
$$a = -\tfrac{1}{2} \log_{10} \left(\frac{\log e}{2\sigma^2}\right)$$
$$k = \text{constant.}$$

In the preceding, constant k determines the ordinate scale expansion, and σ^2 is the variance of a circular or Rayleigh distribution whose density function for the envelope voltage, V, is:

$$p(V) = \left(\frac{V}{\sigma^2}\right) \exp - \frac{V^2}{2\sigma^2} \qquad (2\text{-}11)$$

where the time dependence in $V(t)$ has been suppressed. The cumulative distribution of V is denoted by $P(V > V_o)$. Equation 2-11 is the representation of the band-limited envelope of a noise voltage comprised of normally distributed quadrature components[19] and thus is characteristic of thermal-noise emissions. A thermal-noise APD plots as a straight line in the coordinate system of Equation 2-10 with a slope of $-\frac{1}{2}$, if the ordinate-scale expansion constant k is chosen equal to 1. The resulting coordinate grid is commonly termed a *Rayleigh probability presentation*.

The form and structure of automotive-ignition-noise APD are influenced by the bandwidth of the predetection filter. As the bandwidth narrows, the form of the APD approaches that of thermal noise as a consequence of pulse amplitude rounding and pulse-width extension occurring in the filter. This process is traced in Fig. 2-16 for the same single stationary vehicle, used in Fig. 2-15, with its engine operating at 1500 rpm. Three observation bandwidths of 0.5, 3, and 6 KHz were used with a sampling period of 11.6 min to obtain the

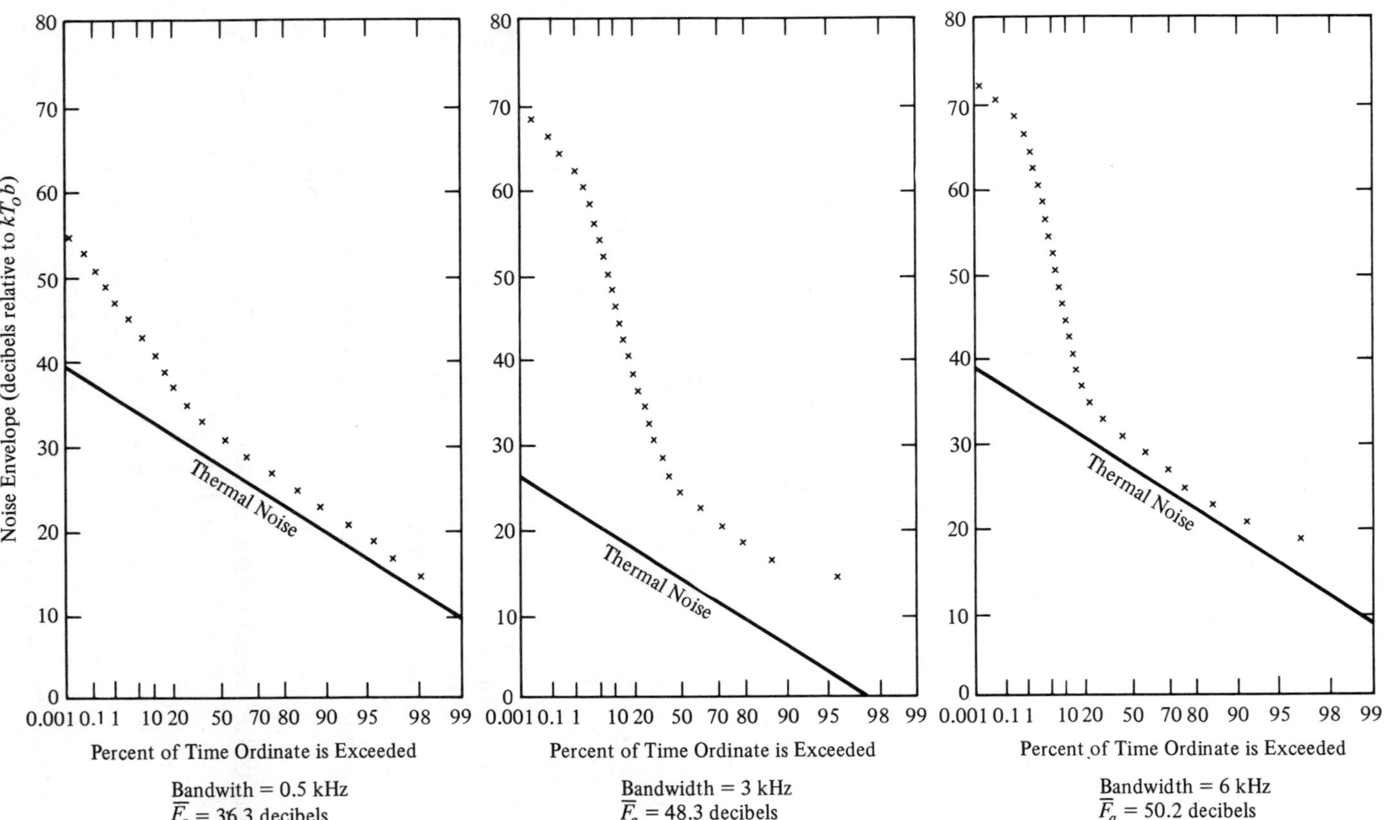

Fig. 2-16. APDs for a single stationary vehicle, 29.91 MHz, engine speed of 1500 rpm. Three detector-noise bandwidths: 0.5, 3, and 6 kHz. (After Shepherd et al., 1975)

vertically polarized ignition noise at a distance of 10 m in front of the vehicle.[18]

Examination of the three APDs obtained at 29.91 MHz reveals the trend toward a thermal-noise distribution as the detection-bandwidth decreases. Note the progressive disappearance in the number and magnitude of the high-amplitude peaks. Notice also the progressive decrease in $\overline{F}_a$ that accompanies a decrease in filter bandwidth. The vehicle used for Figs. 2-15 and 2-16 was studied again in Fig. 2-17, where the effect upon the APDs of changing the observation frequency from 24.11 to 29.91 MHz is presented. Two sampling periods, lasting 22.4 and 11.6 min, corresponding to the lower and higher frequencies, respectively, were recorded. A receiver bandwidth of 6 kHz applies to both observation frequencies as does a separation distance of 10 m between the vertically polarized antenna and the front of the vehicle. The average antenna power is seen to be higher at the lower frequency by 1.6 dB. The gross shape of the APDs is the same.

The general characteristics observed in the form of the APDs for stationary-vehicle ignition noise are present in the APDs determined from measurements of multiple moving vehicles in an urban traffic environment. The impulsive distribution of high-amplitude signals overlying the lower, approximately thermally distributed background is again found. In the HF band at 24.11 MHz, a pair of APD samples for heavy traffic are shown in Fig. 2-18 for two separation distances, 132 and 52 ft, between a vertically polarized antenna and the nearest lane of the freeway. Traffic density varied between 37 and 42 vehicles per minute over the 10.5-min sampling period. The average ignition-system noise power, $\overline{F}_a$, in a 6-kHz detection bandwidth for the two traffic separation distances decreased with range from 33.4 to 28.6 dB at the greater separation but not as rapidly as the inverse square of the spacing.

When traffic density changes, as shown in Fig. 2-19 for an observation frequency of 29.71 MHz, the average observed ignition-emission level increases. For the case of light traffic where an average traffic density of 19 vehicles/min existed for 9.5 min, $\overline{F}_a$ was measured as 28.8 dB at an antenna-to-freeway spacing of 52 ft. $\overline{F}_a$ increased to 33.4 dB as the average traffic density rose to 46 vehicles/min during a 10.5-min observation interval. Each measurement was made with a

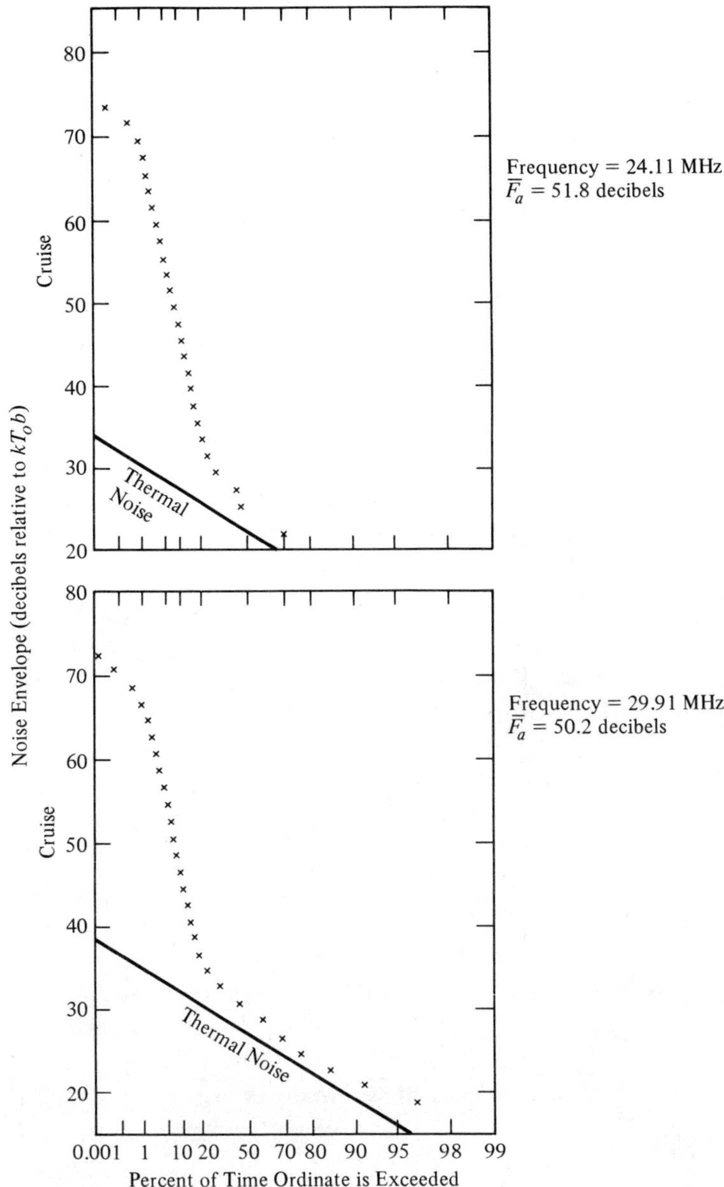

Fig. 2-17. APDs for a single stationary vehicle at two observation frequencies, 24.11 and 29.91 MHz. Engine speed, 1500 rpm; detector-noise bandwidth, 6 kHz. (After Shepherd et al., 1975)

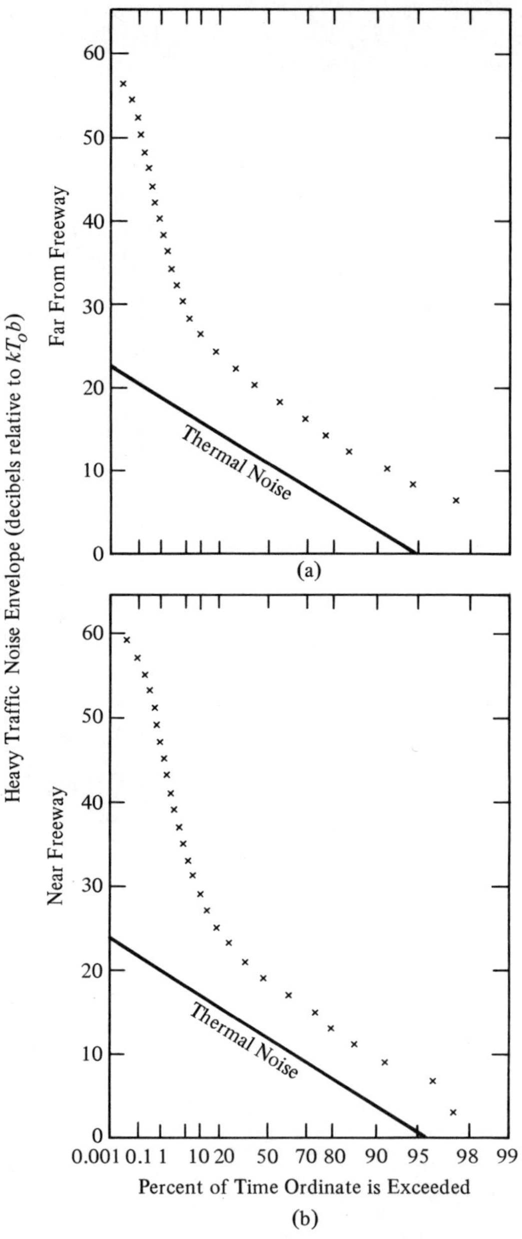

Fig. 2-18. APDs for automotive traffic noise at 24.11 MHz. Vertically polarized antennas; detected noise bandwidth of 6 kHz. Antenna-to-freeway separation distance, 53 ft and 132 ft. Traffic densities: (*a*) 37 vehicles/min, (*b*) 42 vehicles/min. Average power: (*a*) $\overline{F_a}$ = 28.6 dB, (*b*) $\overline{F_a}$ = 33.4 dB. (After Shepherd et al., 1975)

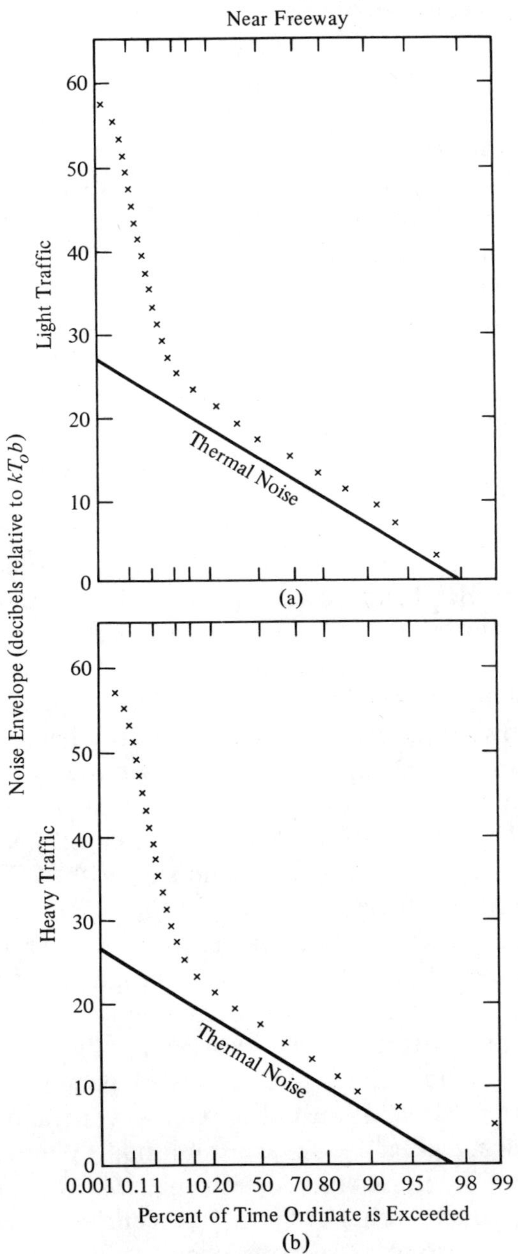

Fig. 2-19. APDs for automotive traffic noise at 29.71 MHz, vertically polarized antenna, detector-noise bandwidth of 6 KHz. Antenna-to-freeway separation distance 52 ft. Traffic density: (*a*) 19 vehicles/min, (*b*) 46 vehicles/min. Average power: (*a*) $\overline{F_a}$ = 26.5 dB, (*b*) $\overline{F_a}$ = 29.3 dB. (After Shepherd et al., 1975)

vertically polarized dipole antenna and a detector-noise bandwidth of 6 kHz. The observed difference in $\overline{F}_a$ is 4.6 dB, which approximately equals the ratio of traffic densities when expressed in decibels, i.e., 3.9 dB. Provided that pulse overlap is very small, the average observed power variation should be proportional to any change in the number of received ignition impulses per unit time, which for identically distributed automotive emissions, is proportional to traffic density.

Figures 2-18(b) and 2-19(b) may be compared for frequency dependence in the HF band. Negligible change in the form of the APDs is evident. Only in the average noise power at the antenna terminals does a frequency dependence effect become evident; i.e., $\overline{F}_a$ decreases from 33.4 dB at 24.11 MHz to 29.3 dB at 29.71 MHz, the same trend as noted for stationary-vehicle emission-noise power.

The APDs of Figs. 2-15 through 2-19 were obtained in a narrow interval of the HF band. At frequencies below those shown, the possibility of contamination of ignition-noise emissions by atmospheric noise exists. As a consequence, few attempts have been made to obtain isolated automotive-ignition-noise data in the lower HF band and below. It is possible to examine the behavior of automotive-ignition-noise APDs at frequencies in and above the VHF band where ionospheric propagation of atmospheric noise is unlikely or nonexistent. The impulsive features of the emissions from automotive-traffic-ignition systems seen previously in the HF band are preserved at higher frequencies. Referring to Fig. 2-20 where the ordinate is referenced to the RMS voltage of the traffic-noise sample expressed in dB, it is evident that the impulsive component of the mixed distribution is superimposed upon a thermal-noise background. A solid line has been drawn through the experimentally observed high- and medium-envelope amplitudes and extended parallel to the thermal-noise distribution (dashed line). The relative vertical displacement of the APDs shown in Fig. 2-20 was established by equating the RMS voltage of the traffic-noise emissions to the RMS voltage of a thermal-noise sample, which occurs at the 37 percentile. A measured value, which may be spurious, is seen to occur at the 98 percentile. The data presented in Fig. 2-20 were obtained during a 5-min period at a well-traveled intersection in Boulder, Colorado, using a vertically polarized, short monopole antenna and a receiver with a 4-kHz noise

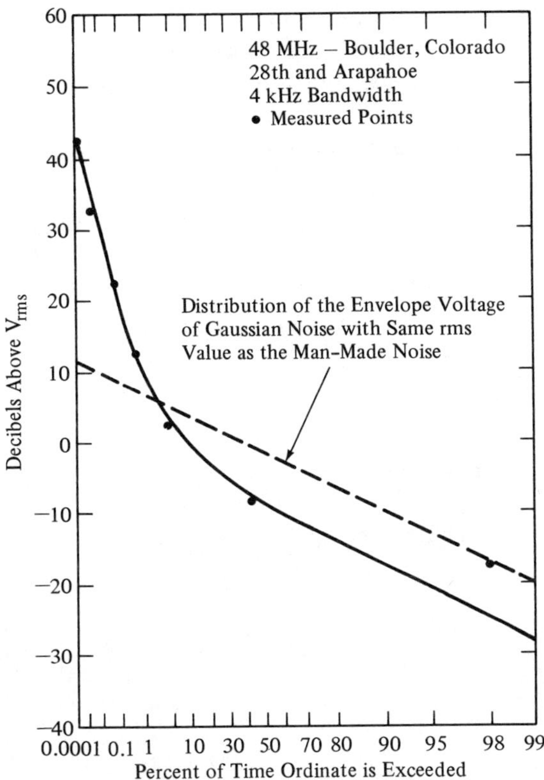

Fig. 2-20. APD of automotive-traffic ignition noise at 48 MHz. Detector-noise bandwidth 4 kHz. $\overline{F_a}$ = 43.5 dB. (After Spaulding, 1974)

bandwidth.[20] The antenna gain was determined separately, yielding a value for F_a = 43.5 dB. The 4-kHz noise bandwidth used for Fig. 2-20 is comparable to the bandwidths used for the HF-band measurements. Comparison of Figs. 2-19 and 2-20 reveal that the data for 48 MHz do not display the reduced slope in the impulsive, high-amplitude portion of the APD evident at 24 to 29 MHz. It is not certain that this difference in slope represents a fundamental change in the character of automotive traffic noise. The instrumentation employed to obtain the APD of Fig. 2-20 possessed a total dynamic range of 85 dB, 50 dB above the 35 dB below the RMS voltage. For the HF-band data, the linear dynamic range of the instrumentation used varied with receiver bandwidth. The receiver with a 6-kHz bandwidth possessed a linear range of 60 dB for the 0.5-kHz bandwidth receiver; the linear range was 40 dB. Consequently, some peak-envelope distortion is present in the narrow bandwidth data of Fig. 2-19.

Figure 2-21 provides a mid–VHF-band sample of automotive traffic noise, likewise using a 4-kHz detecting-system noise bandwidth. The street intersection differed from that represented in Fig. 2-20, but the instrumentation was identical.[20] Contrasting Fig. 2-21 with the preceding traffic-emission APDs, it is apparent that a smaller percentage of high-amplitude impulses occurs in the 102-MHz population, a trend that cannot be presumed to persist into the higher frequencies.

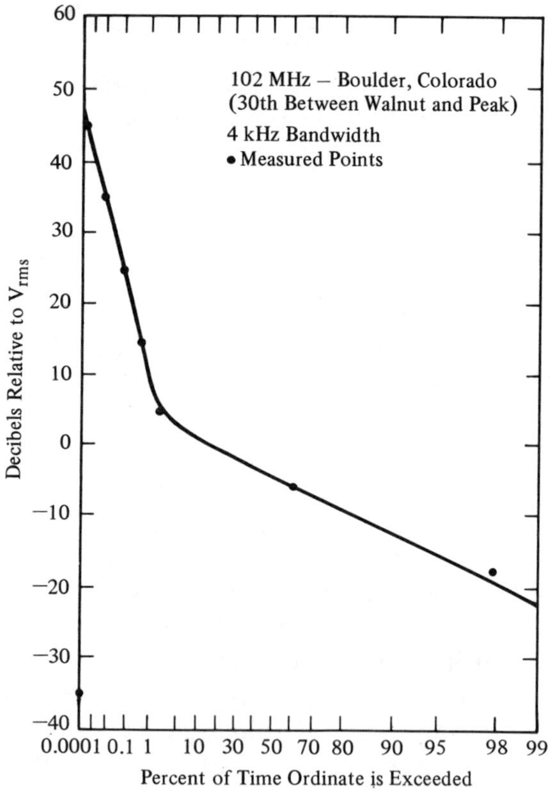

Fig. 2-21. APD of automotive traffic ignition noise at 102 MHz. Detector-noise bandwidth 4 kHz. (After Spaulding, 1974)

Noise-Amplitude-Distribution Measurements. One of the simpler statistical distributions that may be used to characterize unintentionally generated man-made radio noise consists of counting for a sample period the number of envelope-amplitude-exceedances of a set of fixed, ordered threshold levels. Each level exceedance is preceded by a positive crossing of at least one fixed threshold by the detected voltage, and therefore the distribution may alternately be designated as a crossing-rate distribution. Detector bandwidth is important in shaping the distribution; progressively narrower detector bandwidths produce peak-voltage compression, pulse-overlapping and pulse-width enlargement, materially depressing the impulsive portion of the noise-amplitude distribution (NAD). Figure 2-22 provides a representation of three NADs for automotive traffic noise measured by a mobile test laboratory traveling a closed course in the business center of Washington, D.C.[21] The noise bandwidth of the receiver preceding the level setting, envelope detecting, and counting

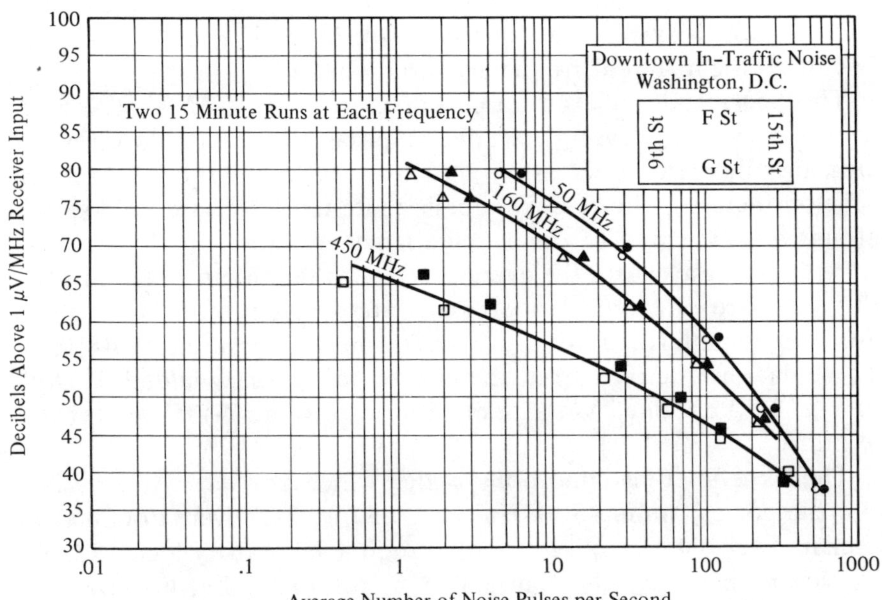

Fig. 2-22. NAD measurements in central Washington, D.C. Receiving antennas vertically polarized. Envelope voltage referenced to antenna terminals. Two 15-min data sets taken at each frequency by mobile laboratory. Bandwidth, 14-kHz.

assembly was 14 kHz. Pulse amplitudes varying in magnitude by 70 dB were countable by adjusting the NAD-level detectors to separately sample the high and low portions of the distribution. The instantaneous dynamic range encomprised 40 dB. An impulse generator was used to calibrate the NAD receiver, yielding a measure of the receiving antenna output voltage. For these data, vertically polarized whip antennas were used with gains of approximately 2 dB above isotropic.

The traffic-noise-envelope amplitude at each average occurrence rate varies inversely with observation frequency. However, the variation of peak-envelope voltage for traffic noise in the range of 50 to 450 MHz presented in Fig. 2-7 displays a positive slope with respect to frequency change; it is most pronounced at the very low amplitudes. Insufficient data are available in Fig. 2-22 at both limiting values of average pulse rate to confirm the trend with frequency for comparison with the peak amplitude distribution of Fig. 2-7. It is possible that at both limits of average pulse rate, the frequency spread may narrow and the curves may transpose. Also, the substantial difference in the instrumentation-detection bandwidth—namely, 300 kHz for Fig. 2-7 and 14 kHz for Fig. 2-22—may contribute to the different frequency variations.

The magnitude of the three curves of Fig. 2-22 progressively decreases with increasing negative slope as the countable threshold crossings increase. Decreasing the lowest threshold of a counting echelon increases the average crossing rate until the threshold setting attains a value equal to the average signal envelope, at which point a further decrease causes the average count to fall. Figure 2-23 displays this sequence of events in a crossing-rate distribution (CRD) for the 5-minute sample of automotive traffic noise measured at 48 MHz. These data are derived from the measurements that yielded the APD of Fig. 2-20 obtained with a detector possessing a 4-kHz noise bandwidth.[22]

The experimental conditions under which the data of Fig. 2-23 and the corresponding 50-MHz curve of Fig. 2-22 were obtained are similar. Each set was observed by a mobile laboratory located in the trafficway of a heavily traveled area; for Fig. 2-23, however, the mobile unit was parked. Vertically polarized antennas were used in each case, and the detectors' bandwidths differed by a factor of less than 4. Data in Fig. 2-23 are normalized to the noise-envelope rms

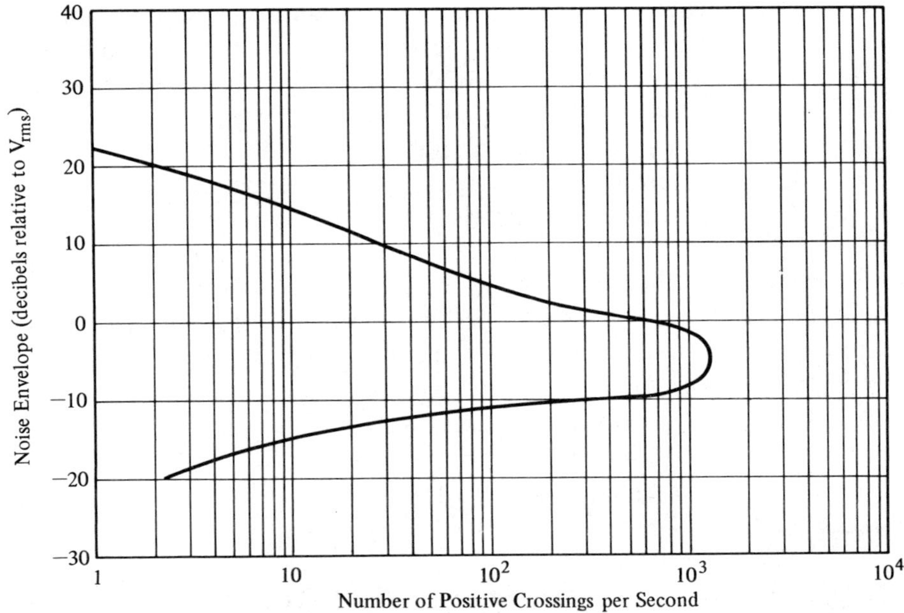

Fig. 2-23. Envelope crossing-rate distribution for automotive traffic noise. Detector-noise bandwidth, 4 kHz. Vertically polarized antenna. Observation frequency. Average noise power F_a = 43.5 dB. (After Spaulding, 1974)

voltage, whereas the data of Fig. 2-22 are referenced to the output of an impulse generator.

Both curves are similar for threshold levels above the onset of crossing-rate saturation, which in Fig. 2-23 appears at −5 dB. However, the inflection point occurring in Fig. 2-23 between 0 and 5 dB is not replicated in Fig. 2-22.

Theory of Automotive Ignition Noise

A description of the radio-noise emission arising from both automotive traffic and stationary gasoline-powered vehicles requires an analytical representation and interpretation based upon the inherent stochastic characteristics of the underlying generation processes; in the case of automotive traffic, the vehicle speed and location with respect to the point of observation are required. The same stochastic foundations are needed to formulate the theoretical models of the primary observables, which for automotive-ignition-system emissions are the APD and the average observed power at the terminals of a reference antenna. Both measurable quantities, the noise envelope APD and the average noise power, depend upon the stochastic-variables ignition-system pulse-height and interpulse period. For situations including moving vehicles, the vehicle separation distance from the observation point and the vehicle speed must be added. The observed APD for both traffic-noise emissions and stationary vehicles may be treated theoretically by examining the mentioned independent random variables and their interactions. Traffic-ignition emissions will be found to represent the more general case, which, by restricting the allowable range for the independent variable, is specialized to the results for stationary-vehicle radiation. Average observed power from either stationary or moving vehicles is analytically representable in terms of the same independent random variables.

Amplitude-Probability Distribution (APD). The amplitude probability distribution of the ignition-noise-envelope represents a mixed random process consisting of two populations. One of the populations distinguishable by its low- and medium-amplitude signal components is distributed in a manner that is approximately representable by Gaussian distributions of its quadrature components or equivalently by a Rayleigh distribution of its envelope amplitude, as observed at the output of the detector filter. The second contributing population, noticeable because of the many large amplitude components that are included, obeys a distribution law very different from that which governs the lower amplitude portion of the mixed distribution. Evidence of each population is seen in Figs. 2-15 through 2-21. The APD of ignition noise, the composite of two distinct populations, may be treated by superposition of the APD descriptive of the Rayleigh distributed population, also known as the *thermally distributed component*, and of the APD applicable to the higher-amplitude components of the ignition-radiation pattern. The higher-amplitude population of the total ignition emission contains many envelope maxima, which are narrow in time duration and appear as impulses under most conditions of observation. From this feature of the emission waveform, the high-amplitude segment of the ignition-radiation distribution obtains the designation "impulsive component." Determination of the APD for the impulsive component of ignition-system radiated noise is the primary problem to be addressed. By examining the phenomenon of automotive traffic noise and obtaining a description of the impulsive portion of the resulting APD in terms of the primary independent random variables, ignition pulse amplitude, sampling interval, and vehicle position relative to the observer, one may by specialization of the results achieve a representation of the APD for stationary-vehicle ignition emissions. The specialization step consists of constraining the range of variation of the vehicle lateral displacement variable, R, of Fig. 2-24.

In Fig. 2-24, an observation point and receiver are located at a distance b measured from the center of vehicle path.[23] The displacement of any vehicle, A, normal to the distance vector of closest approach is indicated by the variable r. The lateral displacement variable r may assume values within the range $-R \leqslant r \leqslant R$. The limit $|R|$ is constrained to be finite in virtue of the inability of any receiver to separately resolve an incoherent source signal–in this case, the vehicle ignition emissions from the total ambient noise. The value of $|R|$ is established in each application by the receiver characteristics, the value of road displacement b, and the transmission loss of the propagation path, which is also frequency-dependent. While b and R are deterministic quantities, r is not; rather it depends upon the

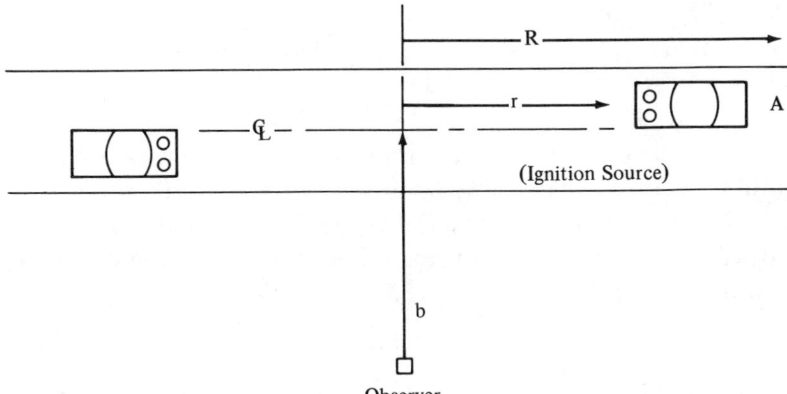

Fig. 2-24. The relative geometry of a stationary observer and vehicle traffic on a roadway.

random lateral displacement of some one vehicle at any time t when an ignition pulse is received by the observer. On multilane roadways, r may assume a value between zero and R for any vehicle in the pattern of traffic flow. The total number of vehicles permitted to reside at identical values of r is limited by the number of road lanes available. All values of r within the range $-R \leqslant r \leqslant R$ are equally likely at any instant during the sampling interval. Therefore, it may be assumed that the random variable r is uniformly distributed over the range $2|R|$ with a density $p(r)$ given by;

$$\begin{aligned} p(r) &= 1/(2|R|); \qquad 0 \leqslant |r| \leqslant |R| \\ &= 0; \qquad |R| < |r| \end{aligned} \tag{2-12}$$

The reciprocal of the separation between the observer and vehicle A is designated s and equals:

$$s = (b^2 + r^2)^{-1/2} \tag{2-13}$$

The probability density $p(s)$ the reciprocal separation distance s has assumed a nonzero value within the interval

$$(b^2 + r^2)^{-1/2} \leqslant s \leqslant b^{-1} \tag{2-14}$$

is obtained from the definition

$$\begin{aligned} p(s) &= \frac{d}{ds}\left[2\int_0^{f(s)} p(r)\,dr\right] \\ &= |R|^{-1} s^{-3} (s^{-2} + b^2)^{1/2} \\ &= 0 \text{ for all other } s. \end{aligned} \tag{2-15}$$

Emitted by each vehicle is an impulsive, radiated waveform with randomly occurring variations in the impulse strength of each envelope peak. The emitted electric field magnitude E may then be represented as the product of a unitless random variable m and a constant peak source strength E'_ℓ

$$E = m\, E'_l \tag{2-16}$$

The source field at any instant in time may assume any value between ϵ and E_k. The range of random variable m is concurrently constrained to

$$\epsilon \leqslant m \leqslant 1$$

As there exists no synchronization between the emission of a pulse pattern from the ignition system of the Kth vehicle and a sampling instant of the observed signal at distance S, all values of m within the indicated range are equally likely, yielding for m a uniform density function $p(m)$

$$p(m) = 1/(1 - \epsilon). \tag{2-17}$$

Among all vehicles that will pass the point of observation during a data-collection period, one will produce a maximum value of E'_ℓ, which may be called E_ℓ where

$$E_\ell \geqslant E'_\ell.$$

Thus, for any vehicle of the set contributing data from which an APD is constructed, the upper limit of $m = M_2 \leqslant 1$ results and permits Equation 2-17 to be generalized as

$$\begin{aligned} p(m) &= 1/(M_2 - \epsilon) \qquad \text{for} \quad \epsilon \leqslant m \leqslant M_2 \\ &= 0 \qquad\qquad\qquad \text{for} \quad M_2 < m < \epsilon. \end{aligned} \tag{2-18}$$

At the observer, the electric-field strength E_o arising from the ignition-system radiation of the Kth vehicle is representable as

$$E_o = m\, E_\ell\, K(b^2 + r^2)^{-1/2} \tag{2-19}$$

Contained in the constant K are factors, for example, the receiving antenna gain and the parameters that remain fixed during a single data-accumulation period. It is presumed in representing the observed field strength as proportional to the inverse vehicle-to-observer separation distance that the contribution of the surface reflected signal to the observed radiated field is much less than the direct-path signal.

The receiver used for detection of vehicle-ignition-system radiation operates upon the waveform appearing at the terminals of the observing antenna. Predetection linear processing of this signal introduces a transfer function $H(w)$, which, for a simple circuit embodiment using a single-pole filter, possesses an impulse response

$$h(t) = a\ell^{-\alpha t}, \qquad t \geqslant 0 \tag{2-20}$$

in which α is approximately equal to the filter bandwidth.

By absorbing the response coefficient, a, of Equation 2-20 into the K, which appears in Equation 2-19, there results for the voltage ρ at the detector output

$$\rho = K\, E_\ell\, m(b^2 + r^2)^{-1/2}\, e^{-\alpha t}. \tag{2-21}$$

The variable t appearing in Equation 2-21 refers to the instant of sampling. For an ignition-noise waveform, the range of t lies between zero, which occurs when one ignition peak appears at the detector output, and a later time τ_{av}, which is the average interpulse spacing for multiple vehicles in roadway traffic. Sampling occurs independently of the noise-generation process, making any value of t equally likely. The distribution function of t is thus uniform, and the associated density $p(t)$ is given by:

$$\begin{aligned} p(t) &= 1/\tau_{av}; \qquad && 0 \leqslant t \leqslant \tau_{av} \\ &= 0; && \tau_{av} < t \text{ and } t < 0. \end{aligned} \tag{2-22}$$

It is necessary to determine the probability density function of the detected voltage ρ and ultimately, by means of Equation 2-9, the APD. To facilitate the necessary computations, a change of variable is introduced.

$$\begin{aligned} j &= mz = msu \\ u &= e^{-\alpha t}. \end{aligned} \tag{2-23}$$

The probability density function, $p(z)$, for quantity z, which is the product of s and u, is determined from the general relationship:

$$p(z) = \int_{\xi_\ell}^{\xi_u} |u|^{-1}\, p(z/u)\, p(u)\, du. \tag{2-24}$$

The range of integration is bounded by finite limits arising from the intervals of nonzero probability occurring for the transverse position measure r, Equation 2-12, and the sampling-time variable t, Equation 2-22. The upper ξ_u and lower ξ_l limits on Equation 2-24, when expressed as functions of z, become:

$$\xi_u = z(b^2 + R^2)^{1/2} \quad \text{for} \quad z(b^2 + R^2)^{1/2} \geqslant 1$$
$$= 1 \text{ otherwise;}$$
$$\xi_\ell = zb \quad \text{for} \quad zb \geqslant e^{-\alpha\tau_{av}}$$
$$= e^{-\alpha\tau_{av}} \text{ otherwise;}$$

which yield for $p(z)$

$$p(z) = (\alpha R\tau_{av})^{-1}\, z^{-2}[(\xi_u^2 - b^2z^2)^{1/2} - (\xi_\ell^2 - b^2z^2)^{1/2}] \tag{2-25}$$

within the range

$$e^{-\alpha\tau_{av}} \cdot (b^2 + R^2)^{-1/2} \leqslant z \leqslant b^{-1},$$

while elsewhere $p(z) \equiv 0$.

The probability density function $p(j)$ for the dependent variable j, related to m, s, and u, as shown in Equation 2-23 may be obtained in a like fashion using the general expression of Equation 2-24 with the following changes made in the variables appearing in the integrand:

$$p(j) = \int_{\zeta_\ell}^{\zeta_u} |z|^{-1}\, p(j/z)\, p(z)\, dz. \tag{2-26}$$

The finite limits of integration are established by the interval within which both $p(m)$ and $p(z)$ are > 0 and when expressed as functions of j they become:

$$\zeta_u = j/\epsilon \text{ for } (j/\epsilon) \leqslant b^{-1}$$
$$= b^{-1} \text{ otherwise.}$$
$$\zeta_\ell = j/M_2 \text{ for } (j/M_2) \geqslant e^{-\alpha\tau_{av}}\,(b^2 + R^2)^{-1/2}$$
$$= e^{-\alpha\tau_{av}}\,(b^2 + R^2)^{-1/2}.$$

The density function $p(j)$ is now defined only over a positive range of the variable z, which permits removing the absolute constraint on the variable in the integrand. Substitution of Equations 2-18 and 2-25 into the expression for $p(j)$ permits rewriting in the following form, which is convenient for evaluation:

$$p(j) = (M_2 - \epsilon)^{-1} (\alpha R \tau_{av})^{-1} \int_{\xi_\ell}^{\xi_u} z^{-3} \cdot [(\xi_u^2 - b^2 z^2)^{1/2} - (\xi_\ell^2 - b^2 z^2)^{1/2}]\, dz \quad (2\text{-}27)$$

within the range

$$\epsilon e^{-\alpha \tau_{av}} (b^2 + R^2)^{-1/2} \leqslant j \leqslant M_2 b^{-1},$$

while $p(j) = 0$ elsewhere.

The APD for the detected output voltage is now obtained using Equations 2-9 with $V(t)$ replaced by $\rho(t)$ and Equation 2-27.

$$APD(\rho) = (M_2 - \epsilon)^{-1} (\alpha R \tau_{av})^{-1} \int_{V}^{M_2 b^{-1}} dj \int_{\xi_\ell}^{\xi_u} z^{-3} \cdot [(\xi_u^2 - b^2 z^2)^{1/2} - (\xi_\ell^2 - b^2 z^2)^{1/2}]\, dz$$

with

$$V = \rho \cdot \rho_{rms}/(KE_\ell)$$

and for

$$\epsilon e^{-\alpha \tau_{av}} (b^2 + R^2)^{-1} \leqq V \leqq M_2 b^{-1}. \quad (2\text{-}28)$$

The rms value of the detected voltage, ρ_{rms}, follows immediately using the basic definition and the density function $p(j)$ for ρ from Equation 2-27

$$\rho_{rms} = KE_\ell \left\{ (M_2 - \epsilon)^{-1} (\alpha R \tau_{av})^{-1} \int_{L_1}^{L_2} j^2 dj \int_{\xi_\ell}^{\xi_u} z^{-3} \cdot [(\xi_u^2 - b^2 z^2)^{1/2} - (\xi_\ell^2 - b^2 z^2)^{1/2}]\, dz \right\}^{1/2}$$

where $L_2 = M_2 b^{-1}$

$$L_1 = \epsilon e^{-\alpha \tau_{av}} (b^2 + R^2)^{-1/2}, \quad (2\text{-}29)$$

in which the limits for the inner integral are given following Equation 2-26.

The stationary-vehicle APD may be obtained from Equations 2-28 and 2-29 by setting $R << b$, which has the effect of removing the dependence of the results upon the transverse displacement, r, of the ignition noise source.

To exemplify the degree to which Equation 2-28 describes the APD of automotive ignition noise, four comparisons have been chosen from a large body of information.[24] The experimental tests upon which these comparisons are based include radiated-noise measurements of single, stationary, idling vehicles and of automotive traffic, performed at frequencies in the HF band between 24 and 30 MHz. Vertically polarized antennas were used to provide signals to a receiver, which incorporated an envelope detector whose output was processed to obtain accumulated time above each threshold of a counting echelon. Three receivers of noise bandwidths–0.5, 3, and 6 kHz–were used.[18] In order to evaluate the APD(ρ) given by Equation 2-28, the additional quantities τ_{av}, ϵ, M_2, b, and R are needed. The average interpulse spacing, τ_{av}, for stationary vehicles is obtainable from a knowledge of the engine revolution rate, the number of cylinders, and the number of vehicles, N. For the dynamic case of automotive traffic, the engine-revolution rate and number of cylinders plus a vehicle-density assessment are required to obtain τ_{av}. The density of contributing vehicles is measured by the average number of automobiles, $\overline{N}$, present at any time between transverse displacement limits $\pm R$ during the period of observation; i.e.,

$$\tau_{av} = \left(\overline{N}\,\frac{\eta\kappa}{120}\right)^{-1} \tag{2-30}$$

where η = the average number of engine cylinders

κ = the average number of engine revolutions per minute.

Equation 2-30 is applicable to either stationary vehicles or automotive traffic, subject only to an alteration in the meaning of $\overline{N}$.

The limits of variation for the impulse amplitude, m, may be assessed from a knowledge of the radiate noise waveform. Extremes of the ratio M_2/ϵ have been observed to exceed 30:1 (i.e., 30 dB);[25] rarely does the ratio approach unity. Typically, M_2/ϵ = 10 for stationary automobiles whose ignition systems are in average operating condition and whose noise-suppression accessories are functioning properly.

Whereas for stationary vehicles it is sufficient in evaluating Equation 2-28 for the APD(ρ) to take $R << b$, to determine R for vehicular traffic moving past an observer necessitates the following considerations. If traffic flow is observed from an unobstructed location on one side of a straight section of a roadway, as shown in Fig. 2-24, the maximum value of the transverse distance r designated by R, is a function of the radio-path propagation loss, the noise level of the receiver, the ambient noise, and average vehicle speed. R may be calculated if the radiated peak-noise field strength of a representative vehicle is known for a range of engine revolution rates κ. R may also be determined experimentally for a specified observation point by measuring the average traffic velocity and average time the radiated noise from each vehicle is present above the ambient noise of the receiving system and test location. The latter procedure was used to arrive at $R = 1000$ ft for the accompanying comparison with automotive traffic measurements.[24]

Figure 2-25 presents two comparisons of measured stationary, idling-vehicle APDs with the theoretical APD (ρ) from Equation 2-28. Both measured APDs shown by circles connected by a solid curve were obtained at a frequency of 24.11 MHz with an antenna placed 10 m in front of a stationary vehicle whose engine speed was 600 rpm. The APD (ρ) computed by Equation 2-28 is shown as small dashes. Detected thermal noise with Gaussian-distributed quadrature components appears in Fig. 2-25 as a broken straight line. The experimental results of Fig. 2-25(a) were obtained with a receiving-system noise bandwidth of 0.5 kHz; the results of Fig. 2-25(b) were obtained with a noise bandwidth of 2 kHz. The divergence between measured APD and computed APD (ρ) is more pronounced for the smaller receiving bandwidth where some ignition-noise-pulse overlapping was observed to occur. One notes in Fig. 2-25(a) that enhanced occurrence of pulse overlapping is manifest as an increase in the percentage of the observed amplitude which is approximately distributed with a thermal noise slope, typically characteristic of the output of a narrowband filter. Both presentations of Fig. 2-25 are representative of the larger set of comparisons available elsewhere.[24]

The value of τ_{av} used for computations of APD (ρ) shown in Fig. 2-25 was 25 msec, obtained using $\overline{N} = 1$, $\eta = 8$, and $\kappa = 600$ rpm in Equation 2-30. Note that R was set at $0.02b$ to conform to the constraint $R << b$. The observed noise power at the receiving antenna

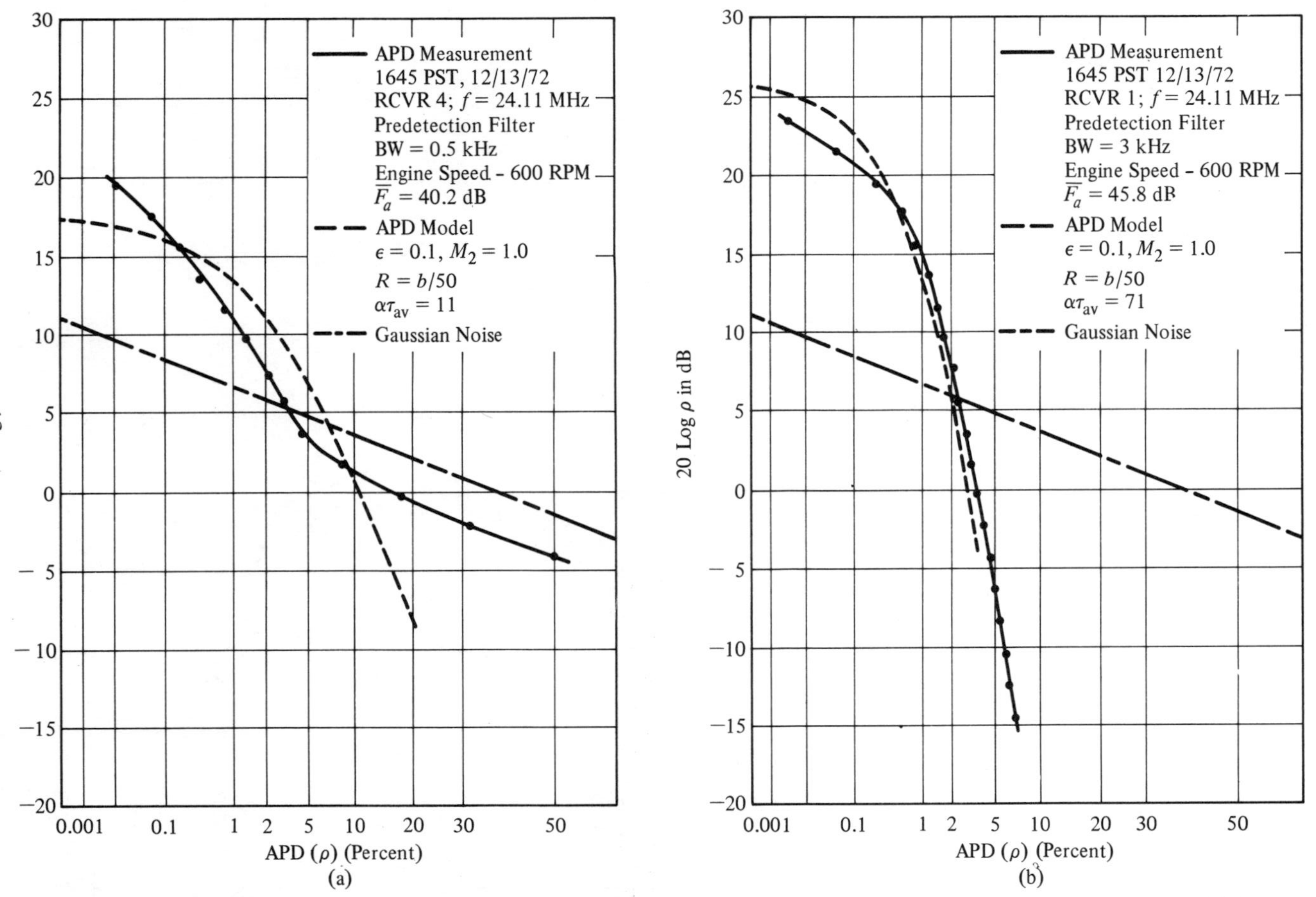

Fig. 2-25. APD for a single stationary vehicle. Measurements —●—. Computed results - - - -. Thermal noise — - —. (a) Detection bandwidth 3 kHz. (After Gillilland, et al.)

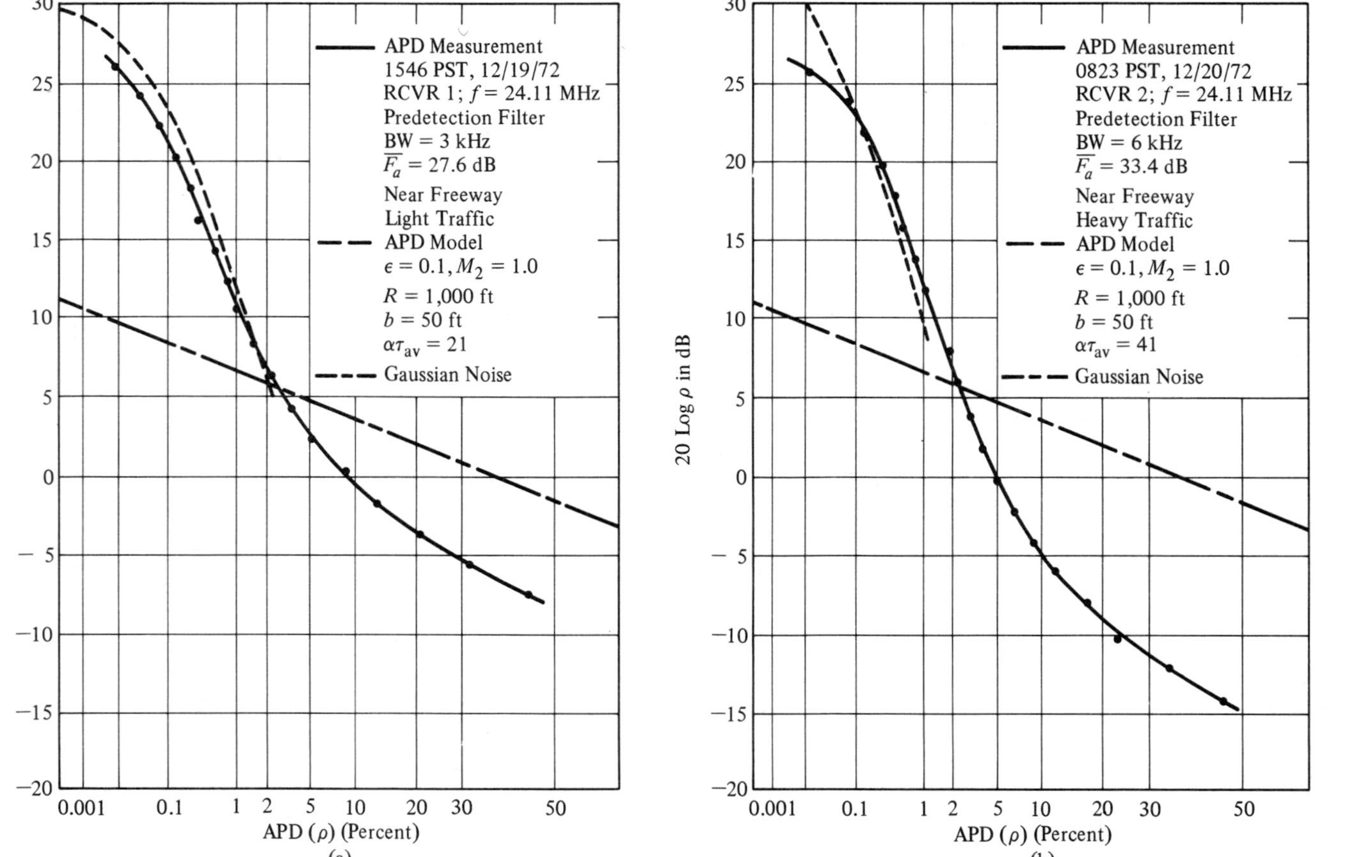

Fig. 2-26. APD for automotive traffic noise measurements at 24.11 MHz —●—. Computed results - - - -. Thermal noise — - —. (a) Detection bandwidth 2 kHz. (b) Detection bandwidth 6 kHz. (After Gillilland, et al.)

terminals, $\overline{F_a}$, is also given on the Fig. 2-25 as 40.2 and 45.8 dB, respectively.

Calculation of APD (ρ) for automotive traffic requires, as noted earlier, a determination of the average number of vehicles $\overline{N}$ present during the measurement period and positioned within the range $-R \leqslant r \leqslant R$. The results of a series of automotive-traffic-noise measurements performed to the side (b = 52 ft) of a surburban freeway under light-traffic conditions, and, in which, on the average, only a single vehicle contributed to the observed noise signal, are compared in Fig. 2-26 with the computed APD (ρ). Receiver noise bands of 3 and 6 kHz are represented. The estimated average values of η = 8 cylinders and κ = 2000 rpm have been used to evaluate Equation 2-28.

The ratio M_2/ϵ has agian been taken equal to 10 as was done for the stationary-vehicle data. The observation frequency for Fig. 2-26 was 24.11 MHz; the road-separation distance of the observation point equaled 50 ft, and the average noise power at the antenna terminals $\overline{F_a}$ observed for the 3- and 6-kHz receiver bandwidths was 27.6 and 33.4 dB, respectively. The experimental results shown by circles interconnected by solid lines are compared to the computed APD (ρ) from Equation 2-28 (drawn with short dashes). Thermal noise is represented by a broken straight line, the slope of which is approached by the measured APD in the low amplitude range.

The computed APD (ρ) for the impulsive portion of the detected envelope distribution, either in the case of stationary vehicles or automotive traffic, shows best conformity with measurements in the high-slope region. The tendency of Equation 2-28 to yield a larger value of APD (ρ) in the high-pulse-amplitude portion of the distribution is most pronounced in the traffic data, although it is evident in the stationary vehicle measurements as well.

Average Traffic-Noise Power. The average noise-power produced by a disposition of vehicles moving in a traffic formation is dependent upon what may be established as the average ignition-noise power of a representative vehicle measured at a fixed separation distance and traffic-flow geometry in relation to an observer. The simplest situation geometry (i.e., linear) is represented by Fig. 2-24, which is, as well, a commonly encountered traffic pattern.

At the observation point □ located at a distance b from the center of the roadway, the observed power p_{1r} radiated by a single vehicle is dependent upon the lateral displacement r. Position variable $r(t)$ is a random function of time, depending upon individual vehicle speed, and characterizable by a probability density function $p(r)$. The greatest value of p_{1r} is recorded when $r = 0$, and is denoted by p_p. For any $r = 0$, p_{1r}, when expressed in terms of b and r, is:

$$p_{1r} = p_p \frac{b^2}{(b^2 + r^2)^{\ell}} = p_p\, b^{2\ell} f(r). \qquad (2\text{-}31)$$

Exponent ℓ lies between the limits $1 \leqslant \ell \leqslant 2$ for cases of general interest, depending upon local geography, frequency of measurement, and antenna height. The average value of p_{1r}—i.e., $E(p_1)$—is expressable in terms of the probability density $p(p_{1r})$ as

$$E(p_1) = \int_0^{p_p} p_{1r}\, p(p_{1r})\, dp_{1r} \qquad (2\text{-}32)$$

since $p(p_{1r}) = 0$ for $p_{1r} > p_p$. The density function $p(p_{1r})$ is related to $p(r)$ by

$$p(p_{1r}) = p(r^{-1}) \left| \frac{\partial r}{\partial p_{1r}} \right|. \qquad (2\text{-}33)$$

using Equation 2-31

$$\left| \frac{\partial r}{\partial p_{1r}} \right| = p_p^{-1} b^{-2\ell} \left| \frac{\partial r}{\partial f(r)} \right|$$

$$= p_p^{-1} b^{-2\ell} \, \{2\ell f(r)^{1+1/\ell} [f(r)^{-1/\ell} - b^2]^{1/2}\}^{-1}$$

$$= p_p^{-1} b^{-2\ell} (2\ell r)^{-1} (r^2 + b^2)^{1+\ell} \qquad (2\text{-}34)$$

The random variable $r(t)$ is defined over a test space $-R \leqslant r(t) \leqslant R$, limited to either side of the point $(0, b)$ of Fig. 2-24 by the sensitivity of the instrumentation, the ambient noise, the traffic pattern, and test geography. Within the test space, vehicle location may be assumed to be uniformly distributed for reasons presented in the preceding discussion of traffic-ignition-noise APD. The density function $p(r)$ is then

$$p(r) = 1/2|R|, \qquad (2\text{-}35)$$

which yields for $p(p_{1r})$

$$p(p_{1r}) = p_p^{-1} b^{-2\ell} (4\ell r |R|)^{-1} (r^2 + b^2)^{1+\ell} \qquad (2\text{-}36)$$

By defining a test space within which the lateral vehicle displacements are confined to values $-R \leqslant r(t) \leqslant R$, an attendant constraint is introduced; namely, at any instant there exists a probability $P(r \leqslant |R|)$ that one vehicle is to be found within the test space. The average observed power (Equation 2-32) must then be altered to include the conditional probability as:

$$E(p_1) = P(r \leqslant |R|) \int_0^{p_p} p_{1r}\, p(p_{1r})\, dp_{1r}. \qquad (2\text{-}37)$$

Using the results of Equation 2-31 and 2-36 and dp_{1r} obtained from Equation 2-31, the expression for $E(p_1)$ becomes:

$$E(p_1) = -\tfrac{1}{2} P(r \leqslant |R|)\, p_p b^{2\ell} |R|^{-1} \int_{\infty}^{0} (r^2 + b^2)^{-\ell}\, dr \qquad (2\text{-}38)$$

For the limiting values of ℓ, the average power per vehicle is obtained as:

$$E(p_1)_1 = P(r \leqslant |R|)\, \tfrac{1}{4}\pi\, p_p b/|R| \text{ for } \ell = 1 \qquad (2\text{-}39)$$

$$E(p_1)_2 = P(r \leqslant |R|)\, \tfrac{1}{2}\pi\, p_p b/|R| \text{ for } \ell = 2. \qquad (2\text{-}40)$$

When the road-separation distance, b, of the traffic observation point differs from the distance d_m at which the average noise power, p_{01} of a single average vehicle has been determined, p_p related to p_{01} by

$$p_p = p_{01} d_m^2\, b^{-2\ell} \qquad (2\text{-}41)$$

may be substituted into Equations 2-39 and 2-40 as required.

The preceding single-vehicle may be generalized to the n vehicle case by summation since individual-vehicle radiated-noise emissions are statistically independent. Equations 2-39 and 2-40 become, using Equation 2-41,

$$E(p_n)_1 = f_{am} kT_o = P(r \leqslant |R|) n \tfrac{1}{4}\pi\, p_{01} d_m^2 /(b|R|) \text{ for } \ell = 1 \qquad (2\text{-}42)$$

$$E(p_n)_2 = f_{am} kT_o = P(r \leqslant |R|) n \tfrac{1}{2}\pi\, p_{01} d_m^4 /(b^3|R|) \text{ for } \ell = 2. \qquad (2\text{-}43)$$

The probability $P(r \leqslant |R|)$ that a single vehicle will be found within the test space $-R \leqslant r \leqslant R$ may be evaluated in terms of the average traffic speed, v, and the average number of vehicles passing the observer per unit of time, D_v, which is also the traffic density:

$$P(r \leqslant |R|) = 2|R|D_v/v \leqslant 1. \tag{2-44}$$

The number of vehicles to be found on the average between the displacement limits $-R \leqslant r \leqslant R$ is related to v and D_v as:

$$1 \leqslant n \leqslant D_v/(44v) \tag{2-45}$$

where the units of D_v and v are vehicles/min and mile/hr, respectively. The lower bound imposed upon n is required to exclude the trivial case of $E(p_n) = 0$.

Experimental data of average noise power for both multiple and single vehicles constrained to a linear traffic pattern have been reported[26] and may be compared with results of the preceding analysis. Using the accumulative distribution of radiated noise power from 958 stationary vehicles measured at a separation distance, $d_m = 50$ ft, and observation frequencies of 20 and 48 MHz, the mean values of p_{01} needed to evaluate $E(p_n)$ were found to be:

for 20 MHz, $p_{01} = 1.82 \times 10^4 \, kT_o$, or
$p_{01} = 42.6$ dB above kT_o;
for 48 MHz, $p_{01} = 2.3 \times 10^3 \, kT_o$, or
$p_{01} = 23.6$ dB above kT_o.

Measurements of average radiated-traffic-noise power at a roadway-separation distance of 100 ft have been reported for 20 and 48 MHz under light traffic-density conditions i.e., 31.6 vehicles/hr and for the observation frequency of 48 MHz when business-hour traffic created a density of 1000 vehicles/hr. From Reference 26, the measured average traffic noise power is shown in column 3 of Table 2-1. In evaluating Equations 2-42 and 2-43, the range limits on the test space must be fixed. For the reported measurements, a road separation distance of 100 ft was used. Since no significant contribution to the maximum observable single-vehicle average power, p_p, is produced when $p_{1r} < p_p/10$, the values of R that effectively define the test space are $R = 1000$ ft for $\ell = 1$ and $R = 316$ ft for $\ell = 2$. The

Table 2-1. Comparison of measured and computed average automotive-traffic-noise power, F_a, for 20 MHz and 48 MHz.

		F_a (dB relative to kT_o)		
			Calculated	
Traffic Density (vehicles/hr)	Frequency (MHz)	Measured Reference 26	Equation 2-42 ($l = 1$)	Equation 2-43 ($l = 2$)
31.6	20	31	21 (26)	18 (28)
	48	18	12 (17)	9 (19)
1000	48	27	27	24

average traffic speed v was estimated in Reference 26 to be approximately 35 mph, which yields for $P(r \leqslant |R|)$, from Equation 2-44:

1. values of 0.34 and 0.11 for $\ell = 1$ and $\ell = 2$, respectively, during the low-traffic-density observation period
2. $P(r \leqslant |R|) = 1$ for the high-traffic-density cases. The calculated results using Equations 2-42 and 2-43 are shown in the right-hand columns of Table 2-1. The four bracketed entries presented under "Calculated" and applying to the low vehicular density are obtained when $P(r \leqslant |R|) = 1$ in Equations 2-42 and 2-43.

Examination of Table 2-1 reveals several facts. First, Equation 2-42 for $\ell = 1$ provides equal or superior agreement with the observed values of F_a at each frequency and traffic density. Second, the observed value of F_a obtained at 20 MHz is appreciably larger than computed results for any ℓ. It is always difficult at frequencies below 30 MHz to observe urban automotive traffic noise free of contributions from power distribution and transmission line s. In Reference 26, local power-line interference was detectable at 10 MHz and possibly contaminated the measurements recorded at 20 MHz as well.

Third, one notices that the experimental results observed at 48 MHz increased by 9 dB while the traffic density increased by a factor of 32. Because the radiations form multiple vehicles in an urban traffic environment are statistically independent, the observed total noise power should be proportional to traffic density and would be expected to increase by 15 dB as D_V varied from 31.6 to 1000 vehicles/hr. The fact that F_a did not increase in proportion to D_V strongly suggests that the low-density traffic measurements are not long-term averages of F_a, but rather were near maximum values recorded when a vehicle was in the vicinity of the point $(0, b)$ of Fig. 2-24 and very likely within the test space $-R \leqslant r \leqslant R$. The bracketed entries included in Table 2-1 were calculated to show the agreement with measurements that exist when this presumption is made.

OTHER AUTOMOTIVE NOISE SOURCES

Although automotive primary and secondary ignition systems generate the most sustained, high-intensity radiated fields originating from modern automobiles, other electrical equipment commonly found on vehicles produces measureable radio signals. Very limited work has been performed on the nonignition automotive noise sources in attempts to characterize their emissions. The published work consists of very little more than source identification, permitting no reliable ranking of their strengths. From the information available, three essential automotive electrical subsystems producing measureable radio-frequency radiation have been identified:

1. Voltage regulators that contain relay contacts capable of generating contact noise, possibly as a result of gas-discharge breakdown stimulated by pitted contact surfaces.
2. Battery-charging circuits containing either a dc generator constructed with a commutator and carbon brushes or an AC alternator rectifier. Worn brush and commutator surfaces are well-known radio-noise sources for all dc electromechanical equipment, including those found in automotive applications. Alternators and associated circuitry are less well understood as incidental noise sources, except for emissions at the fundamental rotation rate and harmonics thereof.
3. Horns that are actuated by either low-power dc electric motors or vibrators may possess worn contacts, which act as the specific noise sources.

Nonessential to vehicle locomotion or safety but included in automobiles for security and convenience purposes are two items that are also radio-noise sources—namely, the ignition key-open door buzzer and the turn-signal light interruptor. Of these, the key buzzer is the greatest source of radio noise. Indications exist that buzzer-radiated noise levels are comparable to ignition-system noise emissions but quantitative evidence remains fragmentary.

Finally, accessory equipment found on custom-equipped vehicles and used to operate windows, adjust seats, and remotely lock doors usually depend upon dc electric motors and relays for activation. From this, a radiated-noise field may arise.

References

1. Ball, A. H., and Nethercot, W. Ignition interference with television reception. *Proc. the Institution of Mechanical Engineers (London)* 19–34 (1952–1953).
2. Burgett, R. R., Massoll, R. E., and Van Um, D. R., Relationship between spark plugs and engine-radiated electromagnetic interference. *IEEE Trans. on Electromagnetic Compatibility* **EMC-16** (3): 160–172 (1974).
3. Schulz, R., Southwick, R., and Smithpeter, C. Measurement of Electromagnetic Emissions Generated by Vehicle Ignition Systems. Southwest Research Institute, 1973.
4. Geselowitz, D. B. Response of ideal radio noise meter to continuous sine wave, recurrent impulses, and random noise. *IRE Trans. on Radio Frequency Interference* **RFI 3**: 2–10 (1961).
5. Automotive Manufacturers Association, Inc. Data Package Radio Interference Tests, South Lyon, Michigan, June 14–17, 1971. Automobile Manufacturers Association, Inc., 320 New Center Bldg., Detroit, Michigan.
6. Ellis, A. G. Site noise and its correlation with vehicular traffic density. *Proceedings of the I. R. E., Australia* 45–52 (1963).
7. Egidi, C., and Nano, E. Measurement and suppression of VHF radio interference caused by motorcycles and motor cars. *IRE Trans. on Radio Frequency Interference* **EMC-3**: 30–39 (1961).
8. Shepherd, R. A., Gaddie, J. C., and Shohara, A. Measurement Parameters for Automotive Ignition Noise. Report MVMA/SRI-75/10. June 1975.
9. Frederick Research Corp. Factor for Predicting Radio Frequency Interference from Vehicular Ignition Systems National Radio Astronomy Observatory Contract RAP-42, 1964.
10. U.S. Navy. Radio Interference Generated by Motor Vehicles. Bureau of Ships, 1953.
11. International Special Committee on Radio Interference (CISPR). Specifications for CISPR Radio Interference Measuring Apparatus for the Frequency Range 0.15 MHz to 30 MHz. Second edition, CISPR Publication 31, 1972.
12. International Special Committee on Radio Interference (CISPR). Specification for CISPR Radio Interference Apparatus for the Frequency Range 25 MHz to 300 MHz. Second edition, CISPR Publication 2, 1975.
13. Society of Automotive Engineers (SAE). Measurement of Electromagnetic Radiation from Motor Vehicles (20–1000 MHz). SAE J551a.

14. Suzuki, H. Characteristics of city noise in the UHF band. *Journal Institute of Electrical Communications Engineers (Japan)* **46**: 186–194 (1963).
15. Shepherd, R. A. Measurements of amplitude probability distributions and power of automobile ignition noise at HF. *IEEE Trans. on Vehicular Technology* **VT-23** (3): 72–83 (1974).
16. Spaulding, A. D., Ahlbeck, W. H., and Espeland, L. R. Urban Residential Man-Made Radio Noise Analysis and Predictions. Institute of Telecommunications Sciences, OT/TRER-14. June 1971.
17. Matheson, R. J. Instrumentation problems encountered making man-made electromagnetic noise measurements for predicting communication system performance. *IEEE Trans. on Electromagnetic Compatibility* **EMC-12** (4): 151–158 (1970).
18. Shepherd, R. A., Gaddie, J. C., Hatfield, V. E., and Hagn, G. H. Measurements of Automobile Ignition Noise at HF. Stanford Research Institute Report Project 2051. February 1973.
19. Rice, S. O. Mathematical analysis of random noise. *Bell System Technical Journal* **23**: 282–332 (1944).
20. Spaulding, A. D., and Disney, R. T. Man-Made Radio Noise, Part 1: Estimates for Business, Residential, and Rural Areas. Office of Telecommunications, Dept. of Commerce, Report OT-74-38. June 1974.
21. Advisory Committee for Land Mobile Radio Services Working Group 3. *Man-Made Noise*. 1966.
22. Spaulding, A. D., Ahlbeck, W. H., and Espeland, L. R. Urban Residential Man-Made Radio Noise Analysis and Predictions. Office of Telecommunications, Dept. of Commerce Report OT/TRE 14. June 1971.
23. Cohen, D. J. A Statistical Ignition Noise Model. Electromagnetic Compatibility Analysis Center, Report ECAC-PR-72-041. October 1972.
24. Gillilland, K. E., and Brewer, T. A. Experimental Verification of Ignition Noise APD Model and Digital-Receiver Bit-Error Probability Model. Electromagnetic Compatibility Analysis Center, Report ESD-TR-73-036. January 1974.
25. Maxam, G. L., Hsu, H. P., and Wood, P. W. Radiated ignition noise due to the individual cylinders of an automobile engine. *IEEE Trans. on Vehicular Technology*. **VT-25** (2): 33–38 (1976).
26. Spaulding, A. D. The Determination of Received Noise Levels from Vehicular Traffic Statistics. *IEEE National Telecommunications Conference Record* 19D-1-7 (December 1972).

3
Electric-Power Generation and Transmission-Line Noise

Radio noise arising in electric-power production, conversion, and transport facilities occurs within the spectral range extending from the fundamental generation frequency, usually either 50 or 60 Hz, into the UHF range. Throughout this frequency interval, the radio-noise intensity in the immediate vicinity of power-transport facilities (e.g., at distances of 100 m or less) arises from one or both of the two types of noise sources, *gap breakdown* and *line conductor corona.* Either source-emission level may be comparable to or greater than the noise levels of other man-made noise sources (Fig. 1-2). Furthermore, the resulting radiation may exceed atmospheric noise levels between sunrise and sunset when the daytime-noise minimum occurs, which in the mid-latitude represents a decrease of 20 dB from the diurnal maxima evident in the middle and lower portion of the HF band (Fig. 1-4). Levels of incidental radiated noise comparable to those observed on transmission and distribution lines and arising from identical causes originate from power-conversion facilities such as local transformer substations.

The distance attenuation of radio-interference field strength from any facility demonstrates a dependence upon separation distance d, which is proportional to d^{-n} where the range of n is $1 \leqslant n \leqslant 3$. The upper limit of n occurs for the smallest values of d resulting in a rapid change of radiated field strength in the facility vicinity. At large values of the factor $2\pi d/\lambda$, where λ is the wavelength of the observation frequency, the exponent n assumes its lower limit of 1, remaining at this value for separation distances comparable to a mile, depending upon the elevation of the noise source and the topography. Although noise radiations from electric-power generation, transformation, and transport facilities are measurable at separation distances in excess of a mile for both fair and foul weather conditions, the resulting levels will usually be masked in metropolitan areas by natural and other man-made incidental-noise sources.

The relative importance of the three types of electric-power facilities as sources of man-made noise is established by their characteristic radiated field intensity and the prevalence of each. The existence of extensive power-transport networks in all industrialized countries places transmission and distribution lines ahead of power generation and transformation facilities as the principal electric-power incidental-radio-noise emitter and not industrial indifference toward improving the technology of power-line noise suppression. Power-transformation facilities, which, because of their numbers and their common occurrence in metropolitan areas, are second only to power transport as radio-noise sources. This ordering is used to arrange the subsequent presentation and interpretation of radio-noise data, which follows a general discussion of gap- and corona-discharge processes, the major sources of radio interference found on all electric-power facilities.

ELECTRIC NOISE PROCESSES

Gap Discharges on Power Facilities

Gap-discharge radio noise is produced by a rapid flow of electric current in the airgap existing between two points of unequal potential occurring on electric-power equipment. As will be identified later in more detail, the points of unequal potential may occur at a myriad of locations—between metal members at interfaces coated with contaminants or partial oxide layers, between ceramic insulators and metal-supporting members, or between metal mounting bolts and wooden members. The conditions for gap-discharge breakdown are created either by induction-coupling or by degradation in the isolation resistance of the line insulators. Mechanical damage, aging fractures, and accumulation of conducting-surface contamination produce a redistribution of the potential drop along the supporting structure between the circuit point of maximum potential and circuit ground. The potential at the base of the insulator rises in an alternating current system at the beginning of each half cycle of line frequency, producing a current flow either through or on a contaminated insulator surface. Between the insulator base and the circuit ground point, mechanically contacting parts, each member of which is intended to function at approximately ground potential, experience a total potential increase and a potential gradient increase as well, if an insulating electrical discontinuity exists at their interface. The insulating discontinuity may take several forms, such as an oil or wood preservative film, an oxide or sulfide layer, resinous inclusions, paint layer, or air. When the discontinuity is air or a form of one of the possible insulating materials, a potential difference is created in an air-filled pocket between the contacting members. In the presence of the potential gradient, free electrons and ions in the air pocket begin to migrate toward the oppositely charged surfaces.

If the potential gradient is sufficiently great, inelastic collisions occur between the charge carriers (primarily the electrons) and the neutral molecules of the air and gap material, resulting in the production of additional ions and electrons, which separate into oppositely charged clouds. As the number of inelastic collisions increases with rising potential gradient, the electron production rate derived from collisions and photon-produced molecular excitations, approaches the rate at which the current carriers are removed from the airspace through the combined processes of attachment to neutral molecules, diffusion beyond the high potential region, and recombination with oppositely charged ions. A further increase in the potential gradient across the gap yields, on the average, more electrons per collision than are lost by attachment, diffusion, and recombination. At this threshold, an avalanche chain reaction is initiated resulting in a current surge across the airgap between the members. Radiation from the current surge is observed as radio noise and gas expansion produced by localized air-heating as the source of the accompanying audible noise. The current-surge deposits charge neutralizing ions on the gap members, which briefly reduces the gap-potential gradient below threshold value. Once the available supply of electrons and gaseous ions has been expended or substantially diminished, the gas-potential gradient commences to rise, and the surge process is repeated.

In alternating-current power facilities, gap-discharge current surges manifest superimposed periodic voltage variations associated with the fundamental frequency of the power-generation system. Polarity reversals occurring across any extant gap change the direction of ion movement and impress upon the gap-discharge radiation a modulation equal to the fundamental frequency and its harmonics that arise from the nonlinearities existing in the ionization process.

The preceding sequence of events is representative of the pattern of occurrences that result in the production of gap-discharge radio noise during conditions when the electric-power facilities are dry. In the event of rain, heavy dew, fog, or melting snow, moisture films form on the supporting structures and penetrate the spaces or gaps between members. The presence of dissolved impurities in the moisture increases its conductivity, which, combined with the large LF permittivity, reduces the gap-potential gradient to a value insufficient to create avalanche ionization. Typically, gap-discharge radio noise undergoes a substantial decrease in intensity when such meterological conditions exist.

The current-surge accompanying avalanche ion production is of very brief duration, consisting of one or several impulses persisting for a few nanoseconds. The radio noise produced by these current surges exhibits RF components of substantial magnitude in the UHF band.

Corona Discharges on Power Facilities

Corona discharge is also a threshold transition process that requires that a minimum potential gradient in the vicinity of a charged object be exceeded before the effect is manifest. The charged object need not be an electrical conductor; dielectric objects are quite capable of producing radiating corona discharges, although the threshold electric field will not be identical. Neither is the occurrence of corona discharge restricted to alternating-current, as opposed to direct-current, power facilities; both will exhibit corona-produced radio noise for either positive or negative polarity. However, the electric field thresholds for initiating noise-generating corona discharges are unequal for the two polarities.

Unlike gap-discharge breakdown, which is always associated with the presence of two oppositely charged surfaces, corona discharge requires but a single charged object at sufficiently high potential, either positive or negative. As the potential of the corona source point increases, the high-mobility free electrons in the vicinity are accelerated by the local electric field either toward or away from the point. When the source point is negatively charged with respect to the zero-potential reference surface, electron movement away from the point affects, through inelastic collisions with the air molecules, the creation of excited molecules, positive ions, and electrons. The molecular excitations emit ionizing photons, which produce additional free electrons, and, together with those created by inelastic collisions, generate an avalanche current if the electron-loss rates from the processes of attachment, diffusion, and recombination are exceeded by the electron-production rate. The net electron-production rate is a direct function of the potential gradient at the source point. Onset of charge avalanching coincides with corona-threshold attainment and initiation of both radio-noise and visible-spectrum emissions. Visible radiation, which is bluish, is confined to the immediate vicinity of the source.[1]

During the avalanche process, positive ions created by electron inelastic impacts and photoionization of the air migrate toward the source point, impinge upon the surface, and create by secondary emission additional electrons and positive ions. The electrons produced by these processes are driven from a negative source, and, through attachment to oxygen molecules, they create a negative-space charge cloud whose center lies somewhat more remote from the source than the average inception point of the electron avalanche. The results are a reduction of the electric field in the vicinity of the source, at approximately 1 mm, and quenching of the current avalanche and all electromagnetic radiation within a few nsec of its inception. Diffusion of the negative-space charge cloud from the vicinity of the source and migration of the positive ions to the source clears the space and permits the potential gradient to increase until it again attains the critical threshold level forming another current avalanche. Electromagnetic pulse durations of a few nanseconds arising therefrom create a radio spectrum extending into the UHF band.[1]

Features of the corona source point that influence the radio emission level are:

1. its dimension, which jointly with the applied potential determines the local electric field strength existing in the air
2. the composition of the source, which determines the secondary emission coefficient for positive ions
3. the source-point surface condition; namely, the level and type of surface contamination that also affects the secondary emission coefficient.

Accumulated experimental evidence has established that negatively charged points on power facilities manifest corona-discharge radio-noise onset at lower operating voltage gradients than do positively potential points. On alternating-current power facilities, corona discharge occurs on the negative half-cycle of the fundamental wave at a lower potential gradient than on the positive half-cycle. The reason that negative corona noise acts as a lesser disturbance to wireless reception than positive corona emissions arises from the appreciably smaller radiated intensity of the former, which is directly ascribable to the confinement of the current avalanche to the immediate vicinity of the source–i.e., within approximately 1 mm. Whenever the source point is elongated and protrudes from the negatively charged surface, negative-corona radio emissions are susceptible to dramatic increases in intensity. Such condition may occur, for example, when the point is formed from foreign material attached to a metal conuctor. This condition is rarely encountered on power facilities leading to the generalization that positive corona is of primary concern and negative corona may be ignored as a cause of radio noise on generators, transformers, and power lines.[2]

The onset of a current avalanche in the vicinity of a positively charged point occurs at potential gradients larger than those associated with negative corona. For clean metallic surfaces, the positive corona-discharge threshold will exceed the negative-threshold electric field by 50 percent.[3] When the state of the metallic surface changes as with the development of localized irregularities, the positive threshold gradient decreases, approaching the value of the negative threshold gradient in the limit of high imperfection density.[2] Free electrons formed by the avalanche are drawn toward the positively charged source and, unlike the situation for negative corona, are not driven from the source into the surrounding air to form a low-mobility negative space charge cloud that quenches the avalanche.[1] An immediate consequence of this difference is a marked increase in (1) the spatial extension of the avalanche discharge, (2) the duration of the avalanche, (3) the number of branching paths that diverge from the source point, and (4) the total avalanche current density.[1,2] Visible radiation (reddish orange) produced by charge recombination and atomic excitations outlines the avalanche traces, revealing a treelike pattern emanating and spreading from the source. Spatial expansion of the positive corona is limited to the region in which the potential gradient is above the avalanche threshold and may extend several inches beyond the surface.

The radiation produced by the discharge is proportional to the current moment–i.e., the product of the current density and the length of the avalanche filament, attaining a value that is several orders of magnitude greater for positive than for negative corona. Positive corona may be classified by the visible appearance of the glow discharge region into plume or streamer and corona glow. Plume or streamer corona whose spatial extension beyond the source point may attain a length of 4 inches on extra-high-voltage (EHV) facilities is, for alternating-current systems, the predominant radio-noise source. Under fair weather conditions, the sources of plumes and streamers may be filamentary dielectrics such as vegetable particles or insects and elongated metallic bosses on various facilities.[2]

When moisture is present in an attached dielectric particle (e.g., an insect), positive corona pluming is inhibited until the moisture has been removed by negative corona excitation.[2] Plume suppression derives from moisture-induced negative-corona generation, which produces a negative-space charge cloud in the vicinity of the source. Low mobility of the negative space charge prevents its dissipation during the positive half-cycle of the voltage, and, unless diffusion is augmented by strong air movement, the negative cloud affects a reduction of the potential gradient about the source point sufficient to prevent electron avalanching. Ultimately, desiccation of the particle or the commencement of air movement with wind currents of 5 to 15 mile/hr enable the formation of positive corona pluming on the power facility.[2]

Elongated metallic protrusions will produce positive corona emission on new, unweathered surfaces. The chemical-physical processes associated with metallic-surface weathering normally stabilize within 6 months of installation and inhibit streamer formation in quiet air. Negative corona thereafter remains as the only noise source unless air movement intensifies, removing the plume-suppressing negative-space charge from the vicinity of the boss.[2]

The lower-intensity positive-corona noise sources in fair weather are not evidenced by visible plumes or streamers but rather by a low-intensity glow. These arise from high-potential-point weathering caused by windblown abrasives, such as dirt and dust particles striking and adhering to the surfaces.

During foul weather, EHVpower facilities may display a large increase in radio noise. This increased level arises from the presence of both negative and positive coronas in alternating-current systems, the latter producing the highest-intensity electromagnetic radiation.

When negative corona occurs independently of positive corona, the general radio-noise level is low. On EHV transmission lines, negative corona during foul weather emits a low-intensity bluish glow visible at night at the conductor surface. Glow discharge occurs in the presence of a small amount of moisture derived from light rain or the presence of ice. The associated radio-noise levels are low for EHV transmission lines, comparable to the corona-noise emissions existing during fair weather conditions on somewhat imperfect lines.[4]

Foul-weather radio-noise maxima always accompany the presence of a positive corona, which is visibly manifest as streamers and plumes emanating from high-potential points on the power facilities. The meteorological conditions productive of high-intensity positive-corona radio noise on EHV lines are in order of decreasing intensity: snow, rain, melting icicles, and sleet. The snowfall-induced radio-noise maxima observed for EHV transmission lines exceed the minimum fair-weather corona level by approximately 40 dB and the maximum fair-weather line corona radio noise by 20 dB for frequencies from 0.5 to 1 MHz.[4] Ambient temperature plays an important role in snow-induced radio-noise emissions. With an air temperature near $-70°C$

plume generation is inhibited, and the radio-noise levels are reduced by crystallization of the water. When the air temperature rises to -30°C to -10°C, in the presence of snowfall, positive plumes occur, resulting in a radio-noise level that attains a maximum value under moderate wind conditions, 10 to 15 mph. Wind velocities within this range are able to remove the negative-corona space-charge cloud from the vicinity of the generating particles, increasing the local potential gradient and thus the noise emission.

Precipitation in the form of rain produces, on EHV transmission-line conductors, corona plumes and concurrently radio noise, which is only slightly less than for snowfall. As rain commences on a dry line conductor, water drops passing through the high-field region near the conductor initiate a gas-discharge breakdown, the threshold of which is reduced by the high dielectric susceptibility of the water. Simultaneously, some drops will impinge upon the cable to produce positive corona plumes and associated radio interference. The resulting peak levels of the radio noise may exceed minimum fair-weather noise by 25 dB, maximum fair-weather noise by 5dB, and maximum fair-weather particle corona noise by from 5 to 10dB. With continuing rainfall, lines become thoroughly wetted, and impingement plumes cease. Gradually, moisture accumulates as a film enclosing the conductors. Forced by gravity, the water film migrates down the catenary to the nadir where it forms droplets and nuclei for spray plumes, which generate intense radio noise approaching levels of 40 dB above the minimum fair-weather corona value.[4] Plume corona and intense radio noise persist in the vicinity of the catenary nadir after rain cessation and remain for as long as water droplets cling to a conductor. After drying, radio noise is often lower than before the rain. This occurrence is attributed to the removal of dirt and dust particles, sources of positive corona, which had accumulated on the conductor.[5] Evidence developed from studies on EHV direct-current test lines presents a behavior distinctly opposite to ac lines. Positive-polarity high-potential gradient conductors, which under dry-line fair-weather conditions display a high level of radiated corona noise, are observed to manifest a substantial drop in interference level in excess of 15 dB, at the onset of rain.[6] When the rain ceases and the line has dried, corona noise levels return to their initial high value, for reasons not presently understood.

If air-temperature changes occurring during or after precipitation follow a profile that produces freezing and then slow-melting (or melting, freezing, and remelting in instances of snowfall), water drops are formed and attach to the tips of short icicles approximately 2 mm in length. The high-potential points initiate intense positive corona plumes and noise emissions approximately 25 dB greater than minimum fair-weather corona noise, if wind speeds of 5 to 15 miles/hr are present to dissipate the space charge.

RADIO-NOISE LEVELS OF LOW-VOLTAGE LINES

Radio-Noise Sources

Low-voltage power-distribution lines provide the hardwire connection between the private consumer, the commercial and small industrial user of electric power, and the transformer substation, which reduces transmission-line voltages of 42 KV and higher to feeder levels of 2.4 to 12.5 KV. The potential gradients that can be achieved by economical designs of transmission-line conductors operating at voltages less than 70 KV are insufficient to produce corona discharges at any time. The noise sources, when they occur on lines of this lower voltage group, are caused solely by gap discharges. Gap discharges that produce the observed radio interference may occur at a variety of points on a power-line pole or its attachments. The drawings in Figs. 3-1 and 3-2 present most of the known types of gap-discharge-noise sources encountered on power lines, poles, and associated hardware. Noise sources such as the crossarm contact surface between the retaining nut and bolthead do not occur when the line-insulator isolation is high and when the contact surface between the metal and crossarm is low impedance. Corrosion or contamination on the hardware components, loosening of the bolt, carbonization of the wood produced by arc-over tracking, and pocket burns increase the electrical resistance at the metal-to-wood contact surface. Insulator leakage current and inductively coupled voltages may then raise the potential across the airgap or dielectric interface to the threshold for arc discharge.

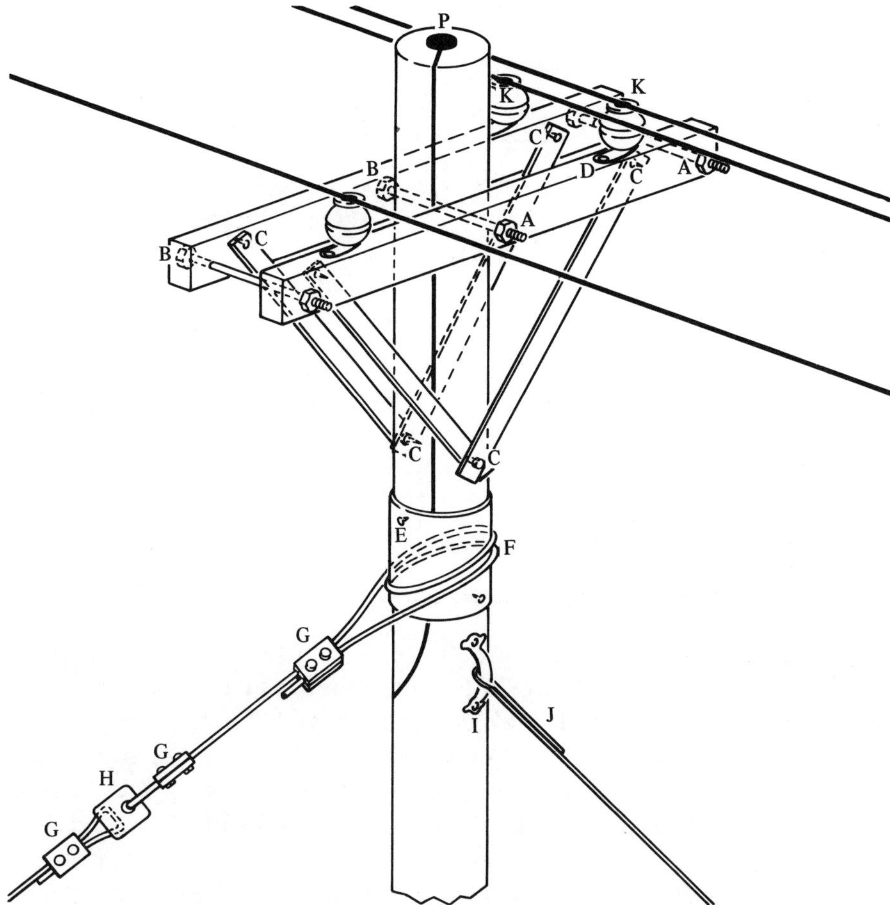

Fig. 3-1. Gap-discharge noise sources Commonly Found on Power Lines:

A Retaining nut on crossarm
B Retaining bolthead on crossarm
C Crossarm brace bolthead and overlap joint
D Insulator mounting post
E Metal-sleeve mounting bolthead
F Guy wire to metal-sleeve contact point
G Guy-wire shackle
H Guy-wire insulator
I Clete bolthead
J Guy-wire weld joint
K Insulator to conductor joint
P Pole cap and grounding wire

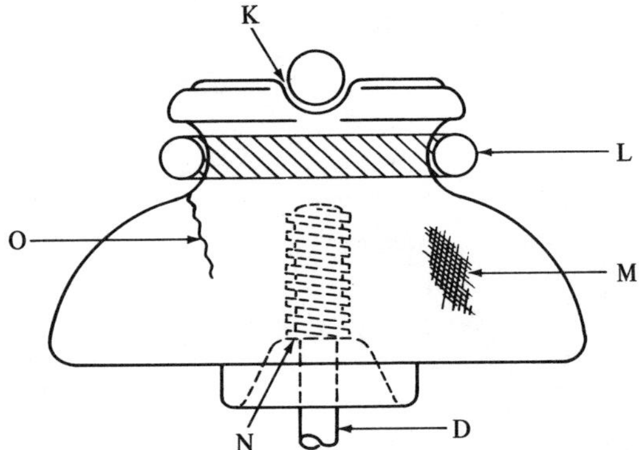

Fig. 3-2. Gap-discharge noise sources Commonly Found on Power-Line Pin Insulators:

D Insulator mounting post
K Insulator to conductor joint
L Tie-wire contact joint
M Insulator contamination
N Insulator-mounting bolt joint
O Insulator fracture

Gap-Discharge Radio-Noise Measurements

Measurements of radiated noise from distribution and transmission lines of 70 KV or less have been performed using several types of detectors. Quasi-peak and peak detectors have been most frequently used by the power-transmission industry. From this work, much data have accumulated characterizing gap-discharge noise levels arising from the numerous pole and line sources shown in Figs. 3-1 and 3-2. A smaller amount of data of very recent origin has become available from experimental work performed using average power instrumentation and envelope detectors supplemented with level quantizing processors permitting the determination of the radiated signal APD.

Quasi-Peak Field-Strength Data. Measurements of power-line radio-noise levels arising from gap discharges occurring at the line-supporting insulators were performed using quasi-peak detectors for a large group of representative 41.6-KV lines in the Detroit area.[7] The data were obtained with radio field-strength meters having 6-dB bandwidths of 6 kHz and charge and discharge time constants of 1 and 600 msec, respectively. Three instruments were used to span the frequency range from 14 to 400 MHz, each possessing a detection bandwidth that varied with frequency. The instrumentation bandwidth dependence

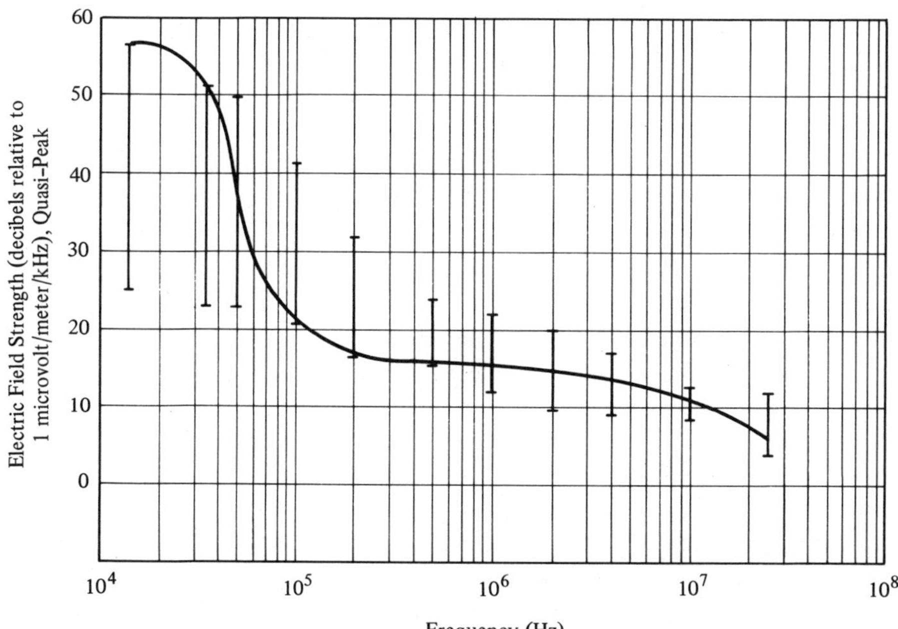

Fig. 3-3. Radio-noise field-strength quasi-peak detector, measured 6 ft above the surface and 25 ft laterally from a 41.6-KV power line. Gap-discharge radio-noise source on an insulator. (After Hinchman Corporation, 1957)

of the incidental noise has been eliminated in the results, which are presented in Fig. 3-3. Therein, data for a 1-kHz bandwidth, a single location, 25 ft laterally displaced from the conductor and a height above the surface of 6 ft are plotted. Appreciable radio interference is observed throughout and below the HF band for the 10 lines studied. For each line, insulator noise was established as the dominant source. The range of variation of the observed quasi-peak field-strength data about the average value curve is shown by vertical bars. The rapid decrease in noise field strength, in excess of −40 dB per decade of frequency variation that occurs above 30 kHz, indicates that insulators located remotely from the test point are the dominant cause of radio noise. Notice that the higher-frequency noise components have been heavily attenuated by the line-to-earth propagating mode.

Peak Field-Strength Data. The results of radio-noise field-strength measurements obtained with peak detecting instrumentation have been reported for several power lines operating below 70 KV.[8] Fig. 3-4 presents the peak detected spectrum for a 4.16-KV distribution line possessing a gap-noise source remotely located from the observation point. The data discontinuity noticeable at 25 MHz, arises from a change in measuring instrumentation. The internal noise levels of the receivers for a 50-ft laterally displaced observer obscured the power-line radiated noise in the portions of the spectrum lying between 8 and 20 MHz and also above 100 MHz. Elsewhere in Fig. 3-4, the gap-discharge noise lies appreciably above the peak residuals of the receivers. The relatively gradual negative slope of the peak noise data, from −20 to −30 dB per decade of frequency change, indicates the noise source to be in the vicinity of the observation point.

Figure 3-5 plots the peak radiated noise field for a 46-KV power line with four gap-discharge noise sources located on one wooden pole.[8] The observation point was positioned 90 ft laterally from the line opposite the pole. The four noise sources that produced the measured signals correspond to items *B*, *F*, *L*, and *P* of Fig. 3-2–i.e., a crossarm bolt-contact point, a guy-wire contact point, a tie-wire

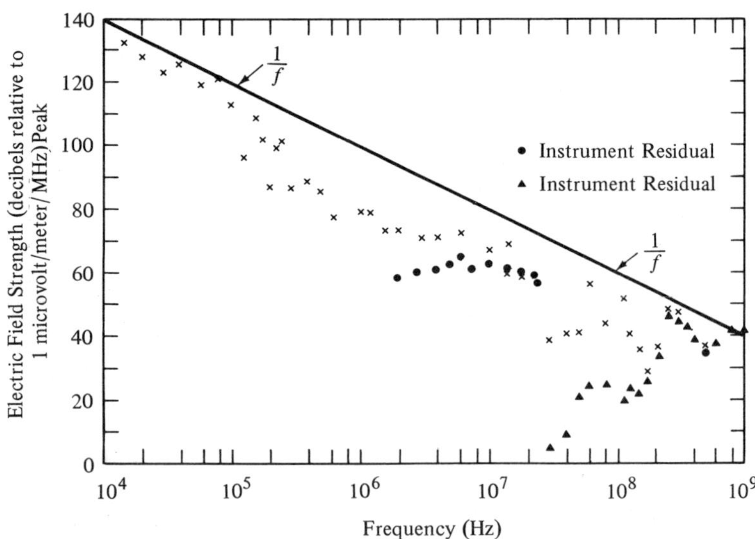

Fig. 3-4. Radio-noise field strength, peak detector, measured 9 ft above the surface and 50 ft laterally from a 4.16-KV distribution line. Gap-discharge noise source on the line. (After Pakala et al., 1967)

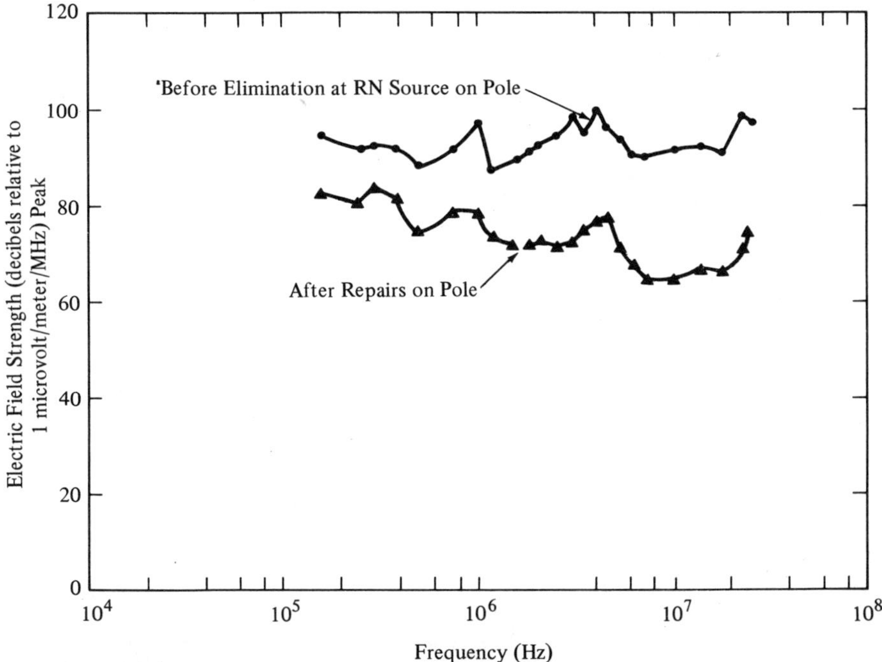

Fig. 3-5. Radio-noise field strength, peak detector, measured 9 ft above the surface and 90 ft laterally from a pole on a 46-KV power line. Four gap-discharge noise sources on pole. (After Pakala et al., 1967)

joint on an insulator, and the pole-cap lightning ground-wire contact surface. The lower curve demonstrates the effectiveness of line maintenance in reducing gap-discharge noise. By repairing the four line and pole defects, the peak-radiated-noise field strength was reduced by amounts varying from 10 to 20 dB.

Figure 3-6 presents peak detected radio-noise field strength for a 69-KV transmission line measured at a point 9 ft above the surface and laterally displaced from the line by 200 ft.[8] A gap-discharge noise source existed on the line at the time of measurement, but its location was at an appreciable distance from the observation point as may be noted by the rapid increase in electric field intensity occurring below 100 kHz. High values of instrument residual noise obscure the line noise levels between 100 kHz and 1 MHz and also those above 400 MHz. The data presented in Fig. 3-6 were obtained using five

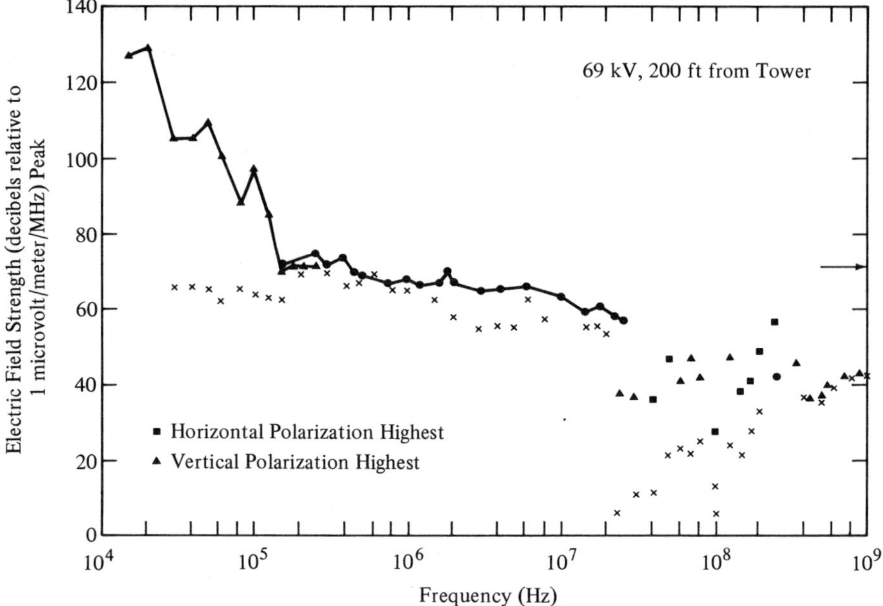

Fig. 3-6. Radio-noise field strength, peak detector, measured 9 ft above the surface and 200 ft laterally from a line tower on a 69-KV transmission line. Gap-discharge noise source on the line. (After Pakala et al., 1967)

overlapping radio-field-intensity meters. Meter changes are noticed in Fig. 3-6 at 150 kHz, 20 MHz, and 375 MHz.

Average Power and APD of Line Noise. A limited number of studies exists on the average radiated-noise power emitted by transmission lines operating at and below 70 KV, the origin of which is the presence of gap discharges on the lines or towers. The experimental average power data derived from these studies are available as a function of either observation frequency or lateral distance from the right of way. A lateral profile for a 12-KV distribution line measured in the vicinity of the wooden supporting pole at a frequency of 3 MHz is shown in Fig. 3-7.[9] Measurements extended from beneath the center of the three horizontally configured phases to 300 ft. Notice that a logarithmic scale has been chosen for the distance variable; consequently, the value of F_a at zero distance is not conveniently representable. Directly under the line, F_a averaged 67 dB above kT_ob with a range of ±1.5 dB. The observed noise power beyond a lateral

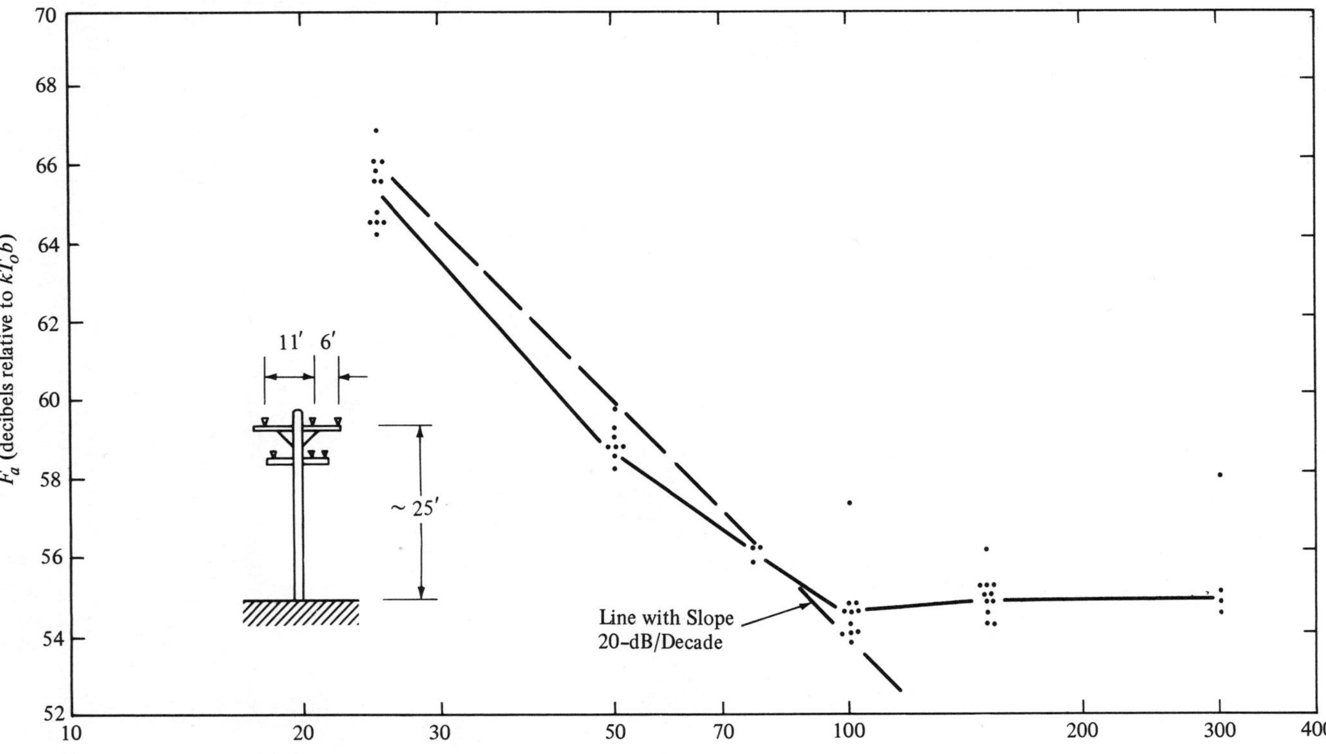

Fig. 3-7. F_a lateral profile on a 12-KV AC distribution line measured from the vicinity of a wooden pole. Gap-discharge noise source present on the line. Fair-weather Conditions. Observation frequency, 3 MHz. (After Shepherd and Gaddie, 1976)

distance of 100 ft represents the ambient noise of the test area, a large portion of which arises from atmospheric noise with some contribution from other man-made noise sources.

The radio-noise-power spectra arising from a gap-discharge source on a 50-Hz, 66-KV transmission line is shown in Fig. 3-8, where vertical bars display the interdecile (10 to 90 percent exceedances) variability of multiple 15-sec data samples.[10] The principal source of the noise is unrecorded, but probably arises from gap discharges along the line or in the transformer substation located within 3000 ft of the observation point. The moderate value of the frequency decrement of approximately −30 dB per decade change in frequency indicates the source to be within a few hundred feet of the observer.

Six examples of APDs generated by gap-discharge sources on dry power-distribution lines are shown in Figs. 3-9 and 3-10.[9] The detector

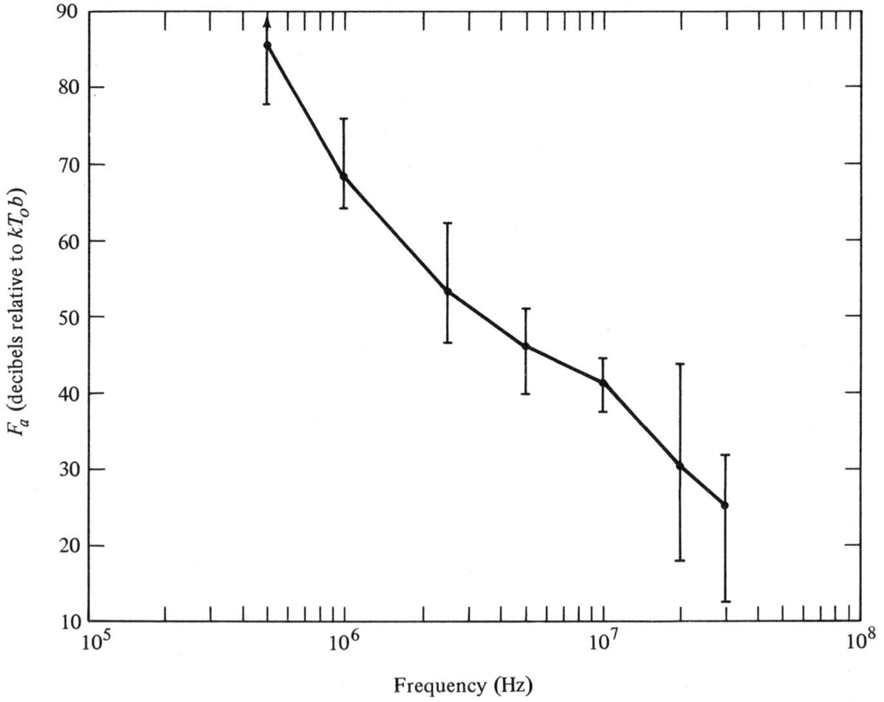

Fig. 3-8. F_a for a 66-KV transmission line operating at a line frequency of 50 Hz. Observer Location: 9 ft above the surface, beneath the line. (After Hagn, 1972)

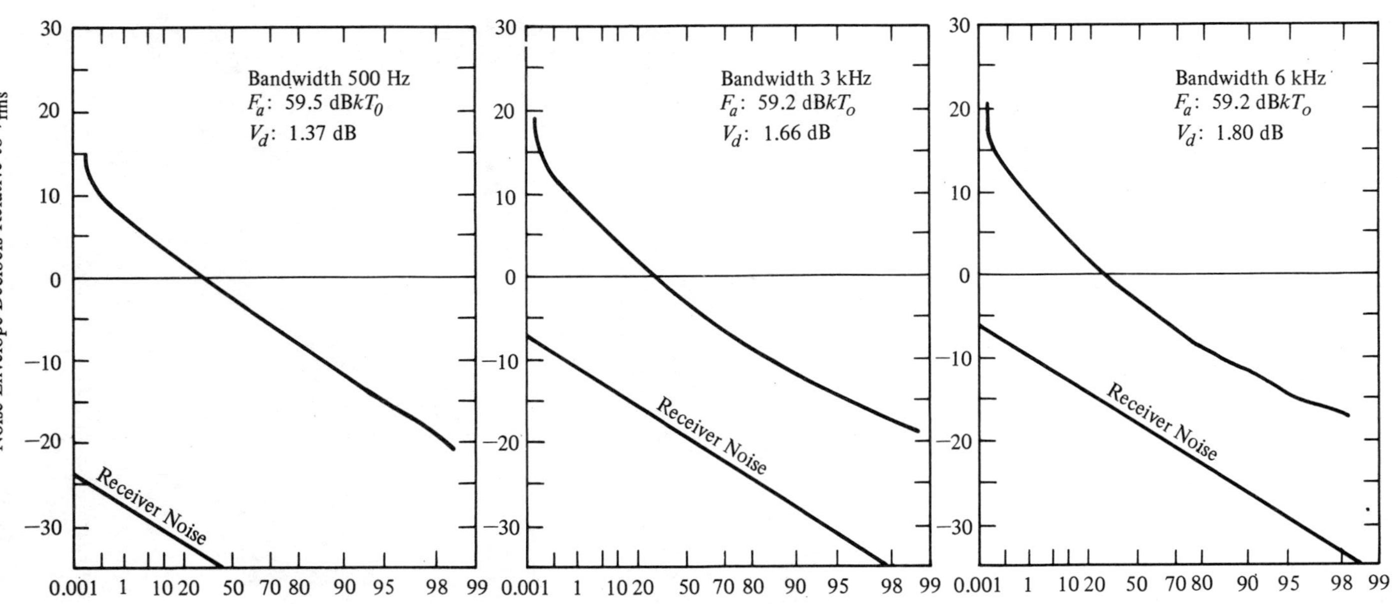

Fig. 3-9. APD of a 12-KV power line at a Frequency of 3 MHz. Observer location: beneath the line. Line condition: dry. Data sample: 5.5-min duration. Three receiver bandwidths of 0.5, 3, and 6 kHz. (After Shepherd and Gaddie, 1976)

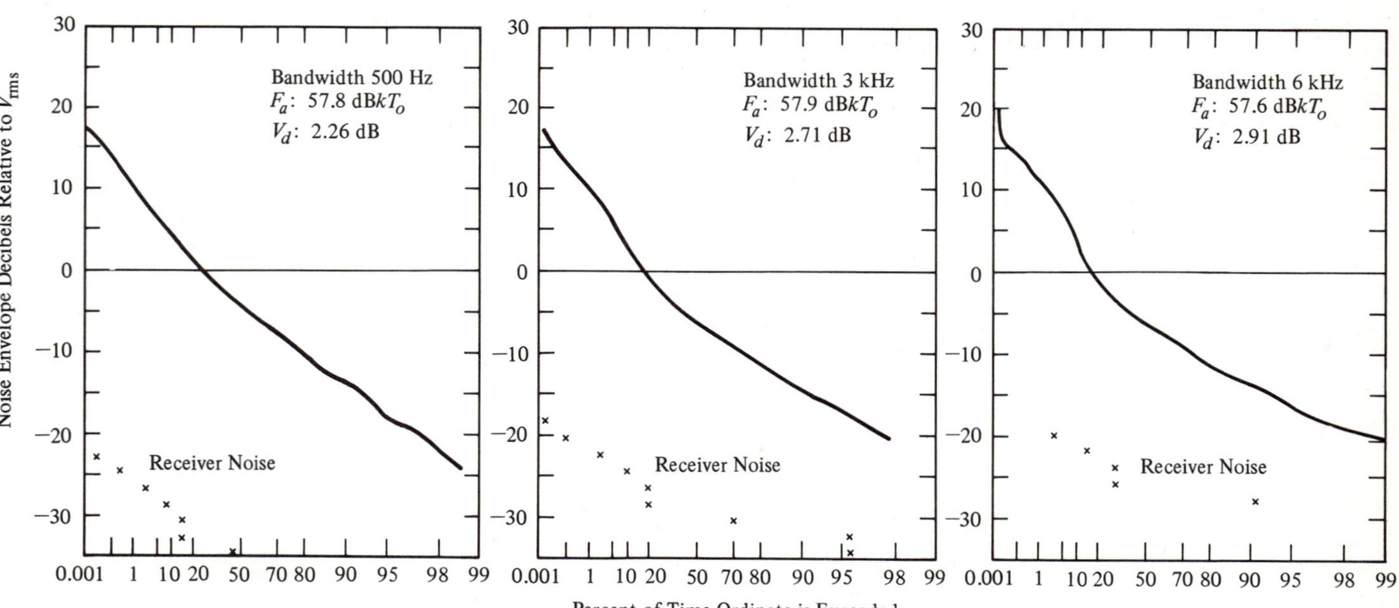

Fig. 3-10. APD of a 24-KV power line at a frequency of 3 MHz. Observer location: 50 ft laterally from line. Line condition: dry. Data sample: 5.4-min duration. Three receiver bandwidths of 0.5, 3, and 6 kHz. (After Shepherd and Gaddie, 1976)

bandwidths range from 0.5 to 6 kHz, yet this variation produces little change in the shape of the distribution or the value of F_a at 3 MHz. The parameter V_d noted on each plot is the ratio of the RMS to average envelope voltage, expressed in decibels, and is used as a measure of the impulsive character of the noise signal. The largest values of V_d correspond to signals containing the greatest proportion of high envelope levels. The thermal-noise levels of each receiver have been included for reference on the plots. Although the line-operating voltages differed by a factor of 2, totaling 12 KV for Fig. 3-9 and 24 KV for Fig. 3-10, the observation point for the 12-KV measurements was beneath the line in contrast to a lateral displacement of 50 ft for the 24-KV study. This range difference equalizes the mean noise levels, which are noted to differ by less than 2 dB. In both studies, 5.5-min data recordings were quantized and digitally processed to derive the distributions. Comparing the noted values of V_d for equal detector bandwidths indicates the gap-discharge noise source impulsiveness was the greatest for the higher-voltage lines.

RADIO-NOISE LEVELS OF HIGH-VOLTAGE TRANSMISSION LINES

Radio-Noise Sources

On high-voltage transmission lines that operate at and above 70 KV, corona and gap-discharge radio-noise sources may coexist. Since line and tower defects that create gap discharges and associated radio emissions are not beyond remedy when proper upkeep procedures are employed, routine preventive maintenance is used in most power-distribution systems to minimize and forestall the development of gap-discharge sources. This technique supplements the trouble-detection and quick-reaction repair procedures that are initiated in response to reported interference or excessive line losses. The consequences of employing these corrective techniques is to substantially reduce or eliminate the contribution of gap-discharge emissions to the total radio noise of high-voltage lines, leaving corona discharges as the only significant cause normally encountered.

Since 1950, the electric power industry has directed its attention toward the reduction of corona noise. Because the existence of corona- and gap-discharge noise emissions has had the greatest deleterious impact on radio broadcast reception, the spectral interval of major interest has been and still is the MF band. Measurement methods and measuring equipment have reflected and continue to emphasize a quantitative representation of the nuisance potential of the interference. In response to a specialized need for an audio disturbance measure, two basic configurations of a modified peak detector were standardized, one each in Europe and the United States, incorporating a rectifying and integrating circuit. Several of the essential electrical specifications of each are listed in Table 3-1. It is evident that the ANSI instrument, because of its greater value of T_d, will provide a larger detected output when excited by non-overlapping impulses of duration less than or equal to the reciprocal bandwidth.

Although quasi-peak field-intensity meters remain the most widely used equipment in obtaining power-line corona-noise data, substantial data have been reported by investigators using receivers incoporating peak-envelope detectors. Most recently, average power detectors and level quantizing circuits have been added to RF receivers and used in transmission-line corona studies to record mean noise power and first-order statistics of the noise emissions, most often the APD.[9,13] In all accumulations of power-line noise data, one of two independent variables—observation frequency or surface distance—is always employed. Surface distance is, in turn, displaced either parallel or perpendicular to the transmission-line path. Movement perpendicular

Table 3-1. Characteristics of CISPR and ANSI quasi-peak noise-field-intensity meters.

		Quasi-Peak		Impulse Bandwidth		6-dB Bandwidth		Predetection Overload Factor	
		Time Constants							
Standard Group	Users	Charge T_C (ms)	Dis-charge T_D (ms)	0.15 to 30 MHz (KHz)	30 MHz to 1 GHz (KHz)	0.15 to 30 MHz (KHz)	30 MHz to 1 GHz (KHz)	0.15 to 30 MHz (dB)	30 MHz to 1 GHz (dB)
CISPR[11]	Europe	1	160 550[a]	9.5	130	9	120	30	43.5
ANSI[12]	North America	1	600	3.2–8.5	130 450[b]	3–8	120 420[b]	10–22	18–22

[a]15 to 300 MHz.
[b]400 MHz to 1 GHz.

(lateral) to the line is of greatest interest since typically noise-field standing waves and longitudinal attenuation along the line produce much smaller variations in noise-field strength than do corresponding lateral displacements. Although not presently amenable to quantitative representation, changes in meteorological conditions (e.g., rainfall, wind velocity) cause appreciable variations in corona-generated noise. These weather-related effects have been studied by several investigators and found to produce line noise-level variations comparable to those arising from longitudinal displacements and full-band frequency changes.

High-voltage transmission lines possess a range of constructional features that have a first-order effect upon the observed radio-noise

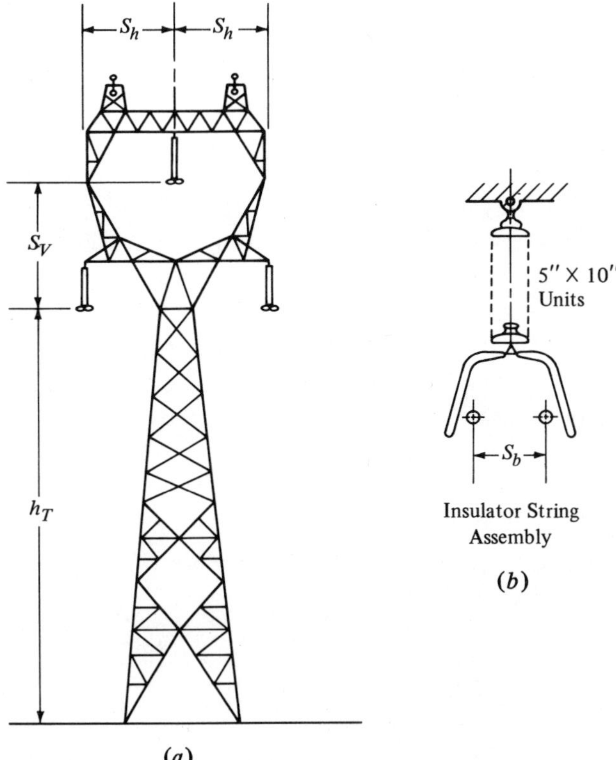

Fig. 3-11. (*a*) Transmission-line tower for single three-phase circuit. (*b*) Insulator string for a two-conductor bundle.

levels. These primary design factors, other than operating voltage, are:

1. The diameter of the conductors
2. The horizontal, S_h, and vertical, S_v, spacing of the conductors
3. The height of the conductors above the ground; at a tower, h_T; and at midspan, h_{ms}
4. The number of conductors, n, and the cable spacing, s_b, in each conductor bundle
5. The number of three-phase circuits supported by each tower.

Figure 3-11 presents a diagram of a representative tower for a single-circuit three-phase transmission line. The tower type is delta. The two-conductor cable bundle shown in (*b*) is the horizontal type suspended vertically by a single insulator string.

Corona Radio-Noise Measurements

Quasi-peak Field-Strength Data. Two elements of information are available for most high-voltage transmission lines—namely, the measured radio-noise level lateral profile at an MF band frequency and the frequency variation of noise level at some point near the line. Occasionally, additional data obtained from long-term monitoring of the noise at a selected point and frequency permit the construction of seasonal cumulative distributions of the radio emissions. A variety of radio-noise measurements for operational transmission lines rated above 100 KV have been reported by investigators using quasi-peak radio-noise meters designed to ANSI specifications. A selection of these data are presented in Figs. 3-12 through 3-15 for AC lines rated from 244 KV to 735 KV.

Measurements of radio-noise level occurring near midspan on a 244-KV transmission line are shown in Fig. 3-12. The radiated noise at six frequencies lying between 154 kHz and 22.3 MHz are shown for lateral displacements to 155 ft. The weather at the time of the study was fair and the line dry.[8]

Ground reflection of some noise signals may occur before they reach a low height-measuring antenna. A partial cancellation in the total field strength results. The effect is noticed in Fig. 3-12 at 80 to 100 ft both east and west of the line center for the 3.05-MHz fre-

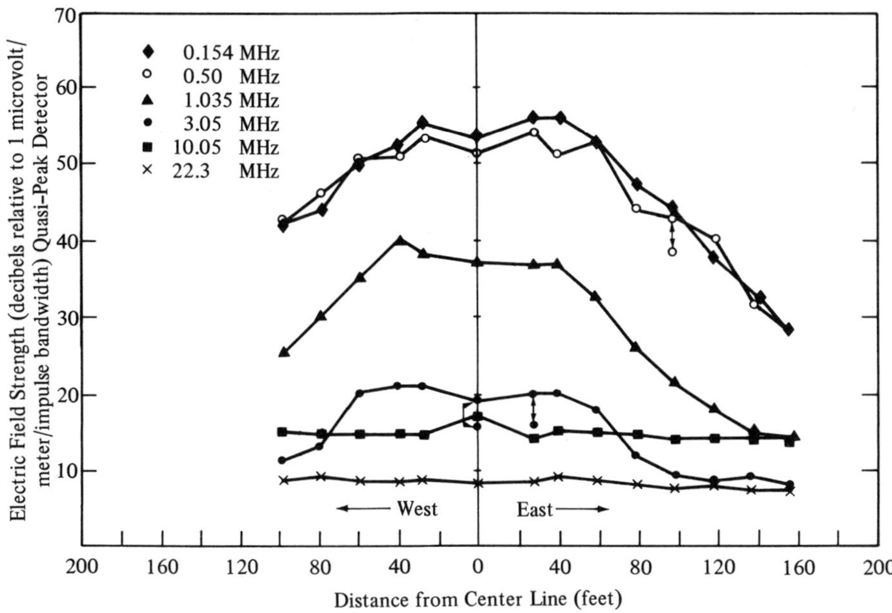

Fig. 3-12. Lateral profiles of quasi-peak radio-noise levels of a 244-KV AC transmission line in fair weather. Measurements made at midspan. (After Pakala et al., 1967)

quency. The line characterized by the data of Fig. 3-12 was a horizontal configuration—that is, each conductor of the three-phase line was suspended at a height of 56.5 ft measured at the steel support towers. The conductors were constructed of single cables of 1.246-in. diameter with horizontal phase spacings of 28 ft. Noise data are reported for the impulse bandwidth of the ANSI detector given in Table 3-1.

A comparable set of radio-noise profiles obtained at midspan on a 345-KV, AC transmission line are presented in Fig. 3-13.[8] Six frequencies were used to record the radio field strength with an ANSI-designed quasi-peak noise meter along profiles to either side of the center of a vertically configured double-circuit three-phase line. Each circuit consisted of three single conductors, the center phase displaced approximately 10 ft outward from the vertical plane containing the top and bottom phases. Conductor diameters were 1.6 in. for the east circuit and 1.75 in. on the west circuit. The height of the lowest

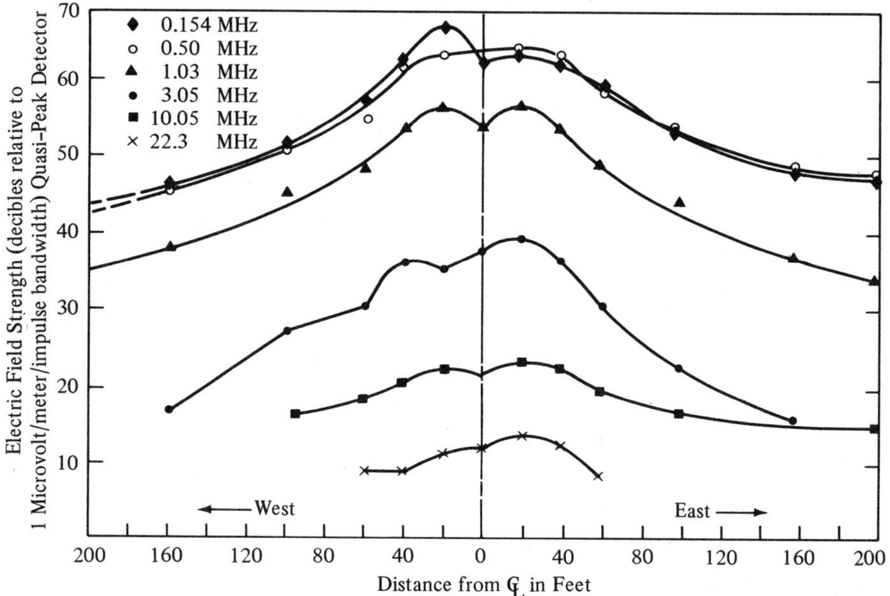

Fig. 3-13. Lateral profiles of quasi-peak radio-noise levels of a 345-KV AC transmission line in fair weather. Measurements made at midspan. (After Pakala et al., 1967)

conductor on each circuit was 108 ft. A vertical phase spacing, S_v, of 21.5 and 23.5 ft existed between the center to lowest and the center to highest phases, respectively. Some asymmetry is noted in most of the lateral profiles near the center line. Lateral-profile asymmetries in noise-field strength are commonly observed and attributed either to the existence of a corona noise source on a horizontally displaced phase or to the influence of a nearby scattering object. Comparison of Figs. 3-12 and 3-13 indicates that the higher voltage line generates a larger radio-noise field strength. It is, however, the conductor-surface-potential gradient, a function in these instances of the conductor diameters, spacings, and line voltages rather than line voltages alone that, to the first-order, produces the evident differences in levels.

Operating at 525 KV, a horizontal-configuration three-phase AC line with two-cable conductor bundles of 1.75-in. diameter yielded the midspan fair-weather lateral profiles shown in Fig. 3-14 for six frequencies from 0.154 to 22 MHz.[14] The conductor heights were

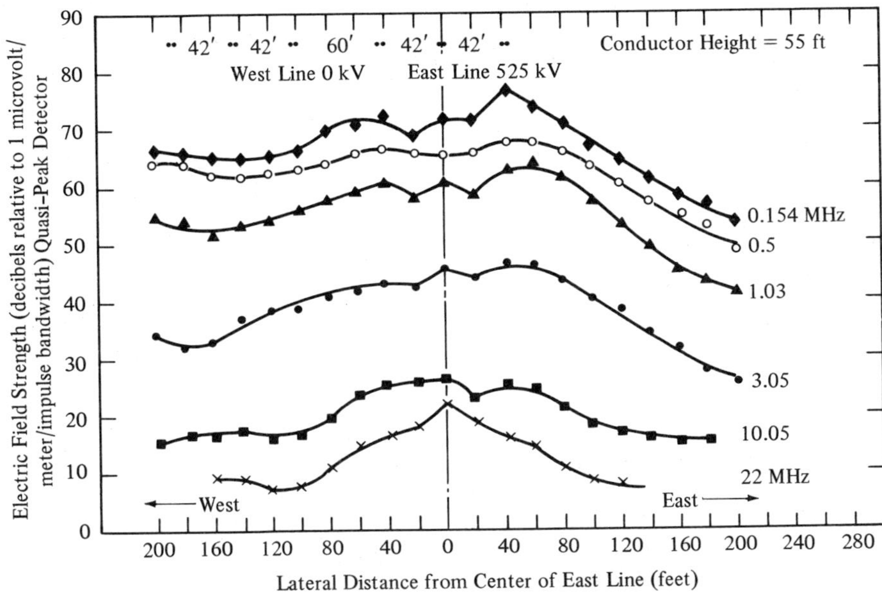

Fig. 3-14. Lateral profiles of quasi-peak radio-noise levels of a 525-KV horizontal configuration, two-conductor bundle AC transmission line in fair weather. A second and identical transmission line is noted to the west but deactivated for these measurements. (After Pakala et al., 1967)

55 ft with S_h = 62 ft. To the west of the activated line lay a second and identical three-phase transmission line, which at the time these data were recorded was deactivated. A portion of the asymmetry in lateral profiles manifest in the data of Fig. 3-14 may be attributed to scattering by the west line of the radio-noise emissions from the eastern line. Measurements of Fig. 3-14 were made with a vertical electric monopole antenna, elevated approximately 4 ft, and a radio-noise meter meeting ANSI specifications. Comparisons of these data with those of the preceding two figures reveal an appreciably enhanced corona noise level but, as noted before, this is primarily attributable to the gradient differential of the conductor.

Data for corona generated noise observed on a 745-KV transmission line with a quasi-peak noise meter meeting ANSI specifications are shown in Fig. 3-15(a) and (b). In Fig. 3-15(a), the laterals obtained at 1 MHz are given for two conductor bundle configurations—namely, four, 1.2-in.-diameter cables per bundle and four, 1.382-in.-diameter

4 × 1.2″
4 × 1.382″
Heavy Rain
80 Percentile
Fair Weather
Electric Field Strength (decibels relative to 1 microvolt/meter/impulse bandwidth) Quasi-Peak Dector
Distance from Center Phase in meters
(a)
Measuring Point
Richelieu
St-Hyacinthe
St-Guillaume
$f^{-1.5}$
Frequency (Hz)
(b)
Maximum Fair Weather RI
Average Fair Weather RI
500 kV
230 kV (100′ from Outside Phase)
735 kV (4 × 1.38″) (Under the Center Phase)
Cumulative Relative Frequency
Electric Field Strength (decibels relative to 1 microvolt/meter/impulse bandwidth) Quasi-Peak Detector
(c)

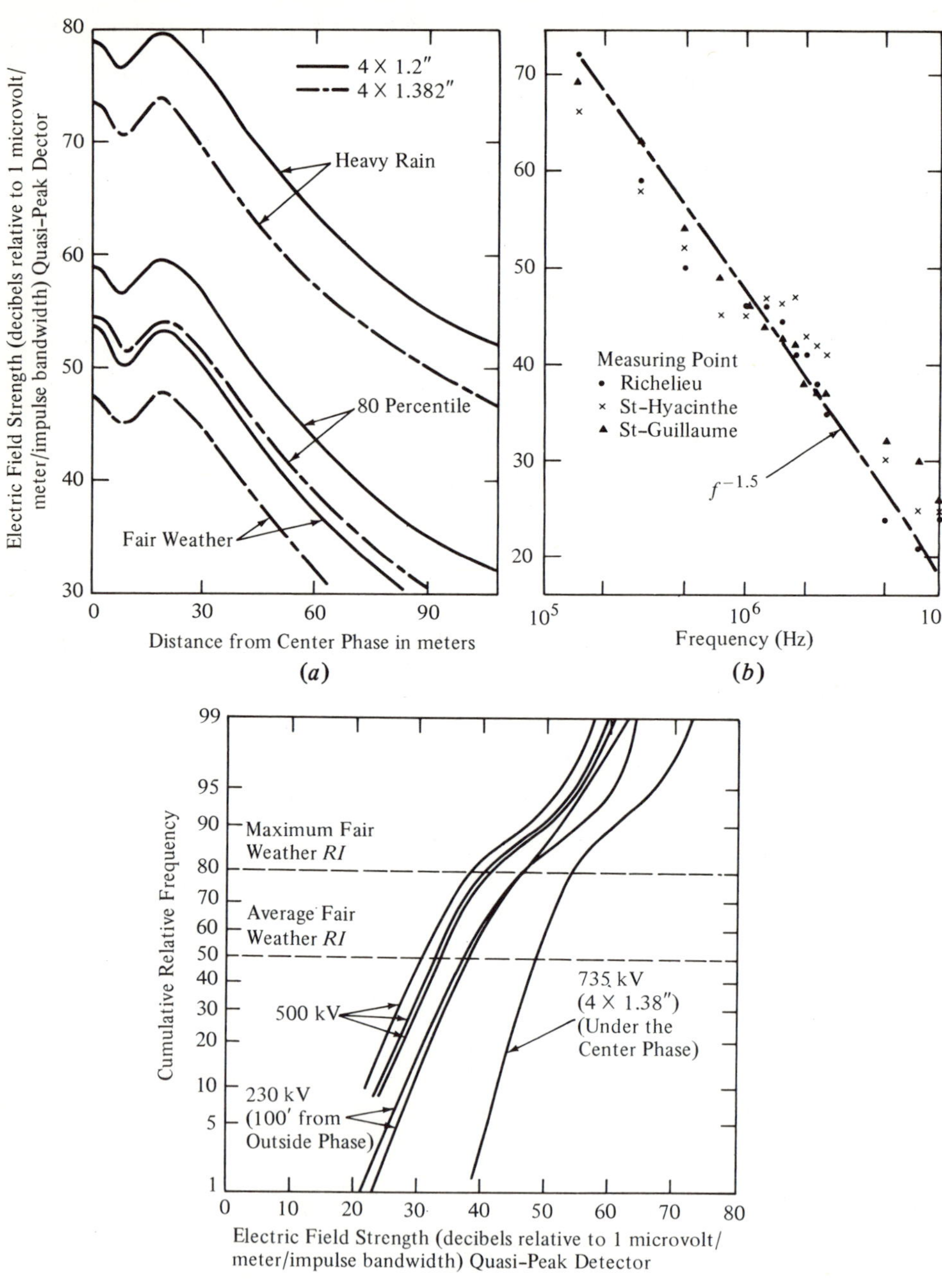

Fig. 3-15.

cables per bundle. Both fair and rainy weather conditions are represented, demonstrating that enhancement of approximately 20 dB above average fair-weather values is possible during heavy rain. The 80-percentile fair-weather value of radio noise is also shown for each conductor size. In Fig. 3-15(b), the fair-weather radio-noise spectrum from 0.15 to 10 MHz is plotted for three observation points along the 735-KV transmission line referenced in Fig. 3-15(a). At each of the three line locations, the lateral displacement of the observation point was 50 ft from the outer phase.

In Fig. 3-15(c), the cumulative distributions of quasi-peak detected, corona radio emissions obtained over a period of 1 year is depicted for the 735-KV lines of Fig. 3-15(a) and (b). These data apply to the 1.382-in. diameter cable design. It may be noted that the distribution is not log-normal in the portion extending above 50 μV/m where the effects of infrequent, heavy rainfall produce excessive radio-noise levels. Cumulative distributions for two other high-voltage AC transmission lines are also shown in Fig. 3-15(c). For a 500-KV line, three distributions are shown obtained over sampling periods of approximately 1 year. The statistical stability of the corona-noise process is qualitatively indicated by the similarity of the separate distributions. Two distributions are also plotted in Fig. 3-15(c) for a 230-KV AC transmission line obtained at a lateral displacement of 100 ft.

Peak Field-Strength Data. Peak measurements of transmission-line corona radio noise have emphasized the acquisition of lateral noise profiles and single-point frequency spectra, much as have quasi-peak studies. Low-height monopole antennas usually raised approximately 4 feet above the surface have served to couple the radiated-noise fields into tunable RF receivers containing peak-envelope detectors. Noise data thus obtained are normally presented as electric-field-strength per unit of bandwidth; the bandwidth conversion factor is

---

Fig. 3-15. High-voltage transmission-line corona radio-noise measurements obtained with a quasi-peak noise meter. (*a*) Lateral profiles for a 735-KV AC transmission line measured at 1 MHz. Four cable conductor bundle. Two values of conductor diameter: 1.2 and 1.382 in. Fair and rain data. (*b*) Quasi-peak radio-noise level as a function of observation frequency during fair weather for the 735-KV AC transmission line in (a). (*c*) Cumulative distribution of quasi-peak noise levels for the 735-KV AC transmission line and similar distributions for 230 KV and 500 KV AC transmission lines. (After Marvada and Trinh, 1975)

derived by calibrating the recording instruments with an impulse generator. In the following sets of peak noise data, 1 MHz has been selected as the unit of bandwidth to permit intercomparisons of the various study results.

Lateral variations of peak-noise field strength arising from corona discharges are of identical form to profiles obtained using quasi-peak instrumentation. The principal difference between quasi-peak and peak corona noise for high-voltage transmission lines occurs in the intensity of the emissions of which the peak-value measurements are the greater.

Figure 3-16 plots peak-electric-noise field strength as a function of frequency for the interval of 10 kHz to 150 MHz obtained at a lateral offset position of 200 ft from a steel tower supporting a single three-phase 244-KV AC transmission line.[8] The line was configured with horizontal conductors containing one cable per conductor; each cable was 1.246 in. in diameter. Line spacings of 28 ft and an elevation of

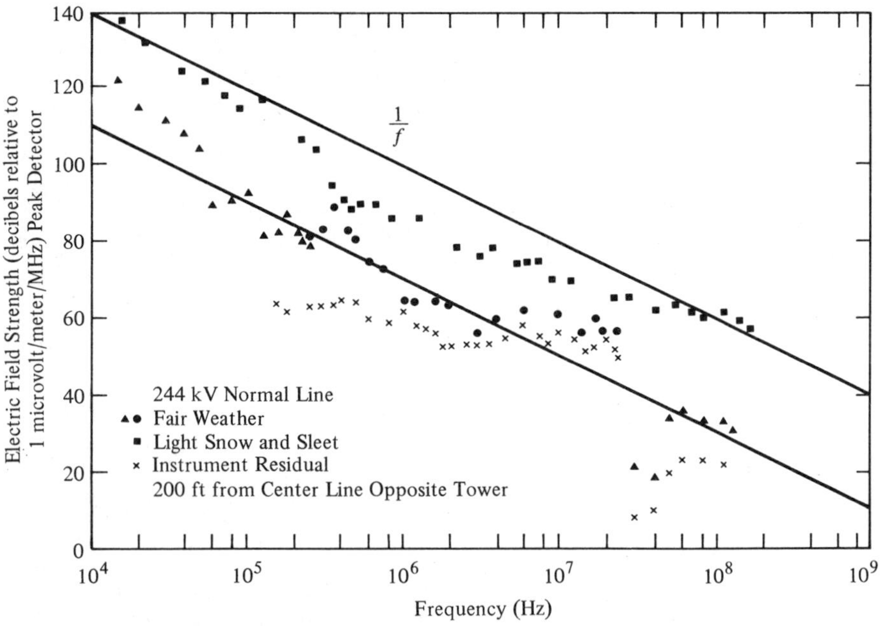

Fig. 3-16. Peak electric noise field strength spectra for a 244-KV AC transmission line with corona noise sources. Measuring location: 200 ft laterally from a steel tower. (After Pakala et al., 1967)

56.5 ft apply. Two sets of fair-weather noise data are shown and may be compared with the higher noise fields created by positive corona streamers during snow and sleet conditions. The solid points of Fig. 3-16 record the internal noise of the receiving equipment, which above 150 MHz equals or exceeds the corona noise and renders the data meaningless. The two solid lines are drawn for reference and have a frequency decrement of −20 dB per frequency decade.

The repeatibility of peak-electric-field strength spectra for a 345-KV double circuit, three-phase AC transmission line is indicated in Fig. 3-17 for dry-line corona noise obtained at one location for two periods separated by approximately 3 years.[8] The double circuit consisted of vertically configured, single-cable conductors of the same geometry and spacing as Fig. 3-13. The observing location used for both sets of measurements was 200 ft perpendicular to the line and opposite a specified steel tower.

Measurements of the dependence of corona-noise field strength

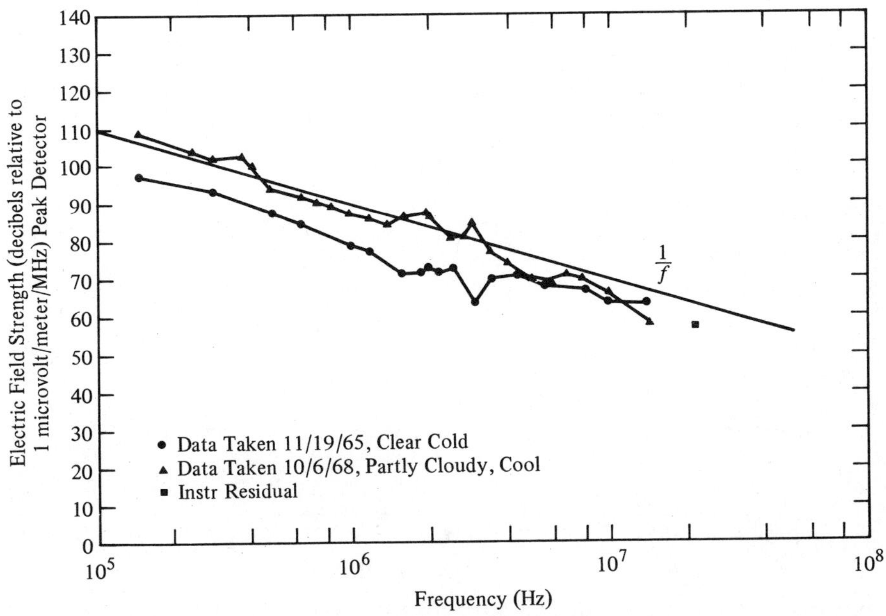

Fig. 3-17. Peak electric noise field strength spectra for a 345-KV AC transmission line with corona noise sources. Measuring location: displaced 200 ft laterally from steel tower. (After Pakala et al., 1967)

upon receiving antenna polarization have shown the near-surface vertical and horizontal electric fields to be identical. Figure 3-18, obtained using both horizontally and vertically polarized antennas, stresses this point.[8] No detectable difference between the results for the two polarizations was observed on the 525-KV AC line throughout the frequency range 10 kHz to 200 MHz. Above 200 MHz, these data are compromised by the high internal noise of the receivers used in the study. Figure 3-18 represents fair-weather dry-line conditions with observations made at a point 50 ft laterally displaced from the outer conductor of the horizontally configured, double-circuit AC transmission line. The tower height of the line was 108 ft; midspan height was 55 ft. Double-cable bundle conductors, consisting of 1.75-in.-diameter cables horizontally arrayed on an 18-in. center-to-center spacing, were supported by steel towers providing phase spacings of 42 ft and a line-center separation of 144 ft.

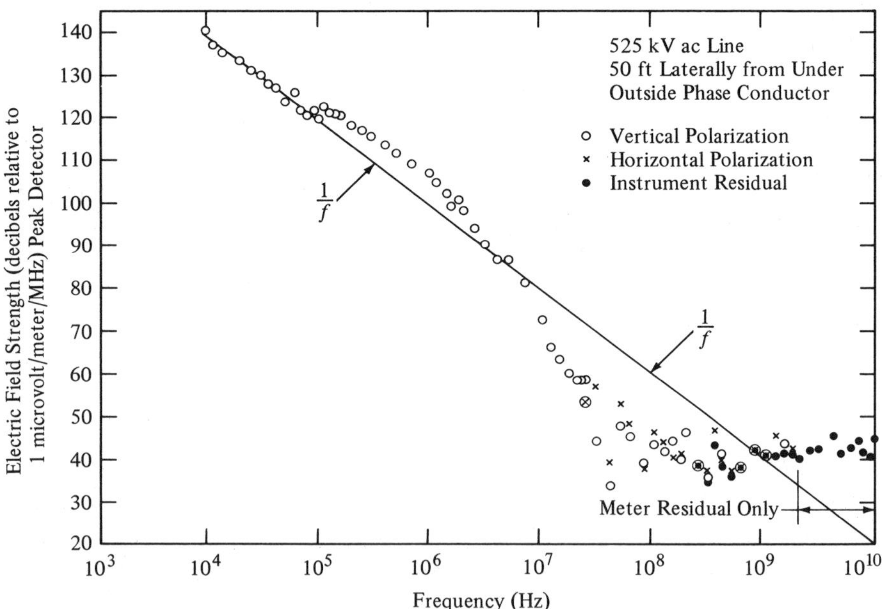

Fig. 3-18. Peak electric noise field strength spectra for a 525-KV AC transmission line with corona noise sources. Measuring location: 50 ft, laterally displaced from the outer conductor at midspan. Horizontally and vertically polarized noise field strength. (After Pakala et al., 1967)

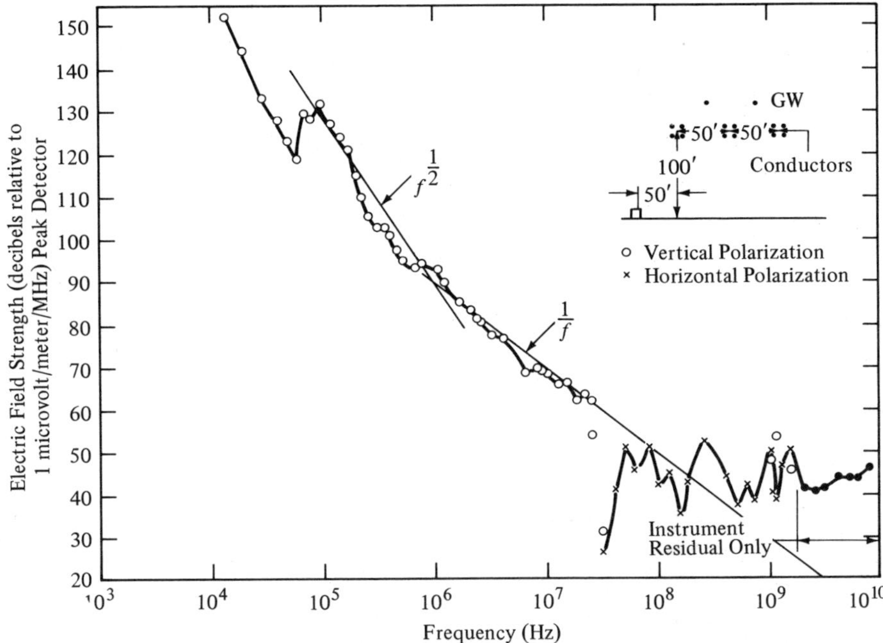

Fig. 3-19. Peak electric noise field strength spectra for a 735-KV AC transmission line with corona noise sources. Measuring location 50 ft laterally displaced from a steel tower. (After Pakala et al., 1967)

Figures 3-19 and 3-20 present fair-weather corona radio-noise spectra for a horizontally configured, four-cable bundled-conductor 735-KV AC transmission line.[8] The line geometry is shown in the insets. The respective bundles are of 1.382-in.–diameter cables on 18-in. center spacings. Two measuring locations of 50 and 200 feet, each laterally displaced from a steel tower, were used. The upper notation in the insets, *GW*, indicates the positioning of lightning-protection ground wires. The greater noise levels evident by comparing Figs. 3-19 and 3-20 with the results shown in Figs. 3-16 to 3-18 for lower-voltage AC lines at equivalent lateral displacements arise from higher-conductor surface-potential gradients rather than line-operating voltages.

Gap-discharge noise sources may arise on any transmission line, irrespective of its rated voltage or construction details. Depending upon the origin of the gap discharge, the resulting emission levels may be either greater or smaller than the fair-weather corona noise.

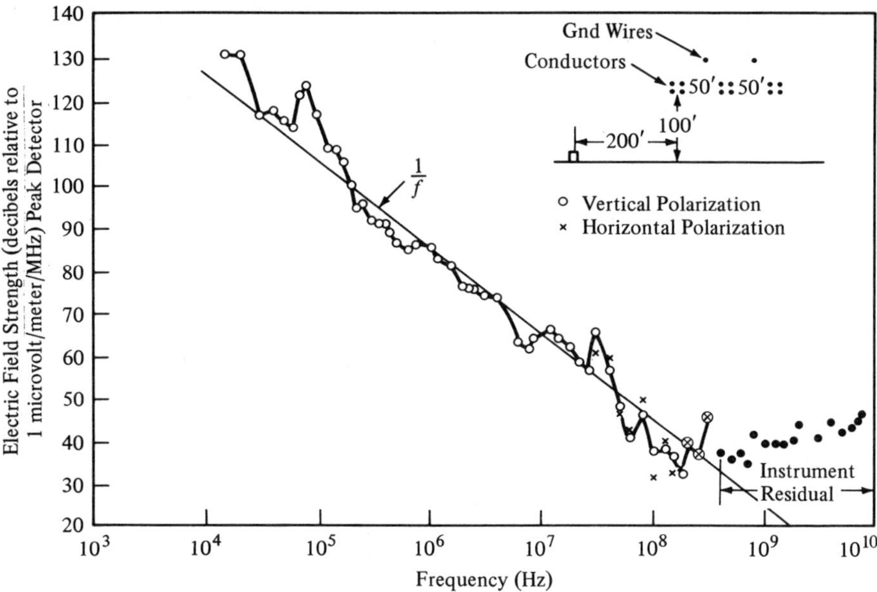

Fig. 3-20. Peak electric noise field strength spectra for a 735-KV AC transmission line with corona noise sources. Measuring location: 200 ft laterally displaced from a steel tower. (After Pakala et al., 1967)

Comparison of the relative intensities of the peak-noise-field spectra produced by a gap-discharge noise source occurring near the observation point on a 735-KV AC transmission line and the fair-weather corona noise sources is performed in Fig. 3-21. The point of observation was approximately 4 ft above the surface, removed 200 ft laterally from the line center. Two features of the spectra are notable. First, gap-discharge emission levels are at least 10 dB greater than the corresponding field intensities of the line corona noise from 0.15 to 7 MHz. Secondly, the frequency decrement for the gap-discharge field strength is characteristic of the spectra of gap discharges observed near their points of occurrence and before line attenuation has selectively diminished the high-frequency components of the arc.

Average Power Data

Supplementing previous and continuing investigations of transmission-line corona radio-noise emissions, which utilize quasi-peak and

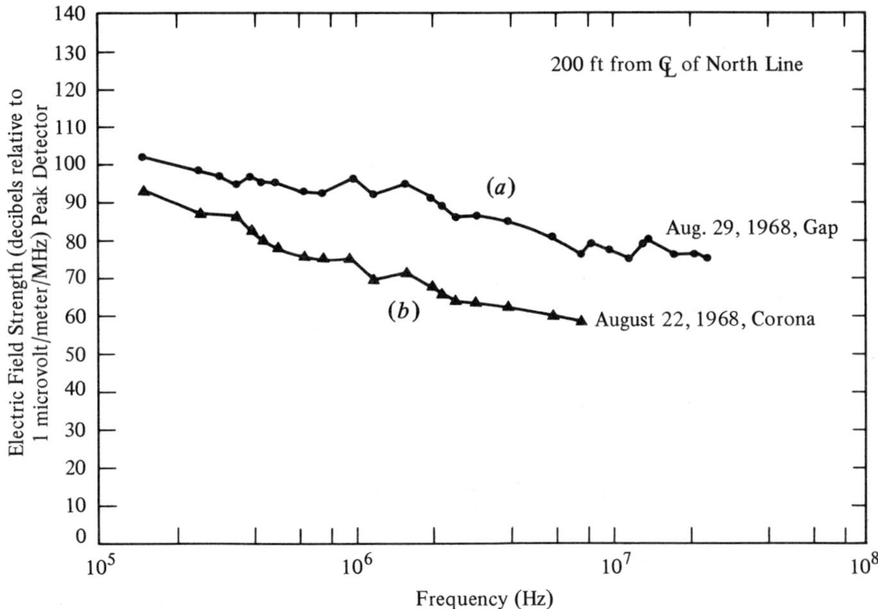

Fig. 3-21. Peak electric noise field strength spectra for a 735-KV AC transmission line. Two radio-noise-source types are shown: (*a*) gap discharge, (*b*) corona discharge. Measuring locations: 200 ft laterally displaced from the center line of the configuration. (After Pakala et al., 1967)

peak detectors, are the more recently initiated studies that exploit average-power-recording instrumentation. Two features of the average-power-measuring equipment that distinguish the newer noise-power instrumentation from the peak and quasi-peak meters are the inclusion of absolute power references and large, continuously tunable, radio-frequency bandwidths.

Data accumulated using average-power detectors provide the lateral attenuation and frequency spectra for a selection of high-voltage transmission lines operating above 100 KV. Lateral profiles of corona radio-noise power have been reported for distances as great as 1500 ft and frequencies within the range of 250 kHz to 250 MHz.

Figure 3-22 presents a set of 3-MHz lateral-profile measurements performed during fair weather on a double-circuit, vertically configured, single-conductor three-phase AC transmission line with a phase spacing of 10 ft and other dimensions as noted in the inset.[9] The observed emissions arose from line corona sources possibly inter-

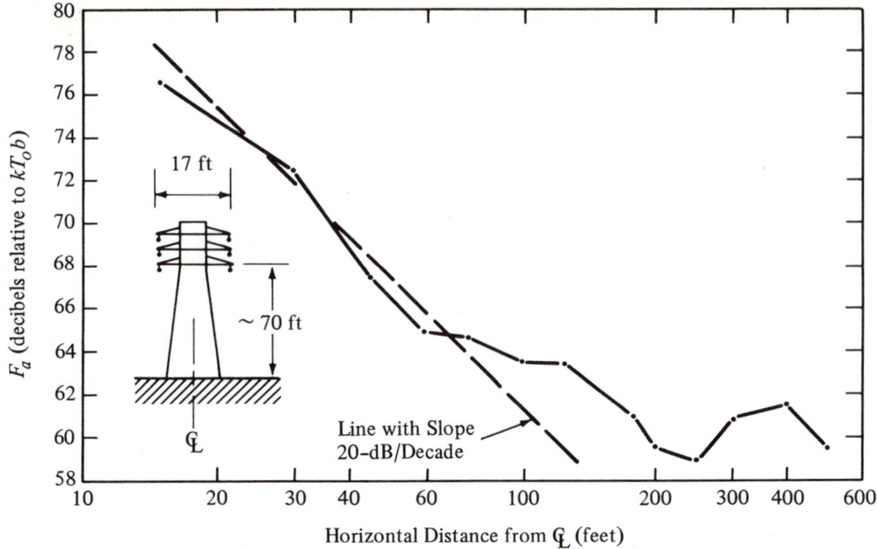

Fig. 3-22. Average power, F_a, fair-weather lateral profile for a 115-KV AC transmission line at 3 MHz. Measurement location at approximately midspan. (After Shepherd and Gaddie, 1976)

mixed with some small amount of gap-discharge interference. Notice that partial cancellation of the direct-wave and ground-reflected noise-field strength occurs approximately at the range of 250 ft. In the execution of these measurements, a vertically polarized monopole antenna was used.

Figure 3-23 presents a fair-weather corona radio-noise lateral profile for a 115-KV AC transmission line determined using an average-power detector and an observation frequency of 102 MHz.[16] The associated average-power spectra for two lateral locations is shown in Fig. 3-24.[16] The power-frequency decrement, ${}_f\delta_P$, for both observation points, lies between −25 and −30 dB per decade change in frequency and is related to the electric-field-strength decrement, ${}_f\delta_E$, as follows:

$$ {}_f\delta_P = 10 \log \frac{p(f_2)}{p(f_1)} = 10 \log \left[\frac{E^2(f_2)\,A(f_2)}{E^2(f_1)\,A(f_1)}\right] = {}_f\delta_E + 10 \log \frac{A(f_2)}{A(f_1)} \tag{3-1} $$

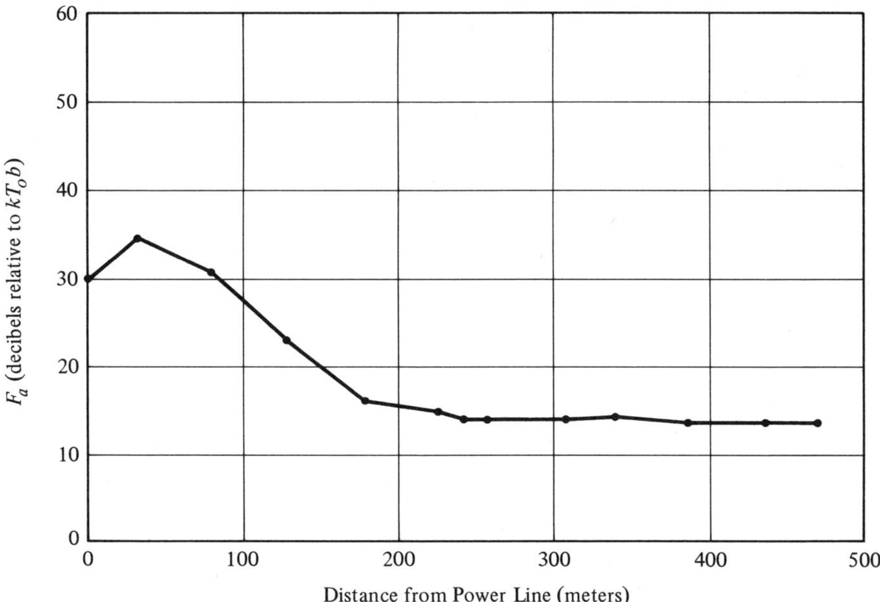

Fig. 3-23. Average power, F_a, fair-weather lateral profile for a 115-KV AC transmission line at 102 MHz. (After Spaulding and Disney, 1974)

where

$$_f\delta_E = 20 \log \frac{E(f_2)}{E(f_1)}$$

and

$$A(f) = G\lambda^2/4\pi = Kf^{-2}.$$

When the frequency increment is chosen to be one decade, $f_2 = 10 f_1$, Equation 3-1 becomes:

$$_f\delta_P = {}_f\delta_E - 20 \text{ dB} \tag{3-2}$$

In the above expressions, $E(f)$ is the noise electric-field intensity at frequency, f, and G is the gain of the detecting antenna, which for tuned monopoles and dipoles is independent of frequency. Using Equation 3-2, the electric-field decrement, $_f\delta_E$, of the power-line corona spectra shown in Fig. 3-24 lies between -5 and -10 dB and is comparable to the decrement of approximately -10 dB occurring for

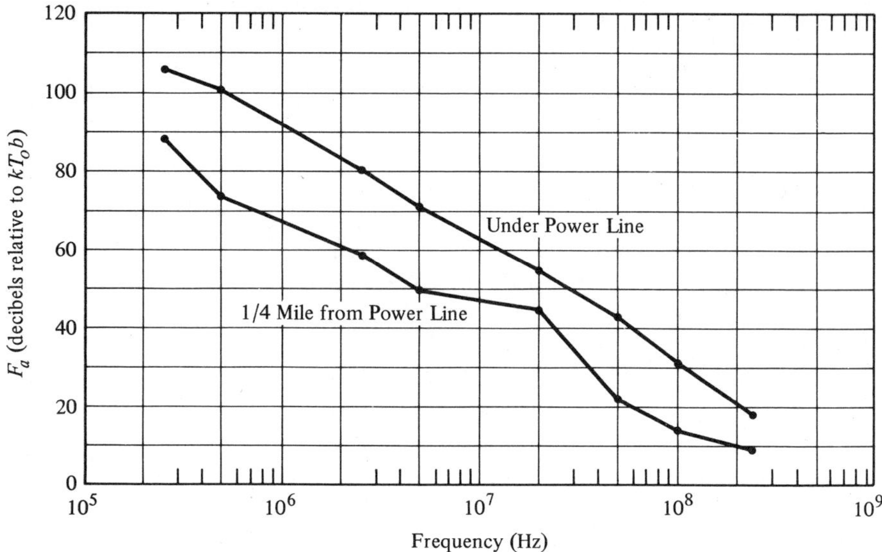

Fig. 3-24. Average power, F_a, fair-weather spectra for the 115-KV AC transmission line of Fig. 3-22 at two locations: under the line and 0.25 miles laterally displaced from the line. (After Spaulding and Disney, 1974)

the peak detected fair-weather corona emissions plotted in Fig. 3-16 for a 244-KV transmission line.

A 250-KV AC transmission-line lateral profile of corona-generated radio noise recorded at a frequency of 500 kHz is shown in Fig. 3-25.[16] The values of F_a at 500 kHz are also shown for the 115-KV line of Figs. 3-23 and 3-24. The background radio noise at 500 kHz arising primarily from atmospheric sources is observed to be sufficiently high to effectively truncate the range measurements at approximately 80 m for the 115-KV line.

Midday corona-noise spectra obtained during dry-line conditions for a 275-KV, 60-Hz transmission are given in Fig. 3-26 for two observing locations—under the line and at a lateral displacement of 300 ft.[17] The internal instrumentation noise level is indicated by a dashed line.

Figure 3-27 extends the average-noise-power spectra for a high-voltage, AC line under dry-weather conditions to an operating voltage of 550 KV. Measurements of the spectra at three lateral locations are shown. Notice the power-frequency decrements for the 275-KV line lies between -40 and -50 dB while for the 550-KV line, it ranges

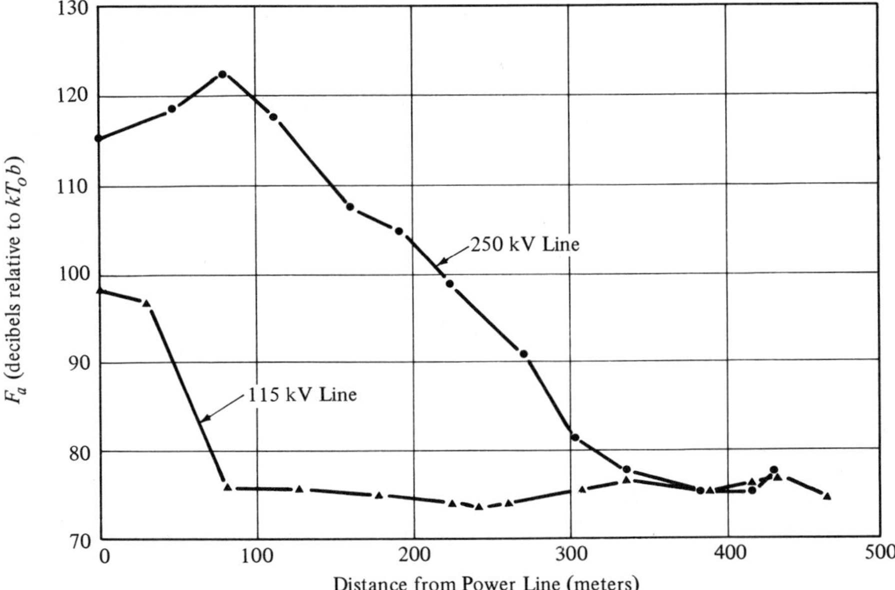

Fig. 3-25. Average power F_a, fair-weather lateral profiles for two high-voltage transmission lines observed at 0.5 MHz. Corona noise sources. (After Spaulding and Disney, 1974)

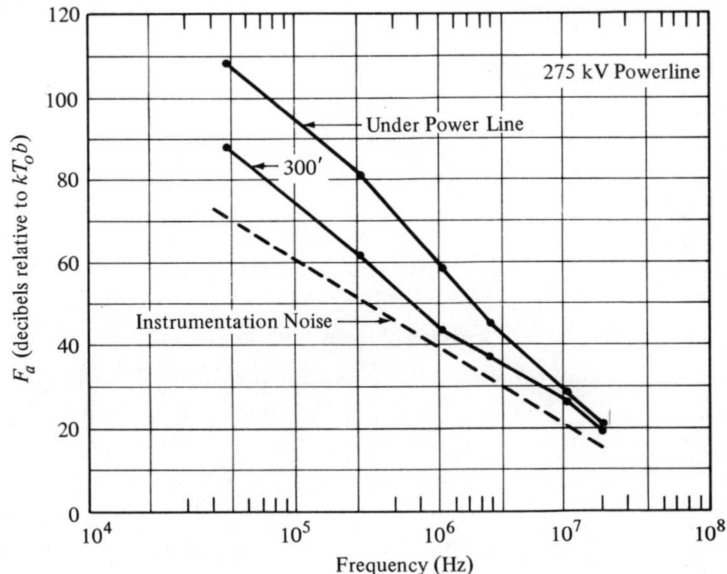

Fig. 3-26. Average power, F_a, fair-weather corona-radio-noise spectra for a 275-KV AC transmission line at two locations. (After Engles, 1970)

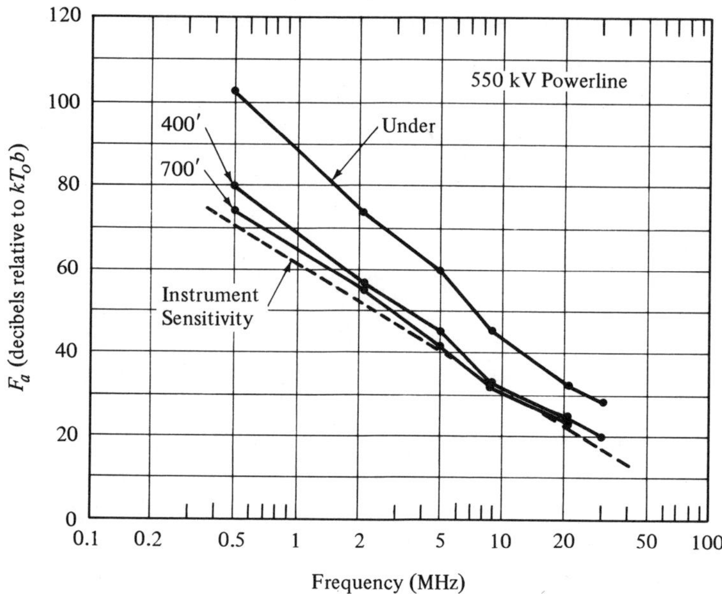

Fig. 3-27. Average power, F_a, fair-weather corona-radio-noise spectra for a 550-KV AC transmission line at three locations. (After Engles, 1970)

between −35 and −45 dB per decade frequency change. In both instances, the slope of F_a versus frequency is appreciably larger than for the spectral data presented in Fig. 3-24 obtained on a 115-KV line.

APD Measurements

High-voltage-line radiated-noise-envelope statistics obtained for conditions during which line-noise sources were predominantly corona discharges are available for a limited selection of transmission lines, frequencies, detection bandwidths, and weather conditions. Tunable receivers with linear dynamic ranges in excess of 40 dB; incorporating linear detectors, envelope sampling, quantizing, and recording assemblies have been used to amass the data from which several measures of the noise distributions have been extracted. The bulk of the statistical reductions have emphasized the preparation of first-order statistical measures—i.e., the APD; RMS envelope voltage, v_{rms}; average envelope voltage, v_{av}; and the ratio of v_{rms} to v_{av} known as V_d when expressed in dB.

Figure 3-28 presents both wet- and dry-weather APDs for a 115-KV transmission line whose lateral profile was plotted in Fig. 3-22.[9] The primary noise source present on the line during these studies was conductor and hardware corona. The recording point at which data were taken was displaced 50 ft laterally from the center of the line. Three receivers with detection bandwidths of 0.5, 3, and 6 kHz, respectively, were colocated and simultaneously used to record noise-voltage-envelope data at a frequency of 3 MHz. The data for three, no-rain conditions were acquired over a 30-min period during which a rain front approached the test site. It could be observed that rain was falling on portions of the line at a distance of several miles when the data (no rain) of Fig. 3-28(b) were measured and that the separation distance between the rain front and the observation point had further diminished when the data of Fig. 3-28(c) in no-rain curves were recorded. The length of the data samples from which these APDs were constructed varied with the case as follows: heavy rain, 5.6 min; no rain (Fig. 3-28[a]), 1.1 min; no rain, 6 min (Fig. 3-28[b]); and no rain, 5.9 min (Fig. 3-28[c]). Evidence of a large increase in rain-induced corona discharge and attendant radio noise is manifest in the comparative values of F_a included in the insets. Also provided in the inset tables are the values of V_d. High-voltage lines demonstrating corona-dominated noise emissions under fair-weather dry-line conditions react to the presence of heavy precipitation by erupting into a plethora of positive streamers emanating from the water droplets on the lower side of the catenary and from drop impingement and proximity passage of the rain enveloping the line. Although the differential increase in V_d between dry and heavy rain conditions is not large, and in Fig. 3-28(c) for the 500-Hz bandwidth receiver is reversed, the trend toward a more impulsive distribution as the transmission line is wetted is distinctive. This trend in V_d with changing weather conditions receives additional support from the data of Fig. 3-29, which were obtained for a 230-KV AC transmission line under three rainfall conditions and for a dry conductor.[9] These data for the 230-KV, double-circuit, vertically configured, single-conductor lines were obtained at a lateral position of 60 ft. The elevation of the lowest phase at the approximately midspan test point was estimated to slightly exceed 100 ft.[9]

A 775-KV, AC, test transmission line, horizontally configured with 45-ft phase spacing and four-cable bundles in 18-in. squares

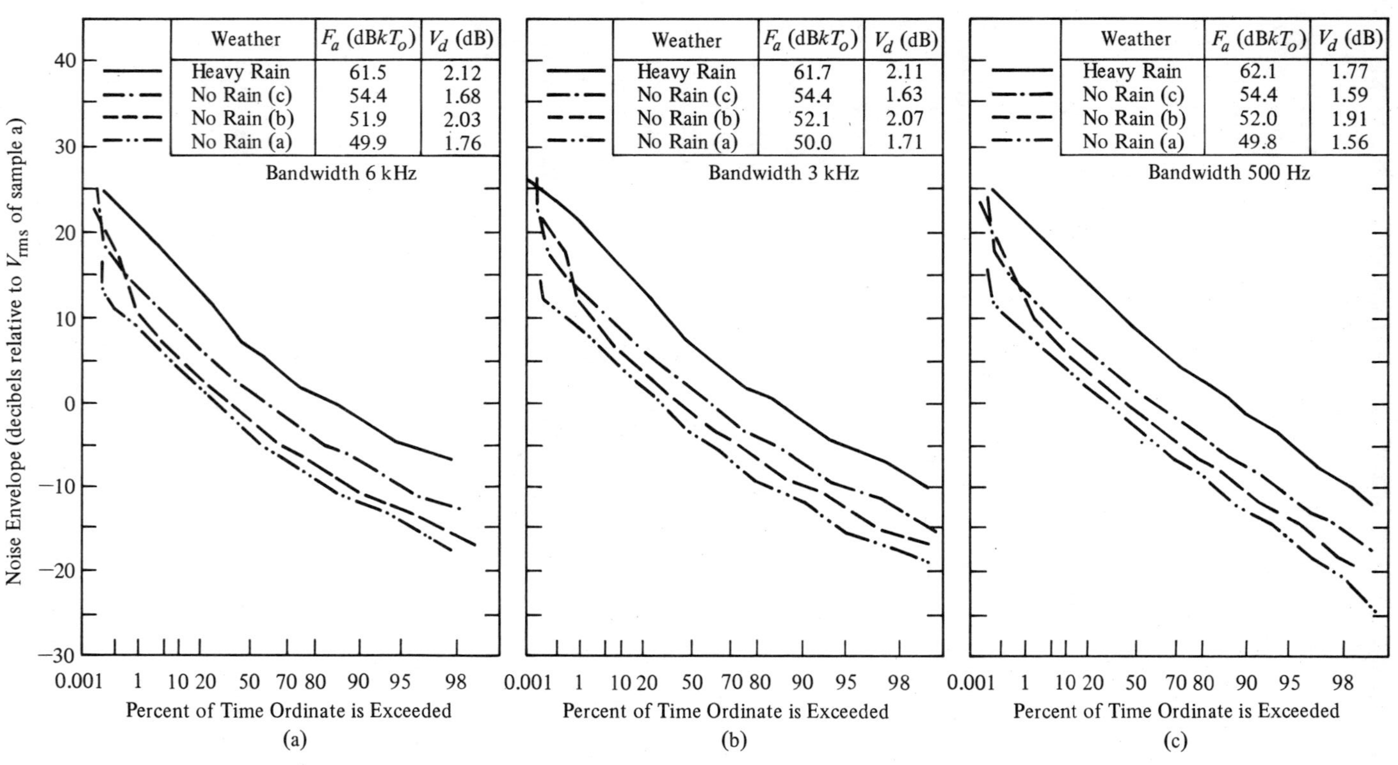

Fig. 3-28. Corona noise APDs for three conditions of dry line and one of heavy rain as measured at 3 MHz for a 115-KV AC transmission line. Test site was laterally displaced 50 ft from the line center. F_a and V_d are noted for each weather condition and each of three values of detector bandwidth 0.5, 3, and 6 kHz. (After Shepherd and Gaddie, 1976)

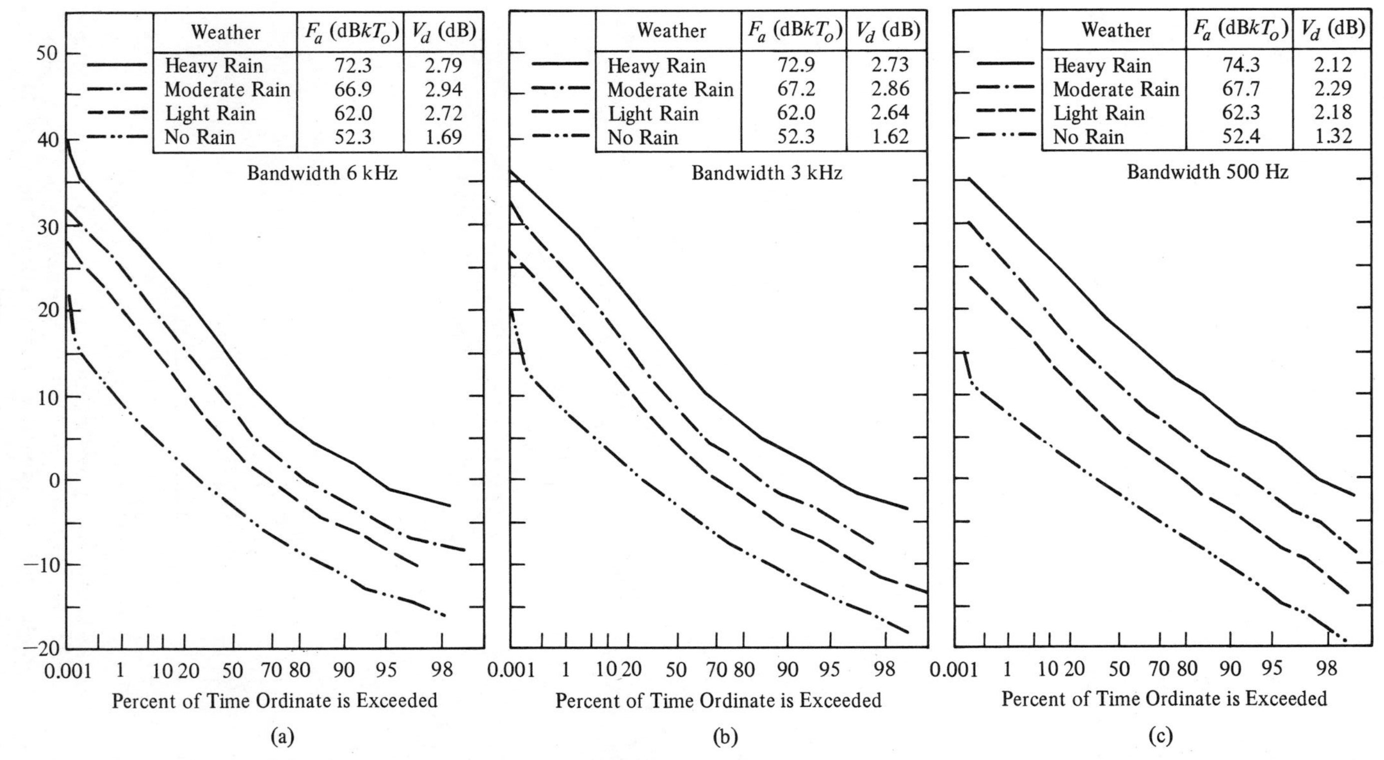

Fig. 3-29. Corona noise APDs for three conditions of wet line and one dry state as measured as 3 MHz for a 230-KV AC transmission line. Measuring location was 60 ft laterally displaced from the line center. F_a and V_d for each weather condition and each of three values of detector bandwidth 0.5, 3, and 6 kHz. (After Shepherd and Gaddie, 1976)

was used to study the statistical distributions of corona-generated noise under several meteorological conditions (Fig. 3-30).[13] With a linear receiver possessing an impulse bandwidth of 5.2 kHz and dynamic range of 90 dB, 50-sec samples of radiated corona noise were digitized and processed to obtain the APD. Independent instrumentation of 50-ohms impedance was used to measure the output voltage from an n-port antenna multicoupler connected to a

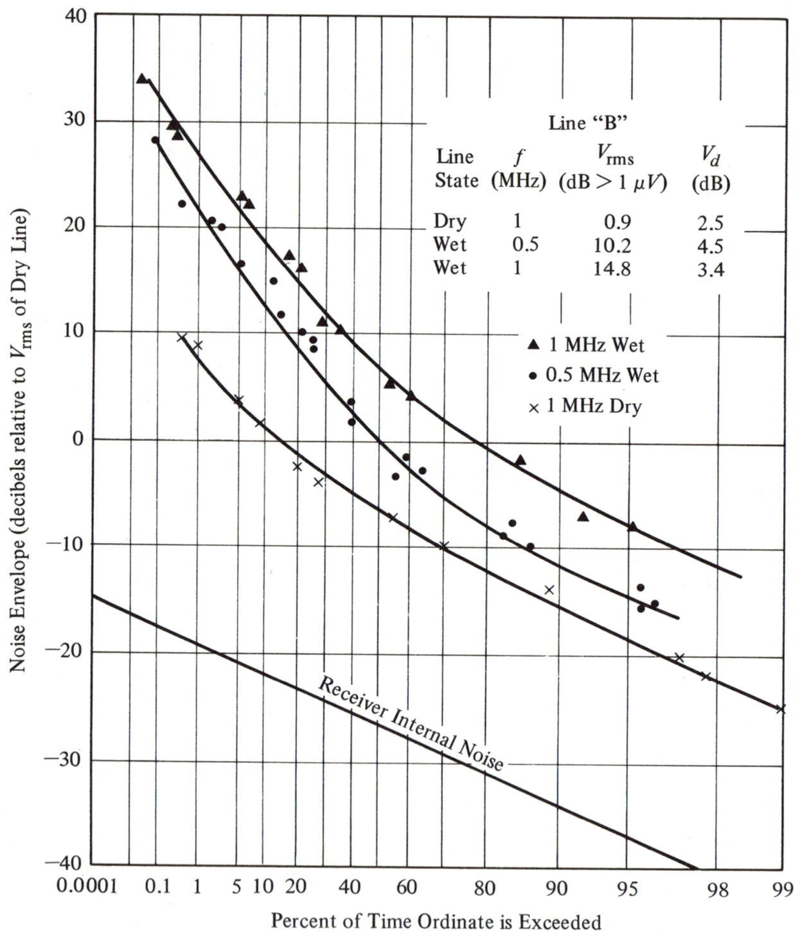

Fig. 3-30. Corona noise APDs for two weather conditions and two observation frequencies obtained on a 775-KV AC test transmission line ("B"). Observation point approximately beneath the outer phase. (After Lauber, 1976)

monopole 10 feet above the surface and beneath the outer phase. Two lines designated B and C, each suspended at a height of 70 ft, were employed. Lines B and C differed primarily in cable diameters, which were respectively, 1.2 and 1 in.

Observations for both wet and dry conditions at three test frequencies (0.5, 1.0, and 1.65 MHz) were used to construct the corona APDs. Wet-line data were obtained by chance, due to the nearly simultaneous occurrence of artificial wetting and a rainstorm. Comparisons of wet and dry conductor data are plotted in Fig. 3-30 for the B line. All APDs, including that of the receiver internal noise, have been normalized with respect to the receiver input RMS voltage, v_{rms}, observed at 1 MHz under dry-line conditions. Shown in the insert are the measured values of V_{rms}, given in decibels relative to

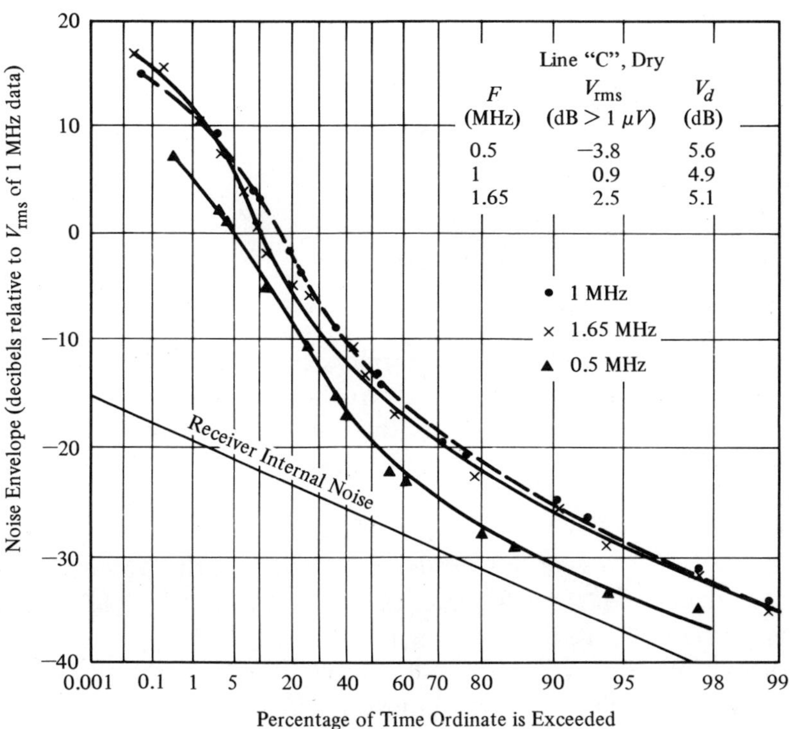

Fig. 3-31. Corona noise APDs for dry-line conditions and three observation frequencies obtained on a 775-KV AC test transmission line ("C"). Observation point approximately beneath the outer phase. (After Lauber, 1976)

1 μV and V_d for each experimental condition. The associated values of $\overline{F}_a$ are relatable to v_{rms} and V_{rms} as

$$\overline{F}_a = 10 \log \frac{v_{\text{rms}}^2}{kT_oR} = V_{\text{rms}} + 67 \text{ dB} \tag{3-3}$$

for R = 50 ohms.

It may be observed in Fig. 3-30 that in addition to increases in the median and rms values of the envelope voltage, impulsive index, V_d, has increased as the line condition changed from dry to wet conductor. Increased impulsiveness of the corona noise was also noted during the investigations yielding Figs. 3-28 and 3-29 when wetted conductor conditions developed.[9]

Figure 3-31 supplements the APD data of Fig. 3-30 by the addition of line *C* results. The APD for an additional frequency, 1.65 MHz, is also provided for comparison with the distributions measured at 0.5 and 1 MHz. All data sets apply to dry-line conditions. The distributions of Fig. 3-31 have been normalized with respect to the 1 MHz value of V_{rms} to facilitate comparisons with Fig. 3-30. It is notable that V_d for dry line *C* is appreciably larger at all observation frequencies than dry-line *B* at 1 MHz.

RADIO NOISE FROM STATIONARY GENERATORS AND SUBSTATIONS

Power substations and line transformers are susceptible to the occurrence of gap discharge and corona breakdown, which are productive of noise radiations comparable to those observed from both low- and high-voltage transmission lines. Loosely grounded conductors, faulty insulators, high-resistance metal contacts, and miscellaneous loose hardware may exist in any substation requiring periodic maintenance for elimination. In medium and large substations, the potentiality for the occurrence of several gap-discharge sources is large, and the likelihood of noise radiation is high.

Pole-mounted power-line transformers are susceptible to fault development similar to that found on substation components. Line transformers normally operating at voltages lower than substation transformers are not free of gap-discharge faults, although the radiated-noise-field strength may be less. Further, corona emissions are comparatively less evident for pole transformers primarily because of their lower operating voltages.

Studies of noise emissions from transformer substations have been infrequently performed largely because substations represent a small fraction of the noise-producing electric power facilities.

Line transformers are only occasionally examined individually for radio-frequency radiations, typically only if they are unusually noisy. Other times, radio emissions arising from pole transformers are difficult to segregate from the noise attributable to the transmission line.

Three examples of substation radio-noise spectra are presented in Figs. 3-32 to 3-34. In none of these studies was the investigation sufficiently detailed to establish the generic source of the noise, although it may be inferred that corona sources are unlikely to have made significant contributions to the peak-noise-field strength of the 46-KV substation (Fig. 3-32).[8] The data of Fig. 3-32 were obtained

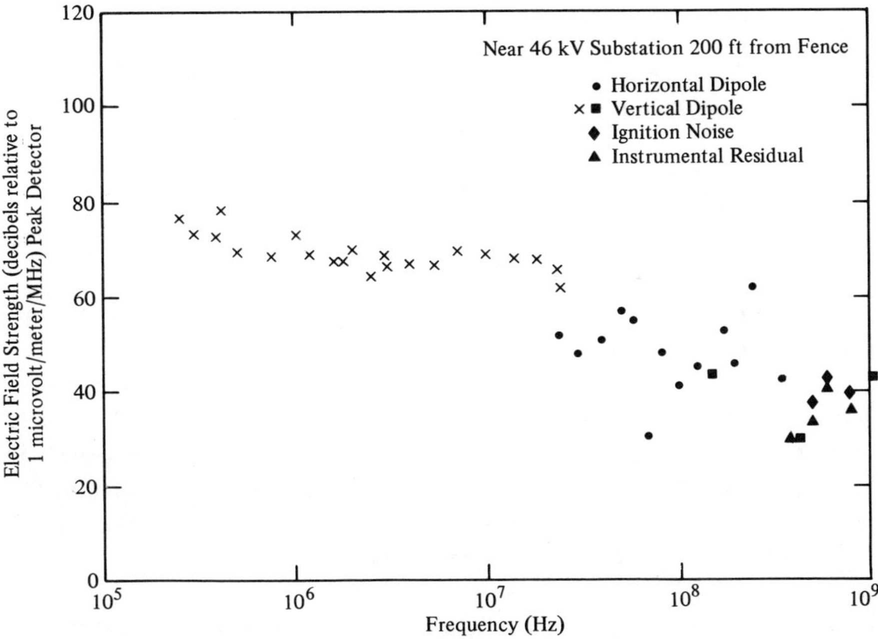

Fig. 3-32. Peak detected radio-noise spectra observed at a distance of 200 ft from a 46-KV transformer substation. (After Pakala et al., 1967)

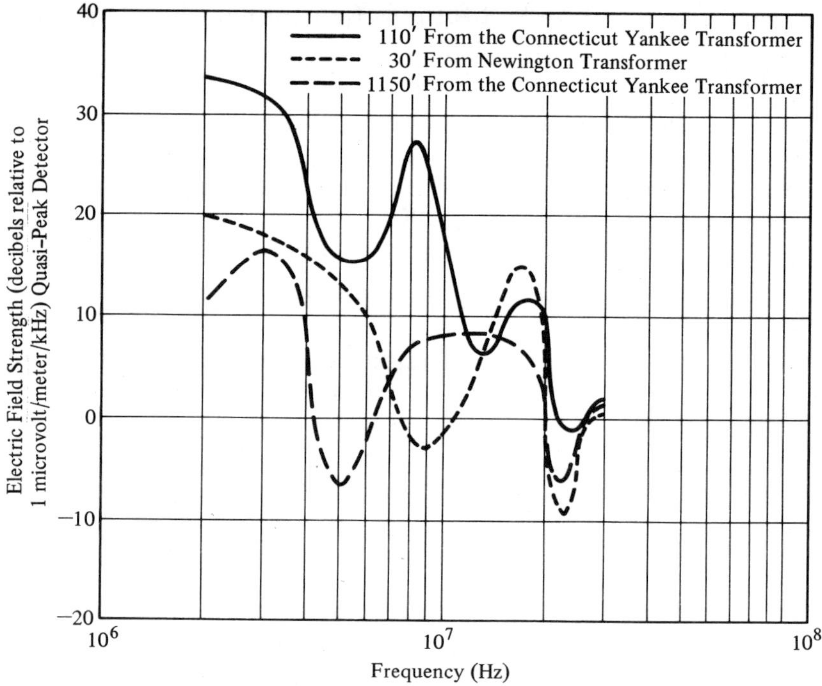

Fig. 3-33. Quasi-peak radio-noise spectra generated by corona noise sources on two 345-KV transformer substations. (After Cook, 1975)

using linearly polarized antennas located approximately 6 ft above the surface and at a range of 200 ft from the station. These fair-weather results may be compared with the lower plot of Fig. 3-5, which was measured at a point 90 ft laterally displaced from the line. Substation interference is seen to be appreciably more intense than normal-line noise radiation when adjustment for the differing separation distances is performed.

In Fig. 3-33, the quasi-peak detected radio-noise spectra obtained at three separation distances from two 345-KV transformer stations are shown. A considerable difference is noticed between the corona-noise levels generated by the two transformer installations. In all cases, the results were obtained during fair weather. From onsite examination of the emitted noise waveform, it was concluded that corona-discharge emissions were the predominant noise-generating mechanism.[18]

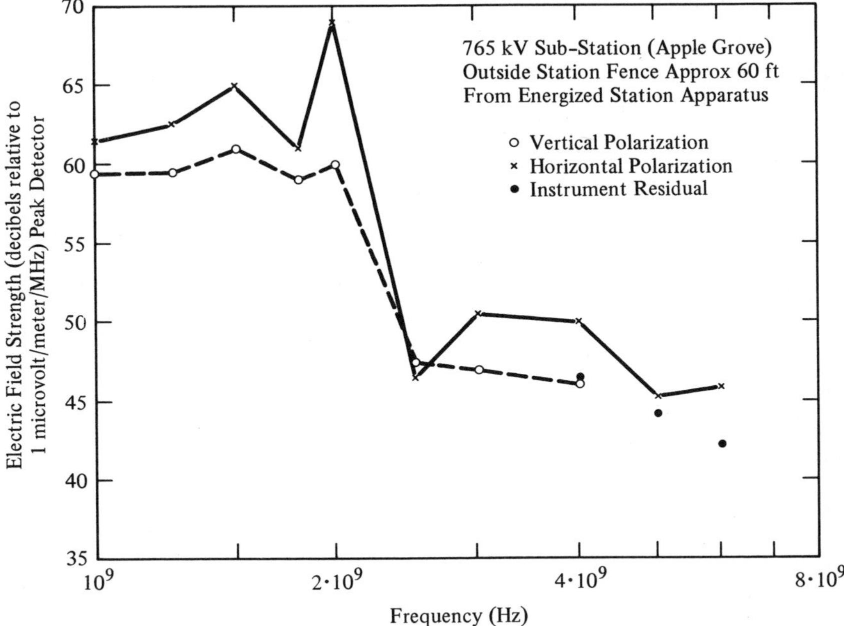

Fig. 3-34. Peak radio-noise field strength spectra for a 765-KV AC transformer substation measured at a distance of approximately 60 ft. (After Pakala et al., 1968)

In Fig. 3-13, lateral profile and quasi-peak spectra data are plotted for the dry-line corona noise arising from a 345-KV transmission line. When these line corona data are compared with the plot of Fig. 3-33, the results indicate that substation noise emissions are approximately 10 dB larger than those for fair-weather lines at frequencies above 8 MHz. Further, at the lower frequencies, the reverse appears to be true. To arrive at this comparison, the assumption is made that corona emissions are approximately thermal in character and that the radiated field strength has Gaussian distributed quadrature components. This is generally found to be a good approximation as is evidenced by the values of the impulsive index, V_d, listed in the insets of Fig. 3-29 for the no-rain condition on a 230-KV line. Recall that the value of V_d for Gaussian-distributed noise is 1.05 dB. Thus, for a quasi-peak detector exposed to thermal noise, the detected voltage is proportional to the square root of the noise bandwidth.[19]

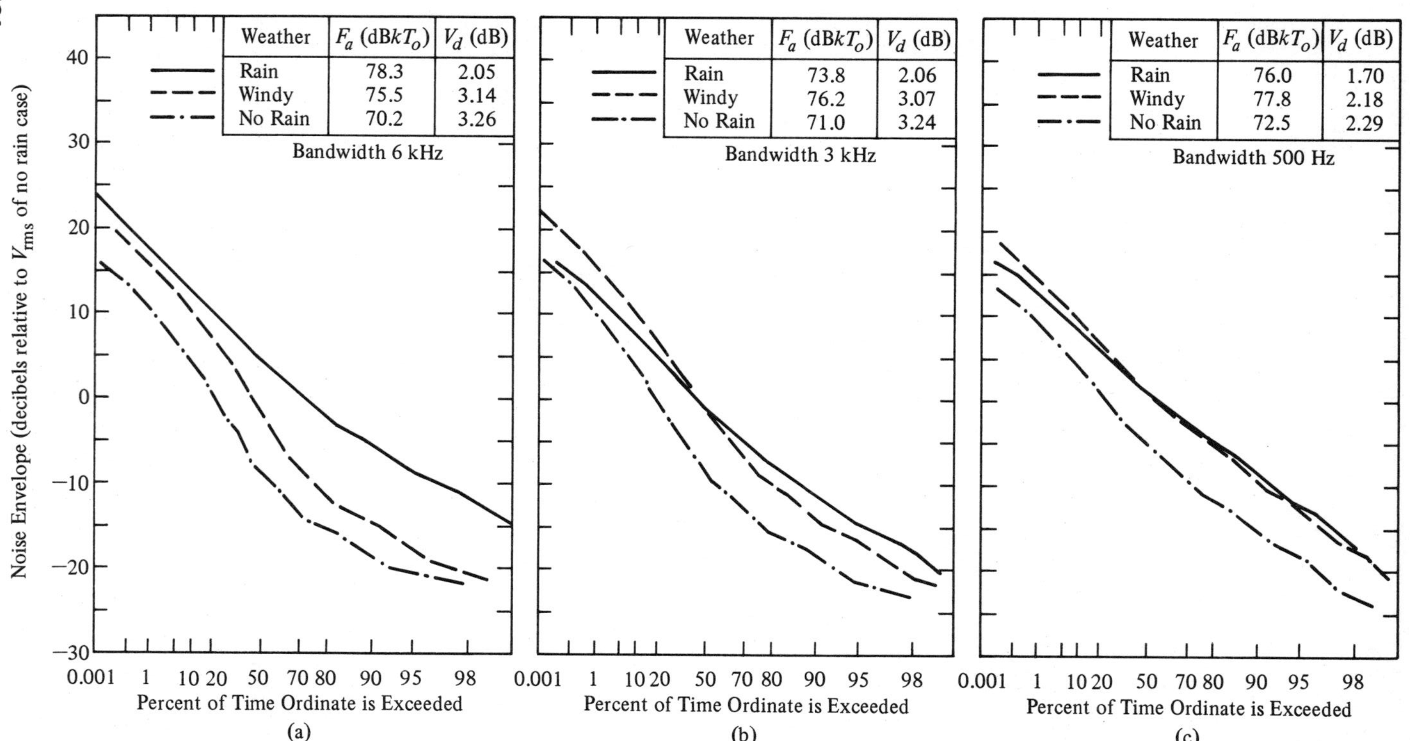

Fig. 3-35. Corona-discharge-generated radio-noise APDs obtained at 3 MHz for a 230-KV transformer substation at a point near the entry of the transmission line into the station. Lateral displacement of the observation point from the line was 50 ft. Three weather conditions: rain, wind following rain, and no rain. (After Shepherd and Gaddie, 1976)

Peak detected radiated-noise-field strength arising from transformer substation corona sources are shown in Fig. 3-34 for the range of frequencies from 1 to 6 GHz.[14] The noise sources producing the large peak-field strengths below 2 GHz were not identified during the study; it is possible they consisted of more than corona emissions. Comparison of the data shown in Fig. 3-34 with peak detected 735-KV transmission-line corona noise in Fig. 3-19 accentuates the point that transformer substations may exist as major incidental-noise emitters. In the frequency interval of 1 to 2 GHz, where comparison is possible, the substation radio interference exceeds normal fair-weather corona noise by a minimum of 10 dB at comparable observer-to-source separation distances.

A set of APD plots for three weather conditions and three receiver detection bandwidths obtained at a frequency of 3 MHz near a 230-KV transformer substation are assembled in Fig. 3-35.[9] These data were recorded at a location near the entry point of the high voltage line into the substation, approximately 50 ft from the line. The sources of the interference were established as corona discharges occurring on the power facilities during the three meteorological states. The condition noted as *Windy* occurred immediately following a rainstorm and while drying was in progress. Each data sampling from which an APD was assembled was accumulated over a period of 5 min or more, with the longest interval equal to 6 min.

Data in Fig. 3-29 may be used for comparison; they represent the incidental noise that is generated by line corona on a 230-KV transmission line for four weather conditions. Both figures represent 3-MHz noise-envelope distributions obtained at approximately equal distances from the noise-producing structures. Identical receiving and data-reduction equipment was employed to obtain and process the data in Figs. 3-29 and 3-35.

There is no general differentiating trend or form notable in the APDs. The most pronounced distinctions found between the transformer substation and line-corona data reside in the values of F_a and V_d. For comparable weather conditions, the transformer substation values of F_a are unusually larger than for the line measurements; e.g., for the no-rain state, the differences in F_a exceeds 15 dB. Notice that the values of mean noise levels differ by 10 dB or less when rain is falling. Also discernible in the respective insets are the greater values of V_d found for the transformer substation during the no-rain conditions. One notices further that the measure of noise impulsiveness observed for the transformer installation decreased during the advent of rain, a direction of change opposite to that encountered in line-corona noise emissions.

THEORY OF TRANSMISSION-LINE-GENERATED RADIO NOISE

Noise field intensities arising from gap discharges occurring between unbonded metal parts, on defective or dirty insulators, between elements of pole or line hardware have been addressed in experimental studies, the finding of which were presented in the preceding section. Theoretical prediction of gap-discharge radiated-noise-field intensity has not shared comparable interest; as a result, there exists no theoretical model for gap-discharge radio-noise-level dependence upon operating-line electrical and mechanical parameters, type or details of the gap sources, line configurations, frequency, observation-point locations, or weather conditions. No generalization of the experimental data that have been assembled from various and diverse transmission line studies is presently supportable theoretically.

Corona-generated radio-noise fields, which encompass the radiation from all discharge and noise sources on lines operating above 70 KV except those identifiable as gap discharges, and improperly operating or isolated electrical apparatus attached to the line have been studied theoretically in an effort to develop a field-intensity prediction model. Concentration of corona noise analytical modeling efforts has persisted within the range of frequencies of commercial AM broadcasting.

Line separation distances to which the analysis is applicable are determined by the ambient noise level (exclusive of line corona emissions), terrain topography, and the minimum detectable sensitivities of commercially available quasi-peak radio-noise meters. These combined restraints fix the maximum observation range and limits of theoretical interest at approximately 1 mile.

Formulations of corona radio-noise models are presently unable to proceed from the physical processes associated with the corona discharge mechanism. In all approaches, there exist a set of undetermined constants and coefficients that require experimental evaluation before the functional dependence of noise level upon the basic electrical and geometrical characteristics of the transmission line, observation frequency, observer location, and meteorological conditions is complete.

APPROXIMATE CALCULATIONS OF CORONA-NOISE-FIELD INTENSITY

Two formulations of the corona-noise-source models are presently in use, each differing with regard to the translation of the primary noise variables into the measurable noise levels. The more recently introduced and less widely used methodology treats radio-noise generation from the viewpoint of the electrical currents flowing on the line, which give rise to the attendant radiated-noise fields. The undetermined constants that appear therein, which presently require experimental determination in laboratory facilities of specialized design, are in effect birth coefficients or excitation functions for the conducted noise currents.[20,21] Once the excitation functions have been experimentally established, the total noise currents at the source point are known. Use of the transmission-line equations permits translation of these currents and the associated potential fields between adjacent conductors and ground to any desired location along a line. At an observation point in the vicinity of the line, the noise field strength is then computed from the electrostatic gradient of the total potential.

The analytical procedure most commonly employed for prediction of corona radio emissions represents the total observed noise-field strength as the product of several terms, each associated with an identifiable aspect of either the generation or propagation of the interference. Individual terms appearing in the expression for observed noise-field strength are treated as independent. Each contains at least one constant or coefficient that requires independent experimental determination. Two particular advantages exist for the latter, often designated as the *comparative method* of analysis. (1) the coefficients and constants appearing in the several correction terms may be determined using operational or test transmission lines rather than specialized laboratory facilities, and (2) the physical process contributory to radio-noise generation and propagation are individually identified, permitting a greater insight into the analytical model and the physics of the corona noise mechanisms.

All comparative-method representations of the noise-field intensity, E, may be expressed in the following form for each phase of an AC line:[22]

$$E = E_o + E_g + E_d + E_n + E_D + E_f + E_{FW} \tag{3-4}$$

which yields E in decibels relative to 1 μV/m in a detection bandwidth established by the measurement of the radio-noise reference value E_o. All terms appearing in Equation 3-4 have been transformed by the operation 20 $\log_{10}$ (e_x/e_{ox}), where the respective reference value of field intensity e_{ox} in V/m are selected for implementational simplicity. Although the comparative method may be applied to field-strength determinations using any signal-detector function, it has become accepted practice in the power industry to employ quasi-peak detectors with receivers conforming to either ANSI or CISPR standards.

The definitions of the remaining terms in Equation 3-4 follow:

- E_g = the conductor gradient factor referenced to a gradient somewhat below corona breakdown
- E_d = the conductor diameter or bundle subconductor (cable) correction factor
- E_n = the correction factor for the number of subconductors (cables) in a conductor bundle
- E_D = the observation-point distance correction factor, a ratio of the radial distance to the observer to a reference distance
- E_f = the observation-frequency correction factor, conventionally referenced to either 0.5 or 1 MHz
- E_{FW} = the foul-weather correction factor, which is referenced to the long-term fair-weather mean-corona-noise level observed at the reference radial-separation distance.

Considerable diversity among reporting investigators exists in both the analytical forms of the last six terms appearing in Equation 3-4 and the values of the constants and coefficients present in the term. In a study to evaluate the ability of the several variant forms of Equation 3-4 to predict corona radio-noise level, 23 AC transmission lines with operating potentials between 220 and 750 KV of known electrical characteristics and line geometries were examined as test cases.[22,23] Fair-and foul-weather radio-noise levels for each transmission line were available for comparison with the theoretical predictions of Equation 3-4.

Results of this exercise, performed for frequencies between 0.5 and 1 MHz, provided for each method the means, variances, and cumulative distributions of the differences between calculated and measured noise field intensity for fair- and foul-weather conditions. Inspection of the mean, variance, and the interdecile differences (the difference between the 90-percentile and 10-percentile values) of the residuals reveals that one comparative method, designated as *400 KV-FG(Germany)*,[24] yields significantly smaller values of the residual error, when averaged over all tested conditions of weather and line voltages, than do the others.

The functional forms of the terms appearing in Equation 3-4 employed by the 400 KV-FG(Germany) comparative method and the attendant constant values are presented for use in the prediction of the corona radio-noise field strength:

E_o = 53.7 dB relative to 1-μV/m, quasi-peak, ANSI noise meter. Frequency of 1 MHz.
 = 52.7 dB relative to 1 μV/m, CISPR noise meter. Frequency of 1 MHz.[22,25]

E_o is the median value of a long-term distribution of fair-weather noise levels. The standard deviation of the distribution is 5 dB. The difference of approximately 1 dB between the ANSI and CISPR median value of E_o is comparable to the average operator-to-operator variation when observing corona radio noise at a fixed displacement from a transmission line.[25]

This gradient factor is representable as:

$$E_g = K_g(g_m - 16.95)$$

where

g_m = the line-surface gradient in KV rms/cm

K_g = 3 for line voltage of 750 KV, or 3.5 for lines of lower voltage whose surface gradient g_m lies between the limits of 15 and 19 KV/cm.

$$E_d = 40 \log_{10} (d/3.93)$$

for conductor diameter d in cm.

$$E_n = 10 \log_{10} (n/4); n \geqq 1$$

where n = the number of subconductors (cables) in each conductor.

$$E_D = 20 K_D \log_{10} (20/D)$$

where D = the radial distance of the observation point from the conductor in meters.

$K_D = 1.6 \pm \sigma_D$ for $0.5 \leqq f \leqq 1$ MHz.
$\sigma_D = 0.1$ dB, the standard deviation of K_D.
$E_f = 0$ for ANSI instrumentation
$E_f = 1.25/(1 + f^2)$ for CISPR instrumentation.

In the preceding, f is the observation frequency in MHz. The latter relationship applies only to CISPR standard equipment.

$E_{FW} = 0$ for fair weather.
$E_{FW} = 17 \pm \sigma_{FW}$ for foul weather.

The value of σ_{FW} is equal to 3 dB and represents the standard deviation of E_{FW}.

The value of noise-electric-field intensity computed using Equation 3-4 and the terms developed by 400 KV-FG(Germany) apply to a single phase of a three-phase transmission line. The procedure is to compute E for each phase of the line and to select the two largest values, E_1 and E_2. These are combined to obtain the line-noise-field intensity, E_{line}, using

$$E_{\text{line}} = \tfrac{1}{2} (E_1 + E_2) + 1.5 \text{ dB} \tag{3-5}$$

relative to 1 μV/m, or, if one value exceeds the others by 3 dB or more, setting it equal to E_{line}. The units of E_1 and E_2 appearing in Equation 3-5 are decibels relative to 1 μV/m.

For many transmission lines, the value of the surface gradient, g_m, is available either from computations performed during the course of design or from measurements. When it becomes necessary to independently compute g_m as a preliminary step to evaluating E_{line}, several basic formulas are available for general use. For three-phase transmission lines with either single or bundle conductors containing one or more subconductors of minimum separation, s, and radius, r, the potential gradient at the conductor is given, per KV of line potential, U, as:[26]

$$\frac{E}{U} = F_1 = \frac{1 + s^{-1}\,[2(n-1)r \sin(\pi/n) \cos\theta]}{nr \ln\{2h/[\sqrt[n]{r}\, s^{1-1/n} \sqrt{1 + (2h/D)^2}]\}} \quad (3\text{-}6)$$

where

E = conductor gradient in KV/cm
U = line operating potential in KV
F_1 = gradient factor in KV/cm/KV
S = minimum separation distance between subconductors in cm
n = number of subconductors in each bundle conductor
h = conductor height above the surface in cm
θ = angle shown in Fig. 3-36 for bundles with $n \geqslant 2$ and is measured relative to the direction of maximum gradient
S = geometrical mean of the subconductor spacing, s, shown in Fig. 3-36 and is given by

$$S = (s_{12} s_{13} \cdots s_{in})^{1/(n-1)} \quad (3\text{-}7)$$

for which

$$n = 2 \text{ and } s = s$$
$$n = 3 \text{ and } s = s$$
$$n = 2 \text{ and } s = s\sqrt[6]{2}$$

D = modified geometric mean spacing of the phases given for the center, D_c, and outer, D_o, phases by

$$D_c = \sqrt{d_{ab} \cdot d_{bc}}$$
$$D_o = \sqrt{d_{ab} \cdot d_{ac}}$$

where the phase spacings are, as shown in Fig. 3-37, measured between the bundle centers.

The approximations represented by Equation 3-6 introduce an error in the center-phase conductor gradient of less than 2 percent if phase spacings are 20 ft or greater and if subconductor diameters range from 0.8 to 2.8 in. A lesser error for the outer phases occurs, which decreases with increasing phase spacings.[26] The indicated range

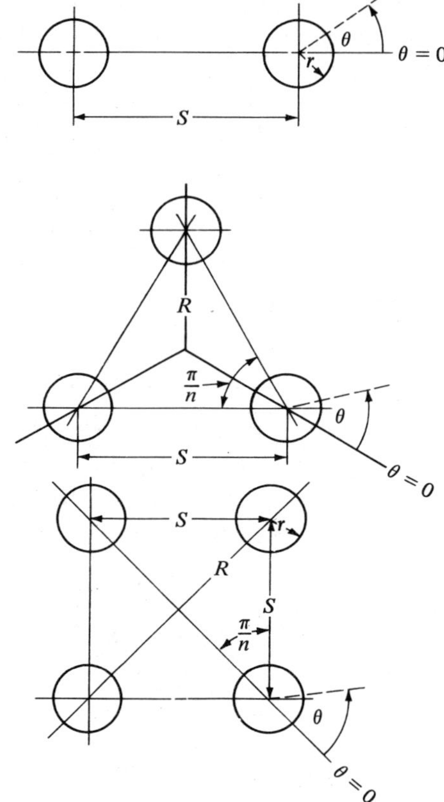

Fig. 3-36. Geometry of a bundle conductor for the cases $2 \leqslant n \leqslant 4$.

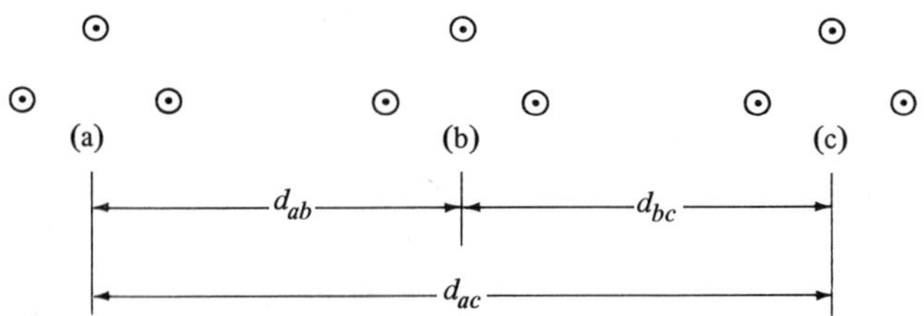

Fig. 3-37. Phase spacing geometry for a bundle conductor line.

of the parameters d and D for which the gradient error is less than 2 percent encompasses most high-voltage line configurations of interest. Plots of F_1 computed for $D < 20$ ft and $d > 2.8$ in. without the noted approximations are available for single[27] and horizontal configuration bundle conductors.[28] Finally, a closed form representation of F_1, applicable, without the aforementioned approximation limitations, to a three-phase AC line with bundle conductors is provided in Appendix 1.[29]

References

1. Loeb, L. B., The Mechanism of the Trichel pulses of short time duration in air. *Physical Review* **86**: 256–257 (1962).
2. Newell, H. H., Liao, T., and Warburton, F. W. Corona and RI caused by particles on or near EHV conductors: I–Fair weather. *IEEE Trans. Power Apparatus and System* **PAS-86** (11) 1375–1383 (1967).
3. Nigol, O. Analysis of radio noise from high-voltage lines: I—Meter responses to corona pulses. *IEEE Trans. Power Apparatus and Systems* 524–533 (1964).
4. Newell, H. H., Liao, T., and Warburton, F. W., Corona and RI caused by particles on or near EHV conductors: II—Foul weather. *IEEE Trans. on Power Apparatus and Systems* **87** (4): 911–927 (1968).
5. La Forest, J. J. Seasonal variation of fair-weather radio noise. *IEEE Trans. Power Apparatus and Systems*. **PAS-87** (4): 928–931 (1968).
6. Gehrig, E. H., Peterson, A. C., Clark, C. F., and Rednour, T. C. Bonneville Power Administrations 1100 KV direct current test project: II—Radio interference and corona loss. *IEEE Trans. Power Apparatus and Systems* **PAS-86** (3): 278–290 (1967).
7. Hinchman Corporation. Report and Criteria of the Method of Measurement and Recommended Allowable Limits of Electromagnetic Interference Voltages from High Voltage Transmission Lines. October 1957.
8. Pakala, W. E., Taylor, E. R., Jr., and Harrold, R. T. High Voltage Power Line Siting Criteria, Vol. 1, High Voltage Power Line Test Results Analyses and Appendices. RADC-TR-66-606, March 1967.
9. Shepherd, R. A., and Gaddie, J. C. Measurements of APD and the Degradation Caused by Power Line Noise at HF. Stanford Research Institute, April 1976.
10. Hagn, G. A. MF and HF Man-Made Radio Noise and Interference Survey-KEFLAVIK, Iceland. Stanford Research Institute Technical Memoranda No. 1, February 1972.
11. Specifications for C.I.S.P.R. Radio Interference Measuring Apparatus for the Frequency Range 0.15 MHz to 30 MHz. Publication No. 1, 1961. Specifications for C.I.S.P.R. Radio Interference Measuring Apparatus for the Frequency Range 25 MHz to 300 MHz, Publication No. 2. International Electrotechnical Commission, Geneva.

12. Specifications for Radio Noise Field Strength Meters 0.15 MHz to 30 MHz, C-63.2-1963. Specifications for Radio Noise Field Strength Meters 20 to 1000 MHz, C-63.3-1964. American National Standards Institute, New York.
13. Lauber, W. R. Amplitude probability distribution measurements at the Apple Grove 775KV project. *IEEE Trans. Power Apparatus and Systems* **PAS-95** (4): 1254–1266 (1976).
14. Pakala, W. E., Chartier, V. L., and Harrold, R. T. High Voltage Power Line Siting Criteria. RADC Technical Report RADC-TR-68-316. November 1968.
15. Maruvada, P. S., and Trinh, N. G., A basis for setting limits to radio interference from high voltage transmission lines. *IEEE Trans. Power Apparatus and Systems* **PAS-94** (5): 1714–1724 (1975).
16. Spaulding, A. D., and Disney, R. T. Man-Made Radio Noise, Part 1: Estimates for Business, Residential and Rural Areas. U.S. Dept. of Commerce, Office of Telecommunications, Report 74-38, June 1974.
17. Engles, J. W. Electromagnetic noise surveys at two sites. *Conference Record National Security Agency Man-Made Noise Conference* 5.1–5.57 (1970).
18. Cook, J. C. The EMC of a Proposed Power Plant and the HF Communication Facility at Westover AFB. Electromagnetic Compatibility Analyses Center, ESD-TR-75-006, May 1975.
19. Geselowitz, D. B. Response of ideal radio noise meter to continuous sine wave, recurrent impulses and random noise. *IRE Trans. Radio Frequency Interference* **RFI-3** 2–10 (1961).
20. Magnein, M., Cladé, J., and Gary, C. La Station Experimentale de l'Electricité de France pour l'Etude de l'Effect Couronne sur les Future Lignes à trés' Hautes Tensions. CIGRE Paper No. 427, 1966.
21. *EHV Transmission Line Reference Book*. New York: Edison Electric Institute, 1968.
22. IEEE Radio Noise Subcommittee. Comparison of radio noise prediction methods with CIRGR/IEEE survey results. *IEEE Trans. Power Apparatus and Systems* **PAS-92** (3): 1029–1042 (1973).
23. CIGRE Study Committee No. 36 and IEEE Radio Noise Sub-Committee. CIGRE/IEEE survey on extra high voltage transmission line radio noise. *IEEE Trans. Power Apparatus and Systems* 1019–1028 (1973).
24. 400-KV-Forschungsgemeinschaft e.v., 69 Heidelberg, 1, West Germany.
25. IEEE Committee. Comparison of various RI meters and reading comparison of RI meter operators on a 735 KV line. *IEEE Trans. Power Apparatus and Systems* **PAS-87** (5): 1249–1259 (1968).
26. Pakala, W. E., and Taylor, E. R., Jr. A method for analyses of radio noise on high-voltage transmission lines. *IEEE Trans. Power Apparatus and Systems* **PAS-87** (2): 334–345 (1968).
27. Adams, G. E. Voltage gradients on high-voltage transmission lines. *AIEE Trans. (Power Apparatus and Systems)* **74**: 5–11 (1944).
28. Temoshok, M. Relative surface voltage gradients of grouped conductors. *AIEE Trans. (Power Apparatus and Systems)* **67**: 1583–1591 (1937).
29. C.I.S.P.R. Publication 1, Second Edition 1972. Specification 2 for C.I.S.P.R. Radio Interference Measuring Apparatus for the Frequency Range 0.15 MHz to 30 MHz. Geneva: International Electrotechnical Commission, 1966.

4

Industrial, Scientific, Medical, Consumer, and Transportation Sources of Radio Noise

Contributors of man-made radio-noise, recognized as either less intense or less frequently encountered in metropolitan areas than automotive or power facility sources, yet capable of adversely affecting wireless receiving equipment, may be assembled into three principal categories. (1) industrial, scientific, and medical (ISM), (2) consumer products, and (3) transportation. These group designations have become accepted as a result of definitions endorsed by the Federal Communications Commission (FCC) and industrial participants.

By associating industrial, scientific, and medical equipment that employ radio waves for purposes other than the communication of information, the FCC has established an equipment nucleus containing most of the intentional-radio-noise generators that are uninvolved in information transfer. These radiation sources have been segregated with respect to their operating frequency from all other radio-frequency equipment and restricted in maximum radiated power within their assigned bands and within adjacent portions of the radio spectrum where potential harmonics, intermodulation products, and sideband spectral components might be present. Seven band assignments have been reserved for ISM equipment, each centered, as indicated in the following, with the spectral occupance widths as noted:[1]

I. 13.56 MHz ± 6.78 kHz
II. 27.12 MHz ± 160 kHz
III. 40.68 MHz ± 20 kHz
IV. 915 MHz ± 13 MHz
V. 2.45 GHz ± 50 MHz
VI. 5.86 GHz ± 75 MHz
VII. 24.125 GHz ± 125 MHz

One category of industrial facilities—namely, industrial heating equipment—is permitted to function throughout the radio spectrum, except for the following, where they are totally excluded:

490 kHz to 510 kHz
2.17 MHz to 2.194 MHz
8.354 MHz to 8.374 MHz.

Elsewhere in the spectrum, industrial heaters that use radio-frequency energy to manufacture articles or process materials must suppress all radiation, except that which fall within the seven ISM bands, to the greatest extent possible. However, they must never exceed 10μV/m at a 1-mile range for all frequencies below 5.725 GHz.

Radio-diathermy equipment, one of several radioactivated medical tools, is specifically regulated by the FCC. While permitted to function without power constraints in any ISM band, the harmonic and spurious emissions of diathermy transmitters may not exceed 25μV/m at or beyond 1000 ft for any unassigned frequency. Diathermy spectral assignment and out-of-band radiation constraints have been applied, with one change, to miscellaneous ISM equipments located in nonresidential areas and generating radio-frequency power P_o, in excess of 500 W. This change admits of an out-of-band emission equal to (1) $(P_o/500) \times 25\mu$V/m at or beyond 1000 ft or (2) the field strength permitted to industrial heating apparatus, whichever is least. Some intermixing of miscellaneous ISM equipment such as the apparatus used in the production of physical, biological, or chemical effects or the acceleration of atomic or nuclear particles may, on occasion, be located in predominantly residential areas. When so located and when operated below 1 GHz, the radiation level at and beyond 1000 ft may not exceed 25μV/m for any frequency outside the ISM bands.

The radiation from radio-frequency–stabilized arc welders is the subject of an FCC Docket Action 11467.

Consumer products and commercially employed apparatus that do not depend upon radio-frequency energy to execute their intended function, yet which nevertheless emit electromagnetic energy, are known as *incidental-radiation devices.* Operators of such equipment assume the responsibility for preventing or eliminating interference that adversely affects the functioning of radio navigation or safety-radio services or that seriously degrades, obstructs, or repeatedly interrupts radio communications services. Although many consumer and commercial products are included in this categorization, e.g., electric appliances, this set is not all-inclusive. There remain consumer and commercial products that require radio energy in their operation, which, may interfere with essential radio services. Designated as *restricted radiation devices,*[2] these apparatuses are assembled into two subgroups. (1) low-power communications devices (LPD), e.g., radio-controlled door openers, and (2) field disturbance sensors (FDS), e.g., radio-frequency intrusion sensors.[1] Any defined member of either subgroup is prohibited from operating if it generates harmful interference to any radio service. Specific radiated field-strength-range limitations are applied in several portions of the radio spectrum to guarantee additional interference protection to certain critical radio services.[2] For each subcategory, selected segments of the spectrum are reserved for implementations requirings high-radiated-electric-field strengths. Typically, regulations constrain high-power applications to frequencies above 470 MHz.

Sources of interference arising from the operation of such electrified vehicles as trains, buses, or automobiles not employing radio control are presently regulated only by extensions of the definition of incidental radiation devices.

Equipment belonging to each of the three principal categories may, even when functioning properly, produce significant levels of radiation that constitute interference for the users of the authorized radio services. In addition, when any equipment malfunctions or operates improperly, the consequence is often an increase in in-band or adjacent channel interference or both.

In the following sections, measured radiation levels and spectral ranges from typical examples of each category of apparatus are presented, compared, and discussed.

INDUSTRIAL EQUIPMENT

The wide variety of industrial equipment that employs radio-frequency energy to perform its intended function includes several classes that have become known as *intense incidental noise sources*. All manufactured examples of each class do not necessarily generate intense radiation fields but sufficient numbers of each do to warrant attention. Dielectric or plastic processing apparatus, electric ovens, induction heating facilities, radio-frequency welding and soldering tools, and electric-discharge metal-machining devices are production-line machines that have received specific experimental examination because of the levels of their radio-frequency emissions.

Dielectric and Plastic Processing Apparatus

The plastic-article manufacturing industry employs radio-frequency apparatus to preheat bulk samples of plastic for various purposes, including molding and encapsulating, and cutting, shaping, sealing, marking, and embossing plastic stock and articles. High-power radio-frequency oscillators provide the source of thermal energy. Electromagnetic fields tightly coupled to the organic materials produce internal power dissipation and heating as a consequence of the large, complex component of permittivity existing in many plastics at and above 1 MHz. Excitation frequencies, methods of coupling the radio energy into the material, drive power, and degree and sophistication of interference suppression vary among the commercially available equipment. Power-line filtering is provided on most presently available apparatus, preventing the impression of the fundamental and its harmonics on the distribution lines. Recently designed apparatus tend to use higher operating frequencies, which manifest greater attenuation coefficients for the power-line conducted leakage.

Harmonics of the fundamental drive frequency are generated in oscillator nonlinearities. When coupling to power line exists, the presence of two or more operating units may produce intermodulation products in one or both assemblies. Harmonics, intermodulation products, and base frequencies can radiate from a unit, either through the leads interconnecting the oscillator circuit to the working tool, which may be a pair of electrodes between which the work is placed, or through imperfections in the oscillator cabinet shield. Use of short-shielded leads, interference suppression filtering, well-designed cabinets, shielding of the high-power oscillator, and minimization of aperture sizes are techniques that have been applied successfully to interference reduction.

The radiated interference from dielectric or plastic processing apparatus is characterized by a line spectrum containing an intense fundamental frequency component, f_o, lying in either the HF or low VHF band plus a harmonic spectrum whose amplitudes remain appreciable for frequencies extending well into the UHF band.

Figure 4-1 provides the line emission spectra for two models of plastic preheaters, one rated at 7.5 kW input power and the other at 13 kW. The noise radiation for the 7.5-kW unit was measured at two separation distances 25 and 100 ft, in each case using a vertically polarized linear antenna.[3] The fundamental oscillator frequency for the 7.5 kW preheaters was 78 MHz. Twelve harmonics are evident (1) below 1 GHz at a range of 25 ft and (2) at approximately half that number when the range is increased to 100 ft. In Fig. 4-1, the first harmonic lying at 155 MHz is masked by the presence of a coherent signal. The maximum linearly polarized electric field strength observed at a distance of 100 ft from a 12.5-kW plastic preheater is shown in Fig. 4-1(c).[4] The fundamental frequency of the preheater was 70 MHz, yielding 13 harmonics below 910 MHz with measurable amplitudes. The higher power preheater in this comparison yields much larger noise spectral components, a trend that is not always found.

Dielectric and plastic processing apparatus, used for bonding, embossing, and cutting manufactured articles and often called *plastic welders*, generate line spectra as shown in Fig. 4-2 for four commercial product examples. In Fig. 4-2(a)–(c), the radiated vertically polarized levels for three identical units operating at 3.5 kW are shown for a measurement station 14 ft from each welder.[5] The model represented by these data is of recent design and was in new condition at the time of study. The variations noticeable in the amplitudes of the emission lines and in the numbers of harmonics present above measuring instrumentation noise reveal the individuality that this apparatus category possesses. Measurements were not pursued above 400 MHz in these investigations, yet large 11th and 12th harmonics of the 26–27-MHz fundamental frequency are seen to occur in each case,

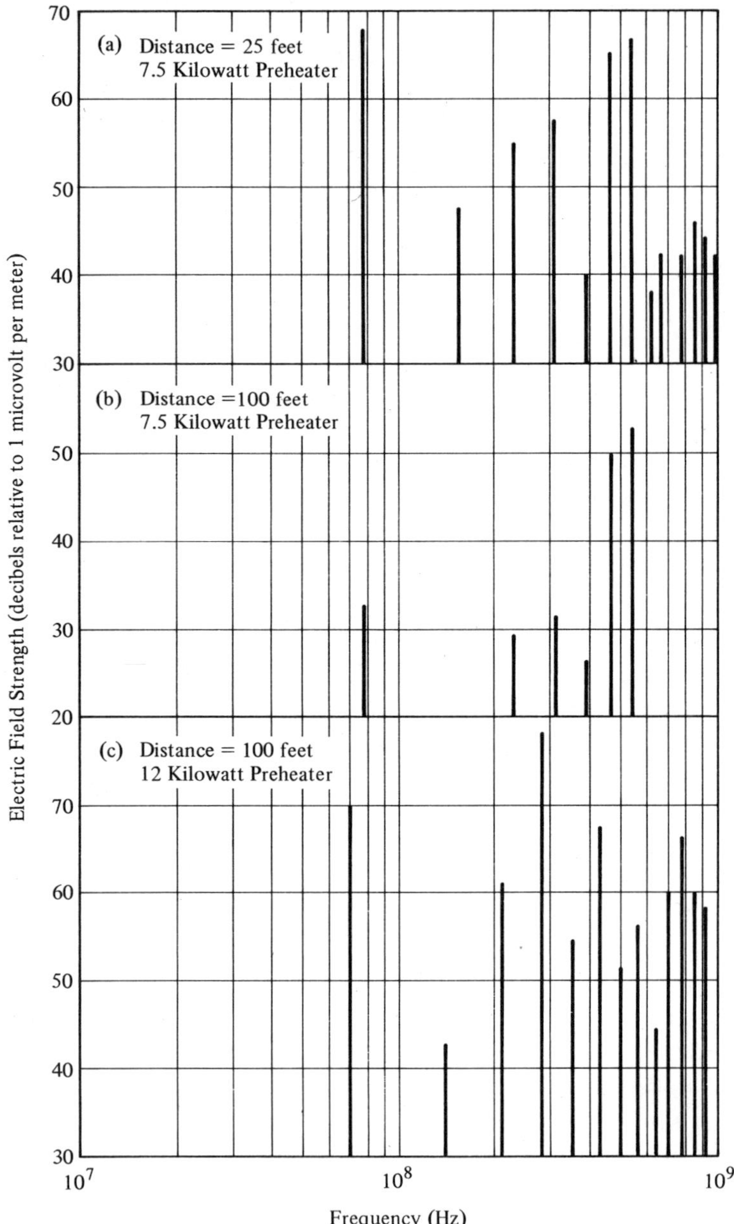

Fig. 4-1. Comparison of the radiated-noise spectra from two models of dielectric preheaters. (*a*) and (*b*) 7.5-kW units observed at distances of 25 and 100 ft, respectively. (After Kernaghan and Ford, 1971) (*c*) 12-kW unit at 100 ft. (After Pearce and Bull, 1964)

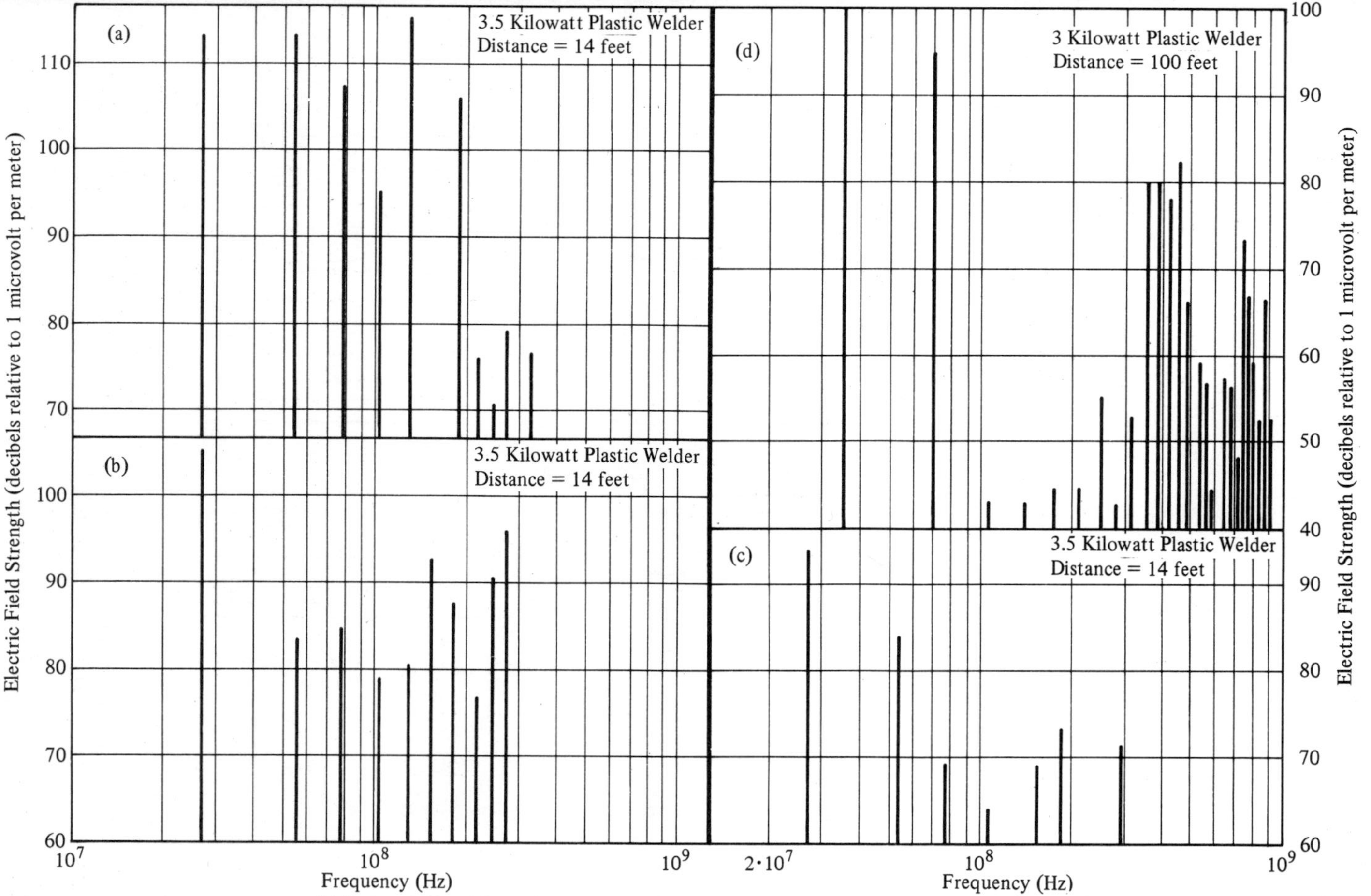

Fig. 4-2. Radiated noise spectra of dielectric processing apparatus (plastic welders). (*a*), (*b*), (*c*) Single unit emissions from three new 3-kW units at a distance of 14 ft. (After Martin and Tabor 1972) (d) Radiated noise from a 3.5-kW unit of different design observed at 100 ft. (After Pearce and Bull, 1964)

implying the presence of significant-level harmonics of even higher order. Figure 4-2(d) provides data for a plastic-processing (welder) unit of a different manufacturer, rated at 3 kW.[4] The greater separation distance of 100 ft is a factor to be considered when comparing these spectral components in Fig. 4-2(d) with those of Fig. 4-2(a)–(c). Design distinctions most probably have substantially contributed to

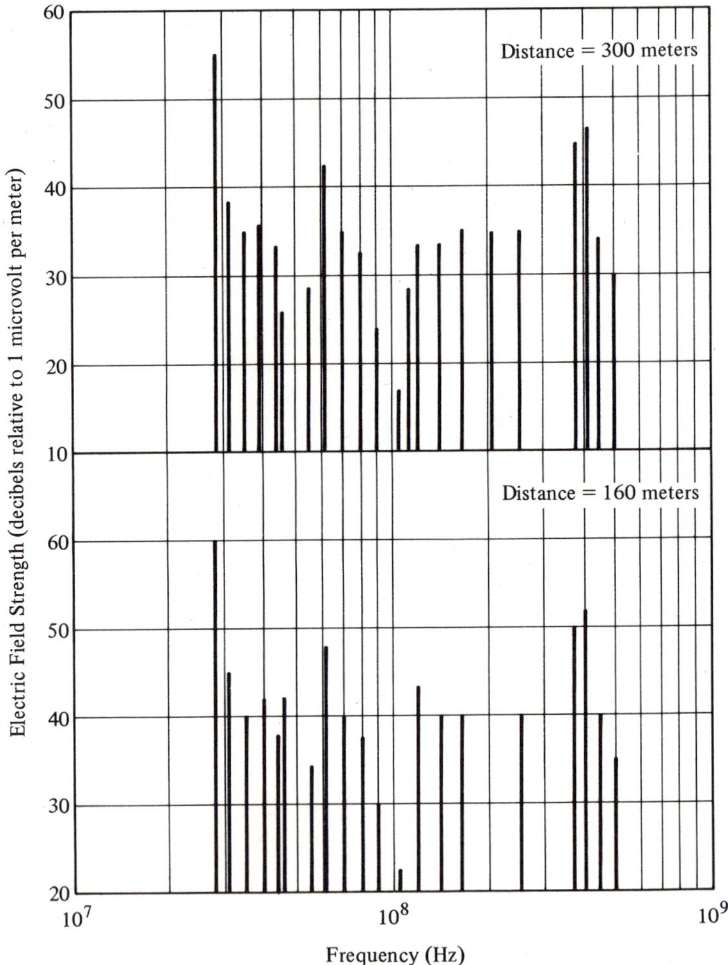

Fig. 4-3. Radiated noise spectra of approximately 400 dielectric processing apparatus (plastic welders) operating in a manufacturing facility. (After Kernaghan and Ford, 1971)

the observed level differences in the higher harmonics. The fundamental frequency for Fig. 4-2(d) lies at 35 MHz, and the highest shown harmonic, the 26th, falls at 910 MHz, which was the frequency limit of the study and not necessarily the highest harmonic which might be resolvable above the instrumentation noise. Linearly polarized antennas were used in the fourth example. These data represent the largest observed values for any antenna orientation.

The individual dieletric processor data given in Fig. 4-2 provide representative examples of emission levels when the apparatus is operated in isolation and the observation point is nearby, unshielded by building facilities and outer walls. Manufacturing installations normally employ several plastic fabrication units, operating concurrently, colocated, and powered from several power buses. Variations in operator techniques, unit age, and condition of repair produce a complicated line spectrum when factories employing several plastic welders are examined for radiated noise level during active production periods. Representative of such conditions are the data of Fig. 4-3, which were recorded at two distances, 160 and 300 m, beyond a factory containing approximately 400 operating units. The measurements were performed using a horizontally polarized antenna situated on the street outside of the factory.[3] The fundamental frequency of most of the units was 28 MHz. Attenuation of the factory walls, which were of light construction, plus scattering from the walls and nearby objects are effects included in the reported data. Range attenuation of the radiated fields is noticeable in the spectra although the amount of added loss with increasing range is seen to be nonuniform across the frequency interval. Although no data were taken above 500 MHz, it is likely this class of dielectric fabricating equipment emits signal components of appreciable energy content at frequencies greater than the upper limits of these measurements.

Nonmetallic-Materials Processing Equipment

Various manufacturing processes often require batch heating of components and assemblies. Handling of the required materials is reduced and simplified by employing conveyer belts that transport items at controlled rates through a programmable heat cycle or heating profile. Equipment satisfying this need are (1) an enclosed, insulated tunnel containing a motor-driven conveyer belt for parts transport and (2) a heat source. When the materials to be heat-treated possess an appreciable radio-frequency complex component of permittivity (dielectric loss tangent), heating may be affected by radio-frequency energy injected into the conveyer oven by properly placed high-frequency coupling fixtures. Oven design is motivated by the need to efficiently provide a heating profile while minimizing the radiation of the high-frequency energy beyond the enclosure. Thus, careful attention is paid to electromagnetically shielding the tunnel and external cabling. However, residual signal leakage may arise from leads, joints, seams, and conveyer-belt openings.

Fig. 4-4 provides the radiated line spectra of one 25-kW radio-frequency-excited conveyer oven measured at a distance of 30 m for three conditions of entrance and exit openings. (a) 50 in. by 12 in., (b) 36 in. by 3 in., and (c) openings closed.[4] The magnitudes of the spectral components for (a) and (b) are shown by solid vertical bars at each emission frequency. The results for both entrance and exit oven openings closed are shown by open circles on the lower plot. At frequencies of 105, 140, and 175 MHz, radiated levels were 28, 23, and 27 dB above 1μV/m for (c). Substantial radiation exists in each of the three cases shown at the 26th harmonic of the 35-MHz fundamental frequency. These data are maximum values for either linearly polarized component. Measurements were not performed above 910 MHz but it may be presumed that substantial harmonic energy persisted above this frequency at the observation range of 30 m.

In the manufacture of wooden furniture and other wooden articles, gluing of joints is accomplished by placing the items to be assembled into a two-surface fixture after application of the glue. The fixture used for this operation is of parallel-plate construction; each plate is connected to a terminal of a radio-frequency circuit. Spacing between the plates is adjusted to the article thickness. Internal work heating is produced by virtue of the high-frequency-complexed permittivity of the wood and glue. Radiation of the drive-frequency and nonlinearity-generated harmonics arises from the capacitive work gap and machine leads. Many radio-frequency-excited gluing units employ less than 10 kW of primary power and fundamental frequencies in the high-frequency band. An example of the line spectra emitted by a 3 kW wood gluer is shown in Fig. 4-5, measured at a distance of 30m using a vertically polarized antenna. When the two-plate work circuit was

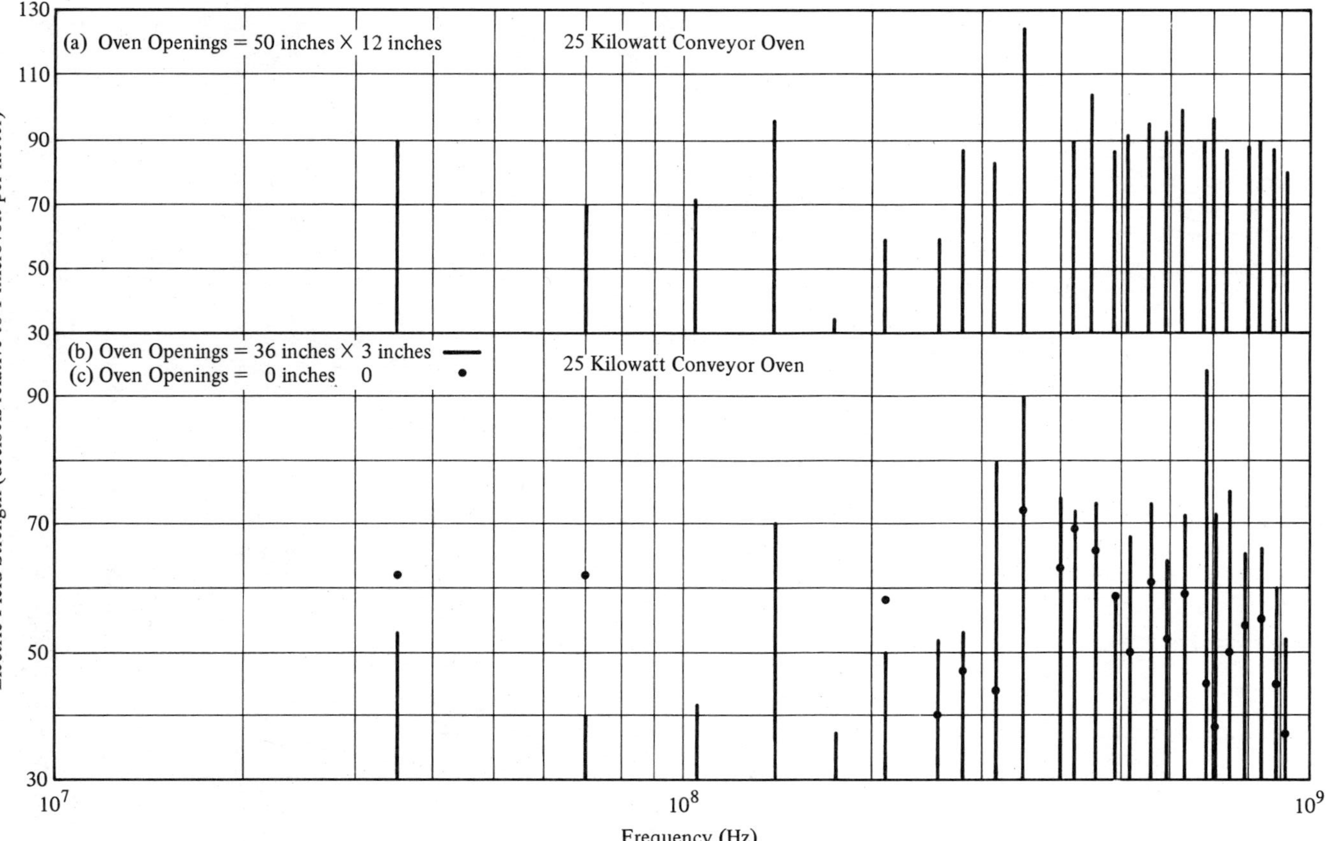

Fig. 4-4. Radiated noise spectra for a 25-kW conveyor oven measured at a distance of 30 m. Three sizes of entrance and exit oven openings are shown: (*a*) 50 in. by 12 in., (*b*) 36 in. by 3 in., and (*c*) zero openings. (After Pearce and Bull, 1964)

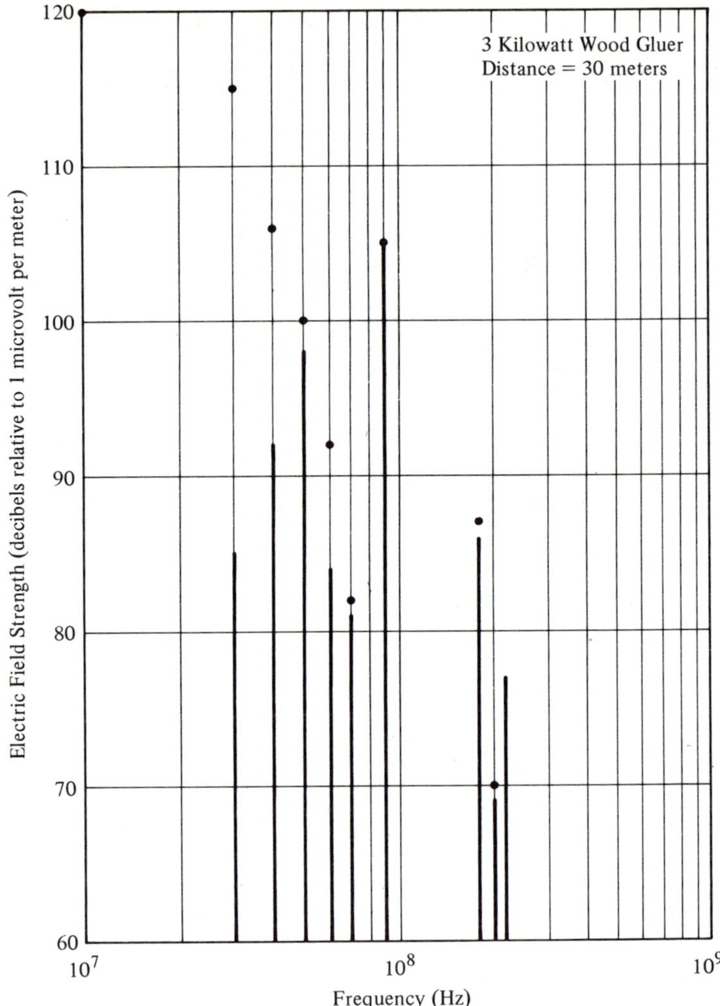

Fig. 4-5. Radiated noise spectra for a 3 kW wood gluer measured at 30 m. Solid lines are measurements with short leads connecting the work fixture. Open circles represent the data when leads of 2-m length are used. (After Pearce and Bull, 1964)

connected to the radio-frequency source by short leads (lengths less than 1m), the radiated noise levels were as shown in the figure by the solid vertical lines. As the leads were increased to 2 m, the observed interference at a distance of 30 m altered to the levels indicated in Fig.

4-5 by the open circles. The fundamental frequency for the unit was 10 MHz. Notice that the harmonics of highest intensity are odd-numbered. Significant radiation was observed at frequencies up to the 22nd harmonic (i.e., 220 MHz), and, although measurements were not extended to higher frequencies, additional spectral components may be presumed to exist above this limit.

Metal-Products Manufacturing Equipment

Radio-frequency induction heating equipment is used in metal-products manufacturing industries as an intense source of localized heat to affect soft soldering of metal seams, curing of high-temperature insulating films, baking of protective layers on metal parts, and annealing of machined metal members. In order to generate the large eddy currents required to quickly accomplish these assembly operations, large inductively coupled electromagnetic fields are necessary. Radio-frequency oscillators producing 10-kW or more of average output power and loop-coupled to the work volume typify the central design characteristics of the machines. The radio-noise emissions of induction heaters manifest a fundamental frequency lying below the AM broadcast band. Analogous to plastic heaters, harmonics are generated in the circuit nonlinearities and radiated by the power unit and interconnecting leads to the work volume.

The radiated line spectra observed at 10 ft is presented in Fig. 4–6 for a 20-kW radio-frequency induction heater used to cure the high-temperature insulation applied to the armatures of fractional horsepower electric motors.[3] Maxima of the radiated noise observed with a linearly polarized antenna are plotted for harmonics through the 28th of the fundamental 500-kHz oscillator frequency. No significant noise-signal power was detected above 14 MHz. Broadcast station interference corrupted the measurements at the harmonic frequencies of 1, 1.5, 5, 9, 12.5, 13, and 13.5 MHz, which have been delected.

An example of the radiated-noise spectra produced by a 30-kW heater used for rapid soft soldering of metal plates is shown in Fig. 4-7 for distance of 30 ft.[3] Electric-field-strength maxima observed with a linearly polarized antenna oriented to detect the largest noise intensities are shown. The radio-frequency oscillator was operated at 4 kW above nominal rating and at a fundamental signal frequency of

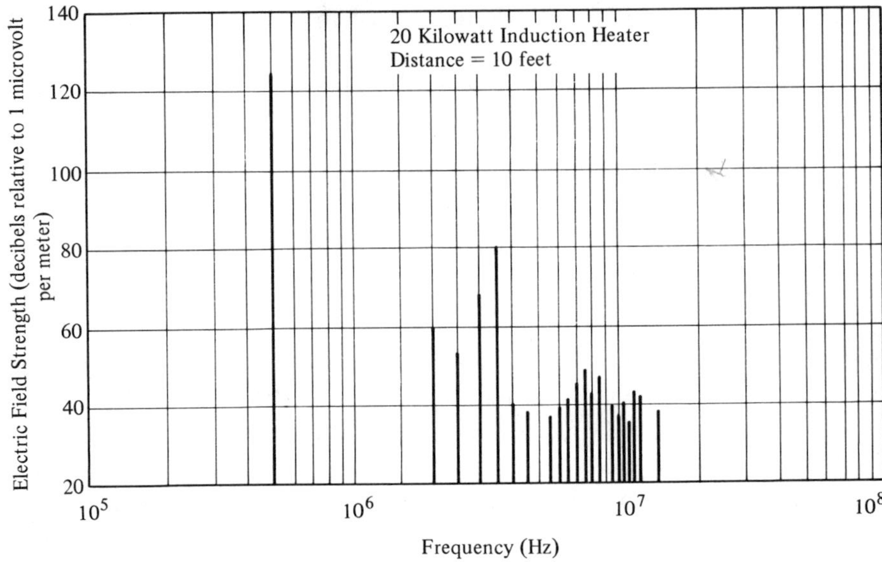

Fig. 4-6. Radiated-noise-line spectra from a 20-kW radio-frequency induction heater observed at 10 ft. Maximum values of linearly polarized, noise-field intensity. (After Kernaghan and Ford, 1971)

378 kHz when these data were obtained. Notice that no appreciable radiation was observed above 220 MHz.

An induction heater rated at 40 kW but operating at the time of measurement at 13.5 kW generated the line spectra plotted in Fig. 4-8. Units of 40-kW rating are commonly employed to anneal and strain-relieve machined metal parts such as those found in internal combustion engines.[3] For the data of Fig. 4-8, the observation distance was 50 ft. Linearly polarized antennas were used and oriented to record the maximum values of the radiated-noise field whose components consisted of the fundamental oscillator frequency, 350 kHz, and harmonics through the 75th.

The radio-noise-field intensities shown in Figs. 4-6–8 typify induction-heater emissions for near-field observer locations. The effect of range attenuation upon the noise intensity for the portion of the spectrum above 1 MHz may be obtained by comparing the induction-heater data shown in Figs. 4-6–8 with the curves of Fig. 4-9 recorded at a distance of 1000 ft. The data in Fig. 4-9 represent line spectra, the maxima for all orientations of a linearly polarized antenna. Peaks

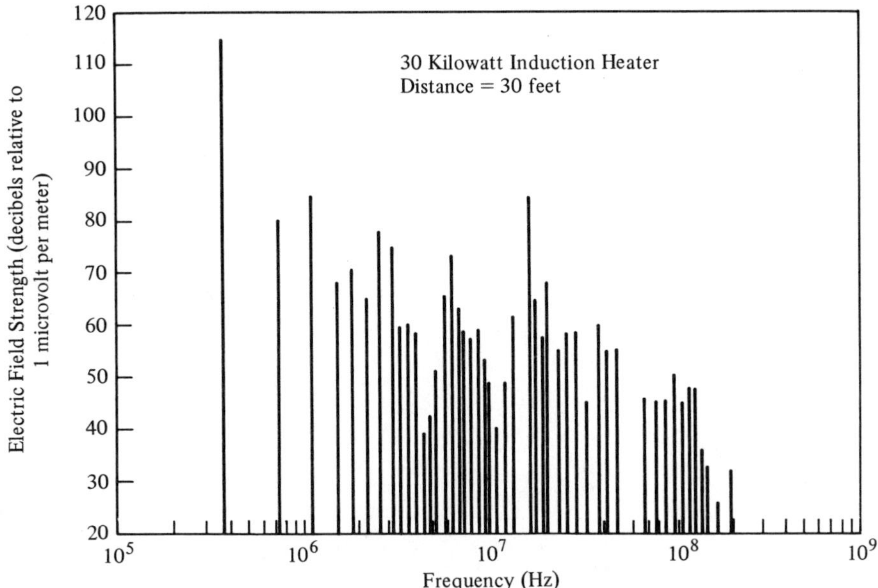

Fig. 4-7. Radiated noise line spectra from a 30-kW radio-frequency induction heater observed at 30 ft. Maximum values of linearly polarized noise-field intensity. (After Kernaghan and Ford, 1971)

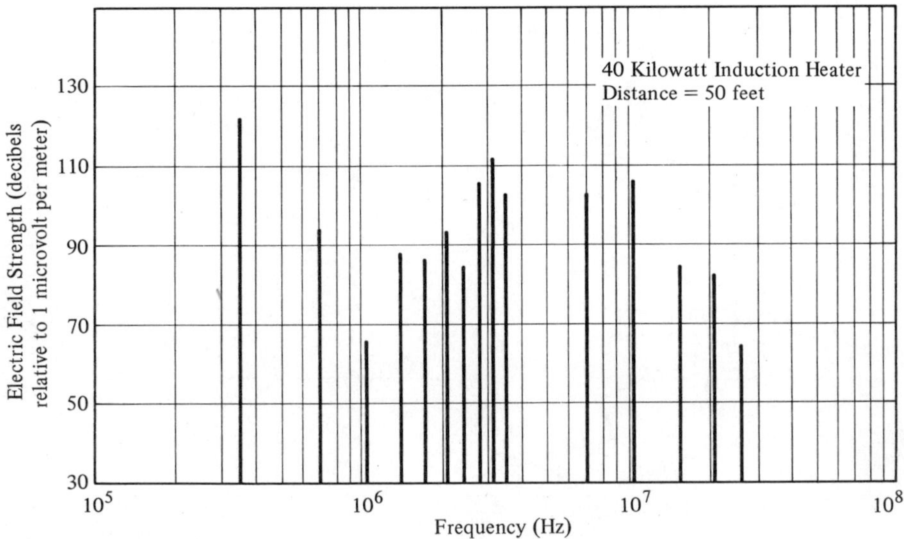

Fig. 4-8. Radiation noise line spectra from a 40-kW radio-frequency induction heater observed at 50 ft. Maximum values of linearly polarized noise-field intensity. (After Kernaghan and Ford, 1971)

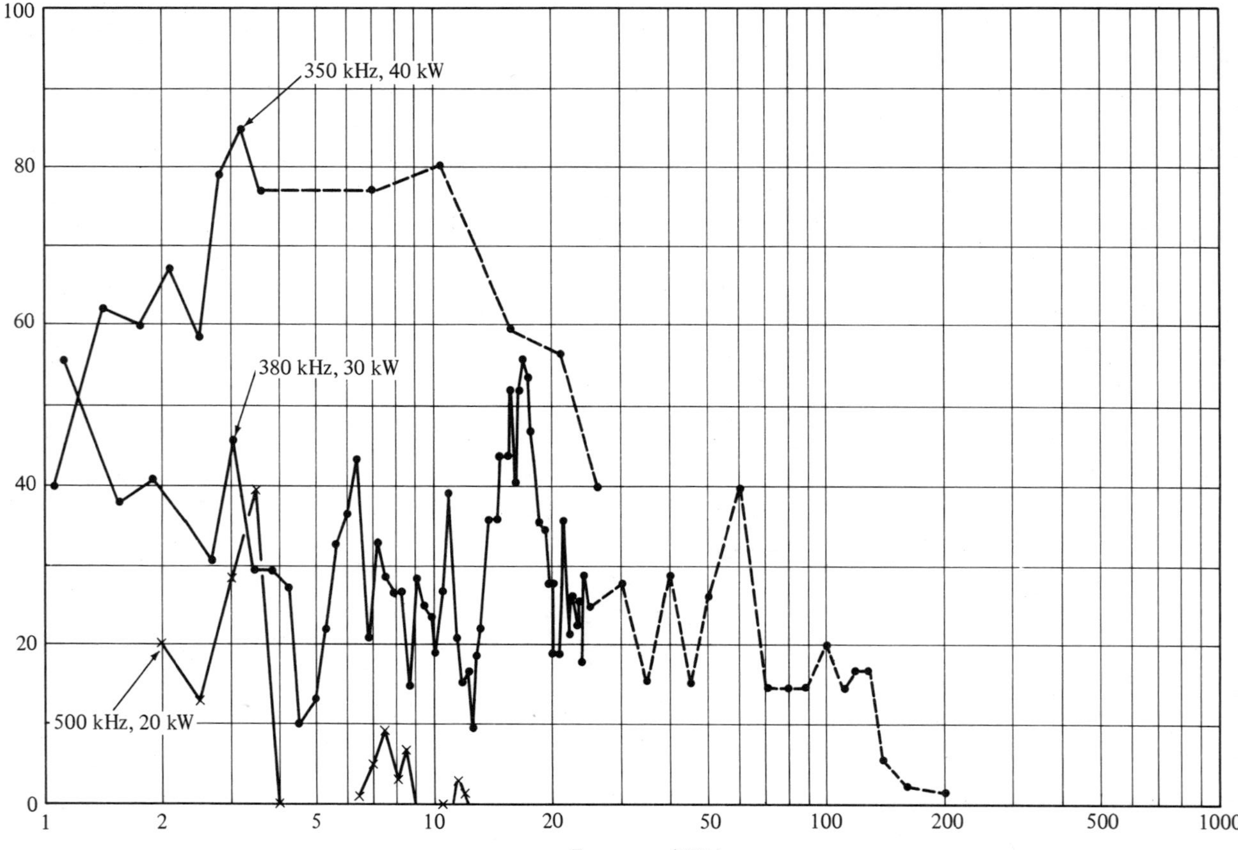

Fig. 4-9. Radiated noise line spectra for three radio-frequency induction heaters observed at 1000 ft. Maximum values of linearly polarized noise-field intensity. (After Kernaghan and Ford, 1971)

of the line spectra have been connected by broken lines to permit easy discrimination of the results for each induction heater. The notation associated with each curve provides the fundamental frequency of the driving oscillator and the rated input power to the heater, in kilowatts.

Welding of metallic parts is commonly accomplished in industrial production by means of either RF-stabilized arc welders or resistance welders. Each device accomplishes intense heat generation in the work by passage of large electric currents through the weld point. The two welding methods differ in one important respect; the proximity of the welding electrodes to the weld. The electrodes of resistance welders rest in physical contact with the weld providing a continuous metal path for current flow. Electrodes of an RF-stabilized arc welder, however, do not make metal-to-metal contact with the work in affecting the passage of large heating currents through the material. In RF-stabilized welders, the addition of a high-frequency medium-current signal to either a direct or 60-Hz high-power source of welding current ionizes a gaseous plasma between the electrodes and weld. The high electrical conductivity of the plasma formed in a chemically inert gas surrounding the weld provides a low impedance path for flow of the large direct or 60-Hz welding current without the necessity for metal-to-metal contact, thereby appreciably reducing the possibility of weld contamination. RF-stabilized arc welding is extensively employed in bonding nonferrous metals as well as steel and steel alloys.

A spark-gap oscillator is used in RF-stabilized welders to produce a high-frequency current of approximately 2 A. The HF-band frequency spectrum generated by the spark gap is superimposed upon the primary welding current. A broadband current modulation is created, which radiates into the surrounding space via the plasma arc, the welding electrodes, the power-supply-connecting leads, and imperfections in the shields and grounds of the spark-gap oscillator.

Figure 4-10 presents a series of measurements performed on a representative RF-stabilized arc welder using a vertically polarized antenna and peak-envelope detector at an observation distance of 100 ft.[5] The resulting spectrum, although continuous, was highly serrated; in plotting Fig. 4-10 only the major features of the data have been reproduced. Mounting, housing, and model design details have a significant effect upon the radiated spectrum of welders of this

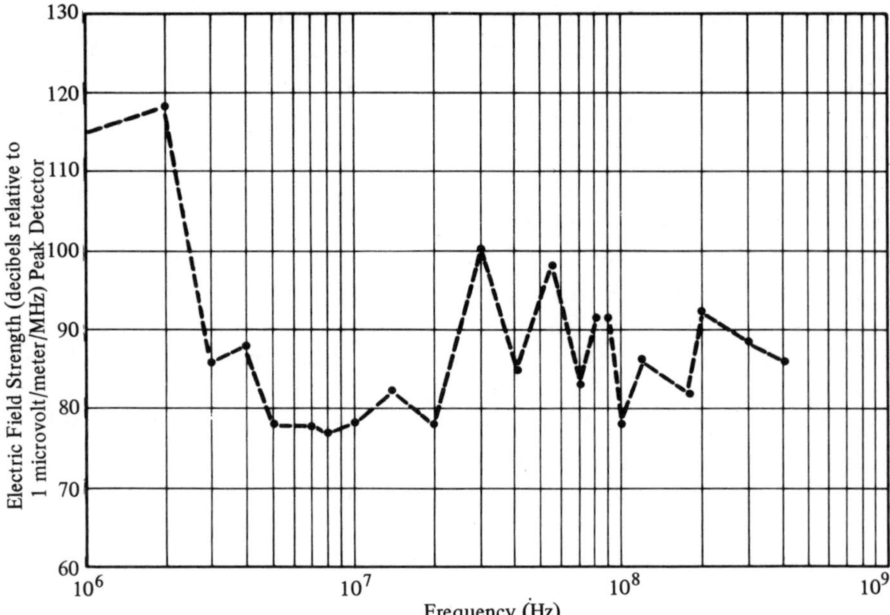

Fig. 4-10. Radiated Noise Spectra of an RF-stabilized arc welder. Distance of 100 ft. Maximum values of linearly polarized radiation. Peak detector. (After Martin and Tabor, 1972)

category. From a series of measurements performed upon 54 RF-stabilized arc-welder models operating in 72 production facilities in the United States, using vertically polarized linear antennas, the plots of Fig. 4-11 have been constructed.[6] The data represent the quasi-peak electric-field strength detected using a field-intensity meter meeting ANSI design specifications, Table 3-1. The vertical bars indicate the range of variation observed in the values of the maxima. Both curves were constructed by plotting the maximum observed quasi-peak electric-field strength values for each welder at the frequency of maximum emission. The resulting graphs for separation distances of 100 and 500 ft sketch the portion of the radio spectrum where RF-stabilized arc welders produce the greatest levels of radio interference. Comparison of Figs. 4-10 and 4-11 reveals several similarities. The primary maxima in all cases occur in the lower portion of the HF band, somewhere between 1 and 2 MHz. Subsidiary maxima lie in the upper portion of the HF band, approximately at

20 to 30 MHz. It has been observed that the construction of the surrounding building walls, the electrical grounding, and the presence of power lines contribute significantly to the propagation and attenuation of the welder-generated radio-frequency noise.[5] Metal building partitions; well-designed, installed, and maintained electrical grounds; and the removal of power lines from the vicinity of the welder are necessary to suppress the noise and to confine emission residuals to the vicinity of the units.

Electric resistance welders produce the necessary current density in the workpiece by applying pressure to the electrodes that is sufficient to puncture the insulating surface film and thereby admit 50 or 60-Hz short-circuit currents exceeding 10 kiloamperes. Several hundred pounds of pressure are typically applied through the electrodes simultaneously with the heating current. Electrode diameter,

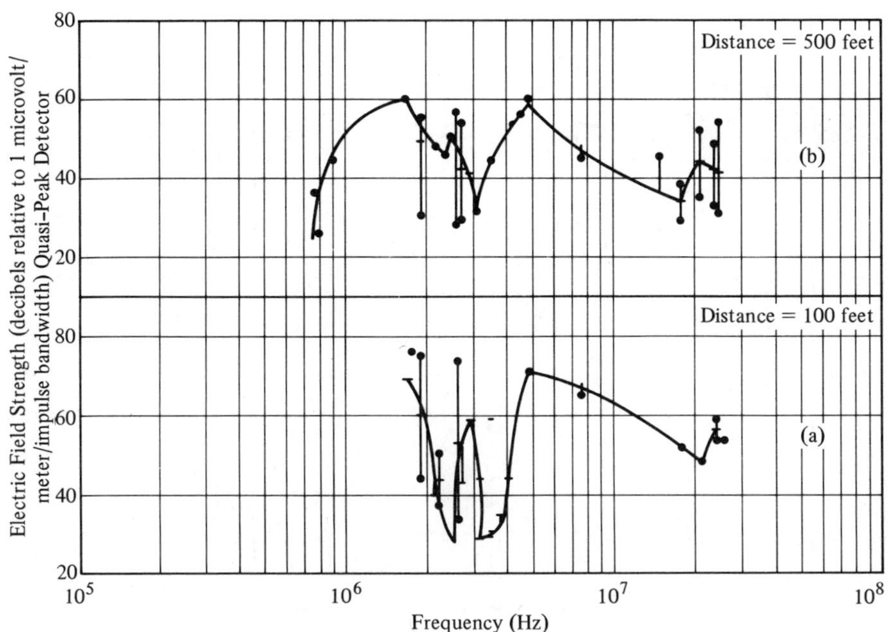

Fig. 4-11. Radiated noise spectra for many RF-stabilized arc welders. Quasi-peak detector with ANSI specifications. Maximum levels for linear radiation. A total of 27 units observed at 100 ft and 34 at 500 ft. (After Garlan and Whipple, 1964)

pressure, current density are increased proportionally with the work thickness.

Linearly polarized, peak-detected noise-field-strength measurements were performed on a representative example of resistance welder in the frequency range from 14 kHz to 1 GHz.[3] The observed radiated-noise spectra is impulsive and thus broadband. Results measured at a distance of 6,30, and 1000 ft from a 150-kW production welder used in automotive vehicle manufacture are plotted in Fig. 4-12 for horizontally oriented, linearly polarized receiving antennas. The resulting noise-field-strength is seen to remain appreciable at least to a range of 1000 ft.

Cutting designs and contours in metal stock for purposes of die manufacture and precision-part machining is performed using an electrical discharge machine whose work tool is supplied with energy from a high-power pulse generator. The metal stock to be machined is electrically connected to the pulse generator anode. The cutting

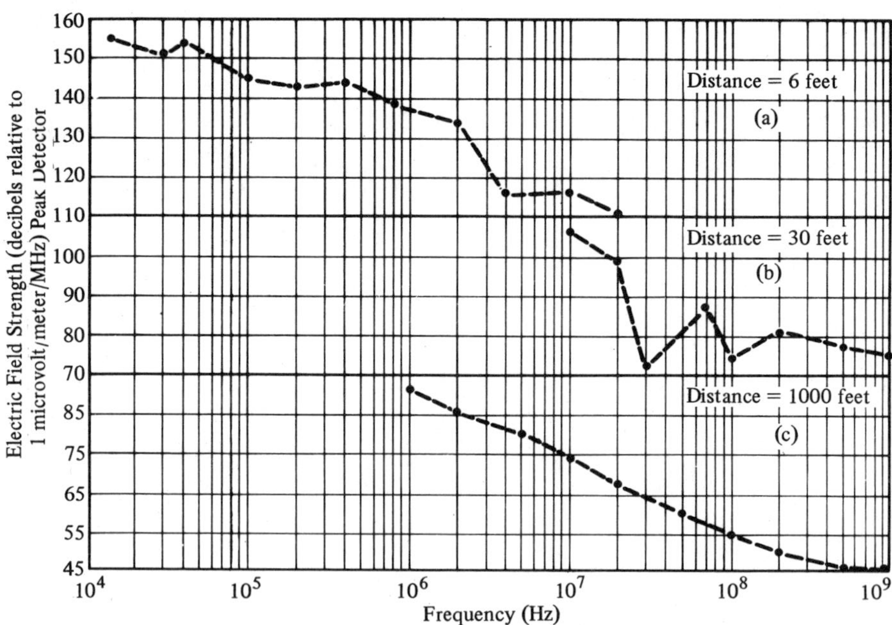

Fig. 4-12. Radiated noise spectra for resistance welder. Horizontal linear polarization, peak detected. Distances of 6, 30, and 1000 ft. (After Kernaghan and Ford, 1971)

tool becomes the cathode, appropriately shaped for the cut. By means of a servo-control circuit, a very narrow gap ($\sim 10^{-3}$ in.) is maintained between the tool and workpiece, across which high-peak pulse currents are supplied by the power unit. Metal particles are eroded from the work by the high-current pulse discharge and swept from the anode region by circulating dielectric fluid in which the work and cathode are submerged. Electrode cooling and uniform impedance loading of the pulse curcuit are additional functions provided by the dielectric bath. The average power rating of electric discharge machines used in production manufacturing is 1 to 2 kW. Units typically operate at pulse rates of from 1500 and 2500 pulse/sec. The resulting radio-noise spectra is broadband, extending from a frequency equal to its pulse repetition rate well into the VHF band. Measurements of the peak-detected-radiated-noise-field intensity for one model of an electric discharge machine are plotted in Fig. 4-13 for two observa-

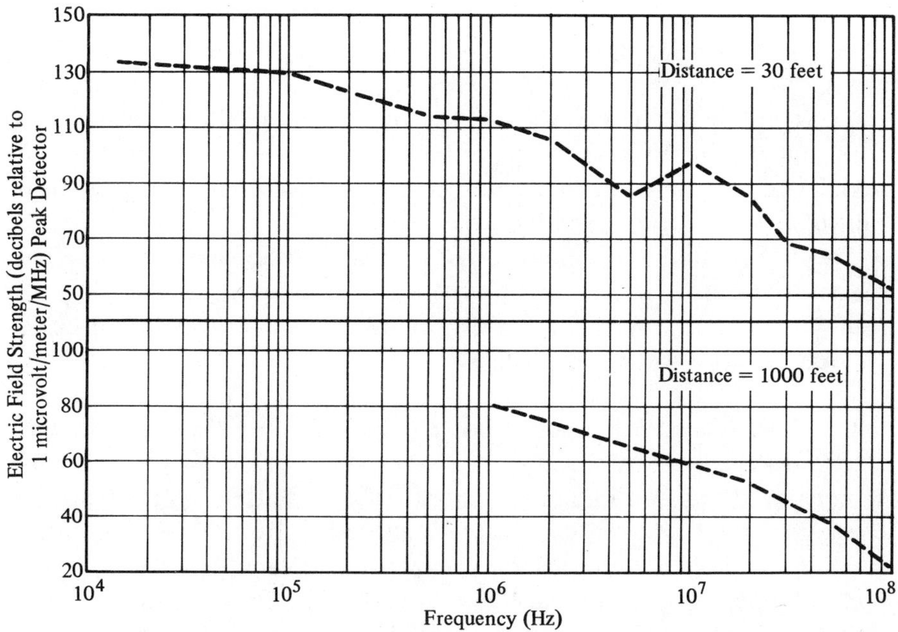

Fig. 4-13. Radiated noise spectra of an electric discharge machine. Peak detected linearly polarized radiation observed at 30 and 1000 ft. (After Kernaghan and Ford, 1971)

tion distances, 30 and 1000 ft.[3] The low-frequency limits shown in Fig. 4-13 were established by the response of the field-intensity meter for the data observed at a separation distance of 30 ft and by decisions of the investigator for the data recorded at 1000 ft. Welder noise emissions are known to remain appreciable to much lower frequencies. The upper frequency limit of these data was dictated by the internal noise of the measuring equipment, equaling the noise emitted by the electric-discharge machine at and slightly above 100 MHz for both distances investigated. Linearly polarized antennas were employed to perform the measurements, during which it was found that the maximum electric-noise field was insensitive to antenna orientation.

Mining Equipment

Subsurface mining operations employ a variety of unique electrical and diesel-driven equipment used in the extraction, haulage, processing, and removal of the coal and ore. Electrical equipment may be powered by either direct or alternating current. When mine equipment requires direct current, available high-voltage three-phase AC sources are often rectified and filtered. Because of imperfect line filtering and rectifier leakage, 60-Hz harmonics, the dominant of which are generally the lower multiples of 360 Hz, frequently appear on the mine electrical circuits.

Diesel-operated equipment, possessing no ignition system, generate electrical interference only by virtue of their auxiliary electrical assemblies, which may include switches, relays, and alternators used to generate battery-charging voltages.

Specialized Measuring Equipment, Data-Reduction Techniques, and Data Formats. Measurements of radio-noise emissions generated by various production mining equipment to be presented were performed using calibrated magnetic loop antennas and either of two configurations of associated amplifiers and magnetic tape recorders.[7] One receiver possessing a linear dynamic range of 62 dB and flat passband response from 40 Hz to greater than 300 kHz was used as a preamplifier for a multichannel AM and FM tape recorder. Data obtained with this assembly were reduced and digitally processed to construct the radio-frequency noise spectra from 1 kHz to 100 kHz.

A second amplifier and magnetic-tape-recording system, providing a response band from 10 kHz to 32 MHz, used a multiple-conversion IF amplifier to down-convert and record the input-signal spectrum. Subsequent digitization and computer processing of the signal-envelope voltages provided the noise-envelope APDs.

The magnetic-loop antennas used in these mine-noise studies were calibrated in MKS units* of magnetic field strength. Resulting spectral plots present magnetic field strength in decibels relative to one of two references, depending upon whether the spectral signature may be considered narrowband or wideband. When the noise signature is narrowband, as in the case of a CW tone superimposed upon a thermal noise background, the decibel reference is chosen as 1 μA/m. For broadband continuous-spectrum noise, the reference used in computing the noise magnetic field intensity is 1 $\mu A/m\sqrt{B}$ where B (in Hz), represents the noise-resolution (observation) bandwidth employed in digitally processing the magnetic field signatures. Amplitudes of the broadband noise spectra may be transformed to a 1-Hz resolution bandwidth by reducing all ordinate values by $10 \log B$.

Mine-Machinery-Noise Spectra and APDs. Two types of coal extraction equipment are commonly used at the mine face in modern operations, a continuous mining machine and a shear machine. Continuous miners cut and gauge the coal seam while advancing into the strata. They leave supporting coal pillars in their wake while extracting the available coal from the surrounding areas in what is known as *room-and-pillar operations*. Shear-extraction machines cut and gauge the exposed coal while moving parallel to the face in a long, continuous movement, which reversed direction at the end of a longwall run.

Continuous miners may be designed for operation on either alternating or direct current. When supplied by direct current, power is obtained from a three-phase 60-Hz high-voltage distribution system, which is full-wave-rectified and filtered at a substation usually located outside the mine. A 600-V direct-current line provides the drive power for a continuous miner whose radio-frequency magnetic-field noise spectra observed 10 m to the rear is presented in Fig. 4-14.[7] The lower curve represents the noise level of the amplifying,

*MKS: meters, Kilograms, seconds.

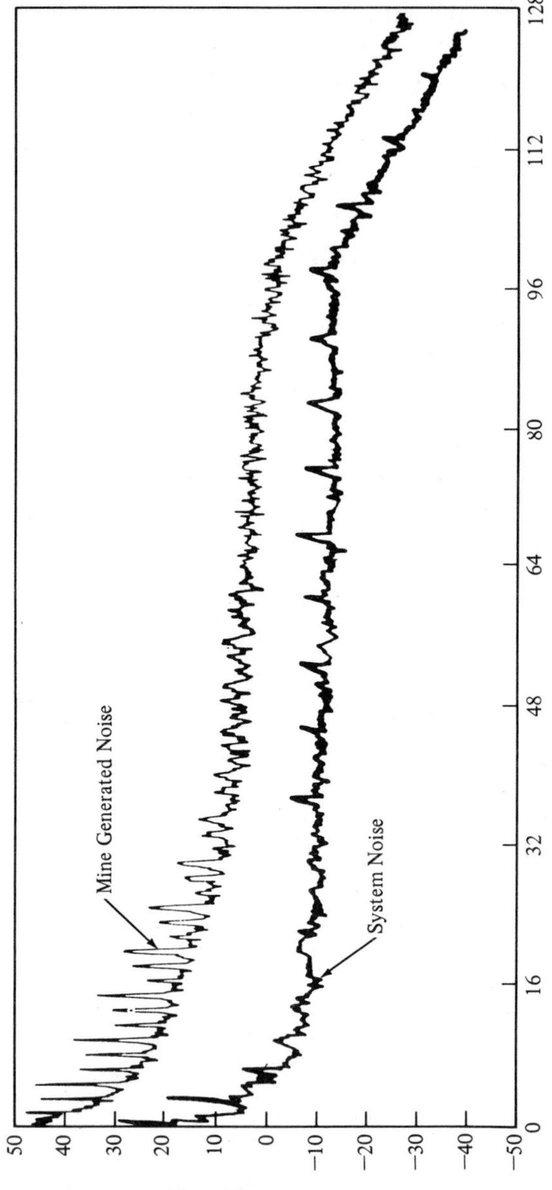

Fig. 4-14. Radiated noise spectra of a direct-current continuous-mining machine observed at a distance of 10 m. Vertical component of the magnetic-field intensity. (After Bensema et al., April 1974)

recording, and data-processing equipment. The single ordinate scale is applicable to each of the ordinate captions. The inner ordinate caption should be applied to those portions of each noise spectra that contain no apparent signal peaks and that are therefore interpretable as broadband emissions. The outer caption should be applied to the sharp peaks, which represent the line spectral components superimposed upon a continuous noise background. The continuous miner-noise recordings shown in Fig. 4-14 were not calibrated below 1 kHz nor above 100 kHz; thus, they are only of qualitative value beyond these limits.

Radiated noise spectra observed at a distance of 3 m from an operating alternating-current continuous miner are shown in Figs. 4-15 and 4-16 for the vertically polarized magnetic field component. Figure 4-16 provides an expanded presentation of the spectrum below 20 kHz using a resolution bandwidth of 19.5-Hz.[8] In Fig. 4-15, the large signal appearing at 100 kHz was produced by the mine telephone system. The strong continuous-wave signal peaks evident in Fig. 4-16 arise as multiples of 360 Hz generated by full-wave rectification of the 60-Hz primary power source. Imperfect filtering of the direct-current output allowed harmonic currents to appear at the alternating-current load where they were radiated and detected by a loop antenna located within 3 m of the operating continuous miner. An associated APD for this miner is presented in Fig. 4-17, measured in a predetection bandwidth of 1 kHz.[8] These data were recorded at an observation frequency of 70 kHz, using a vertically polarized magnetic-loop antenna displaced from the continuous miner by approximately 3 m. The data sample from which the APD of Fig. 4-17 was prepared extended for a period of 23 min. The increase in the negative slope of the APD at and below the 45 percent exceedance value confirms the existence of a large component of impulsive noise radiated from the continuous miner. When comparing the APD with the spectral plot of Fig. 4-15, the inner ordinate caption on Fig. 4-15 must be used, and compensation for the differing resolution bandwidths must be incorporated, using the relationship:

$$H_1 = H_2 + 10 \log\left(\frac{B_2}{B_1}\right) \tag{4-1}$$

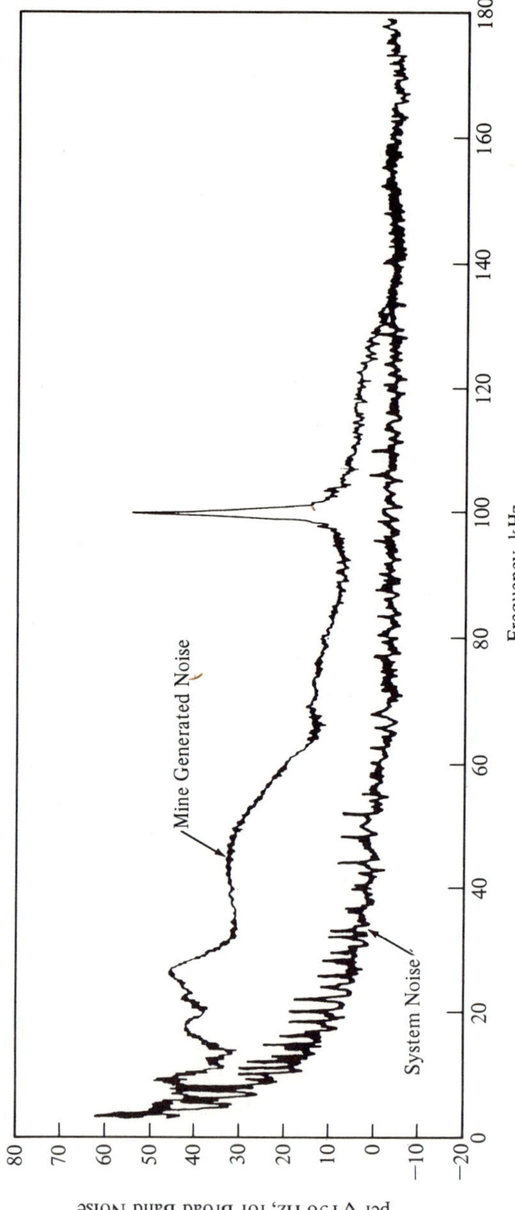

Fig. 4-15. Radiated noise spectra of an alternating-current continuous-mining machine observed at a distance of 3 m. Vertical component of the magnetic field intensity. Spectral range: 1 to 100 kHz. (After Kanda et al., 1974)

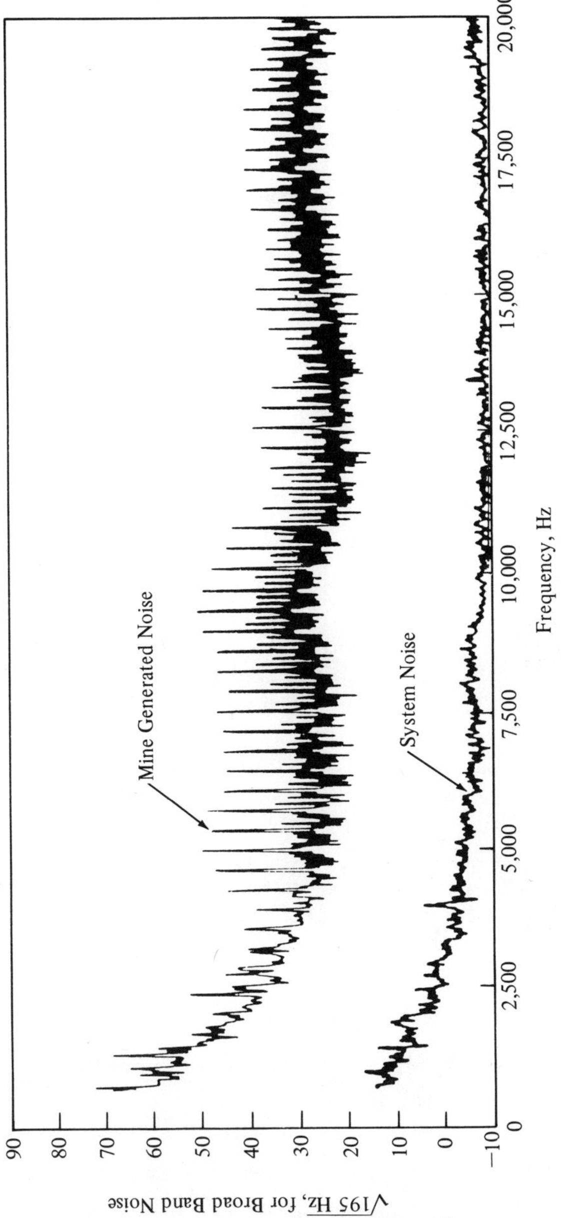

Fig. 4-16. Radiated noise spectra of an alternating-current continuous-mining machine observed at a distance of 3 m. Vertical component of the magnetic-field intensity. Spectral range: 1 to 20 kHz. (After Kanda et al., 1974)

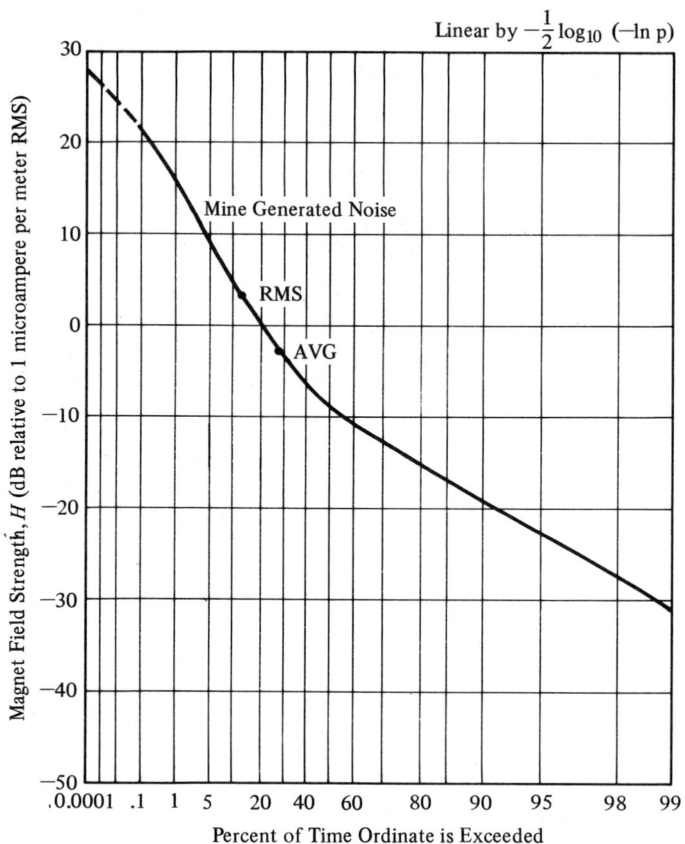

Fig. 4-17. APD of the noise envelope produced by an alternating-current continuous-mining machine observed at a distance of 3 m. Vertical component of the magnetic field intensity sampled for 23 min. and presented for a resolution bandwidth of 1 kHz. (After Kanda et al., 1974)

where

H_1 and H_2 = the rms values of the magnetic field, expressed in dB relative to 1 μA/m/$\sqrt{B}$

B_1 and B_2 = the resolution bandwidths in Hz.

Shear-mining machines, used for the removal of coal along a laterally exposed face, cut and discharge the material into a conveyer belt as the machine moves parallel to the exposed surface. Shear machines may be powered by either direct or alternating current. In the latter case, harmonics of the 60-Hz power frequency, arising from imperfect filtering, often are radiated into the surrounding space, as is evident in Fig. 4-18. Here, the spectra of an alternating-current shear machine is shown for a recording station removed a distance of 3 m.[9] The vertical component of the magnetic field intensity is shown in Fig. 4-18. Notice it contains a few strong single-frequency signals; the dominant portion of the spectra is composed of broadband noise presented for a processing bandwidth of 19.5 Hz.

Continuous mining and shear machines deposit their extract on either conveyer belts or haulage trains, which remove the coal from the vicinity of the mine face. Conveyer belts may be driven by high-voltage electric motors, either direct or alternating current. The vertical magnetic noise field generated during the start-up of a 950-V longwall conveyor belt is shown in Fig. 4-19 for an observer displaced 10 m from the conveyer end.[9] The periodicity appearing in the noise-emission spectra at intervals of approximately 20 kHz is attributed to the high start-up currents required by the conveyer induction motors. An alternate method for transporting coal from the mine face to the main trolley area, prior to removal from the mine, makes use of haulage trains consisting of several shuttle cars. Electric motors used to power the shuttle cars may be sources of appreciable radio-frequency noise. Figure 4-20 reproduces the magnetic field noise spectra for an empty shuttle car train that is moving slowly past a magnetic detecting loop approximately 3 m distant.[8] The noise signature in Fig. 4-20 represents the vertical magnetic-field intensity processed for the broadband component in an observation bandwidth of 3.91 Hz. Several CW noise signals are present, each harmonically related to the 60-Hz primary power source. The largest-amplitude harmonic components are found to be 360 Hz and integer multiples thereof, caused by full-wave rectification of the 60-Hz supply.

Where passage dimensions permit, mines employ rubber-tired diesel power equipment for the transport of mine products from the active faces to the rear areas. There the materials may be temporarily stored, partially processed, or removed to the surface. These vehicles, called *lift-haul-and-dump machines (LHD)*, produce no ignition noise but may mount auxiliary electrical equipment, which generate electromagnetic interference. An example of the noise magnetic-field in-

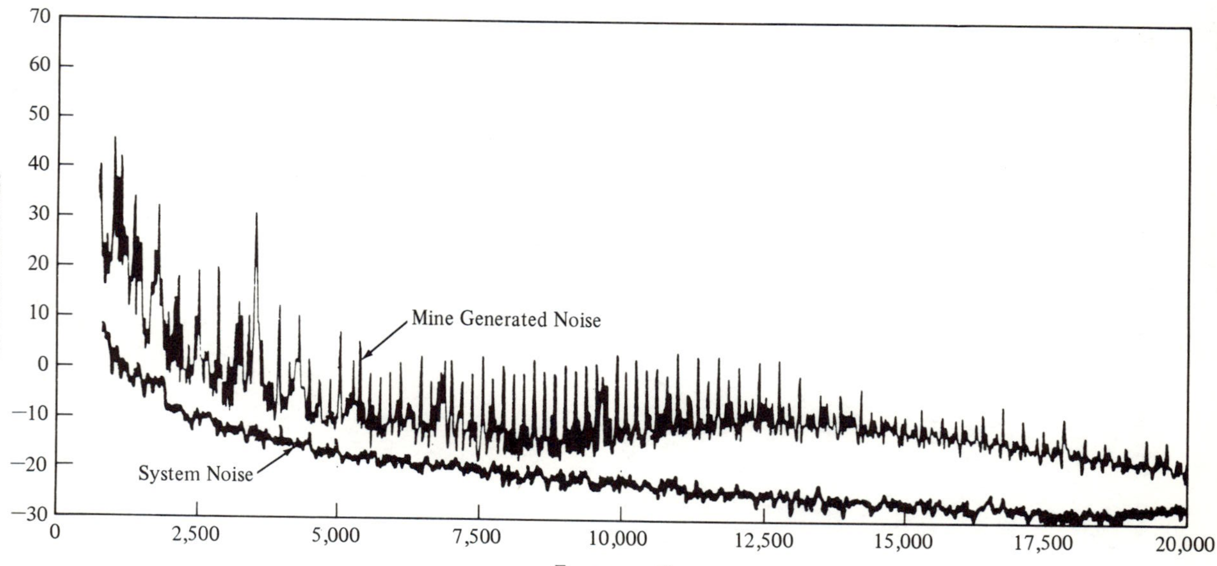

Fig. 4-18. Radiated noise spectra of a shear-wall mining machine at a point 3 m distant. Vertical magnetic-field component. Detection bandwidth: 19.5 Hz. (After Bensema et al., June 1974)

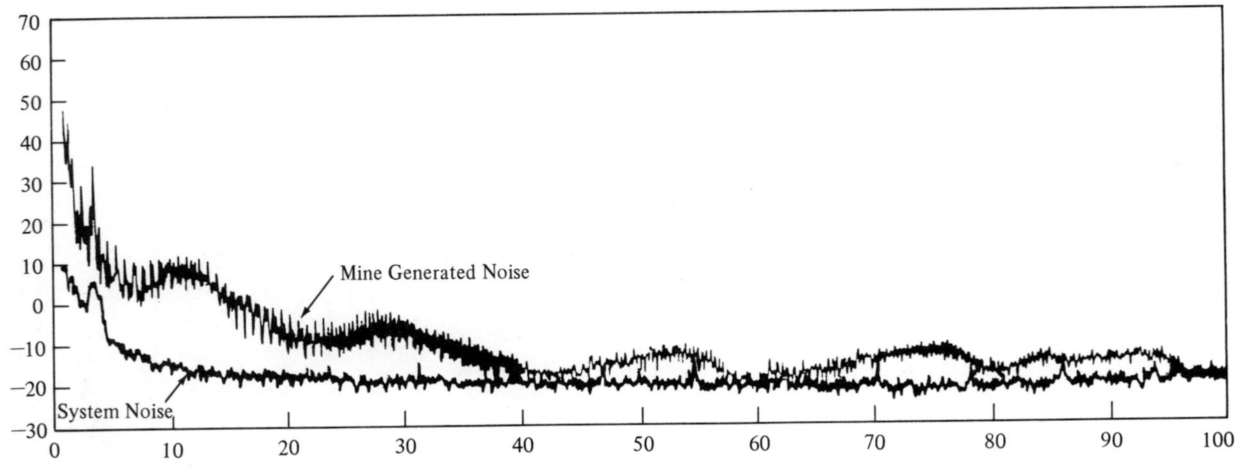

Fig. 4-19. Radiated noise spectra of a longwall mine, conveyor-belt coal transport driven by 950-V alternating-current induction motors. Observation distance-10 m. Vertical component of the noise magnetic field: 1 to 100 kHz. (After Bensema et al., June 1974)

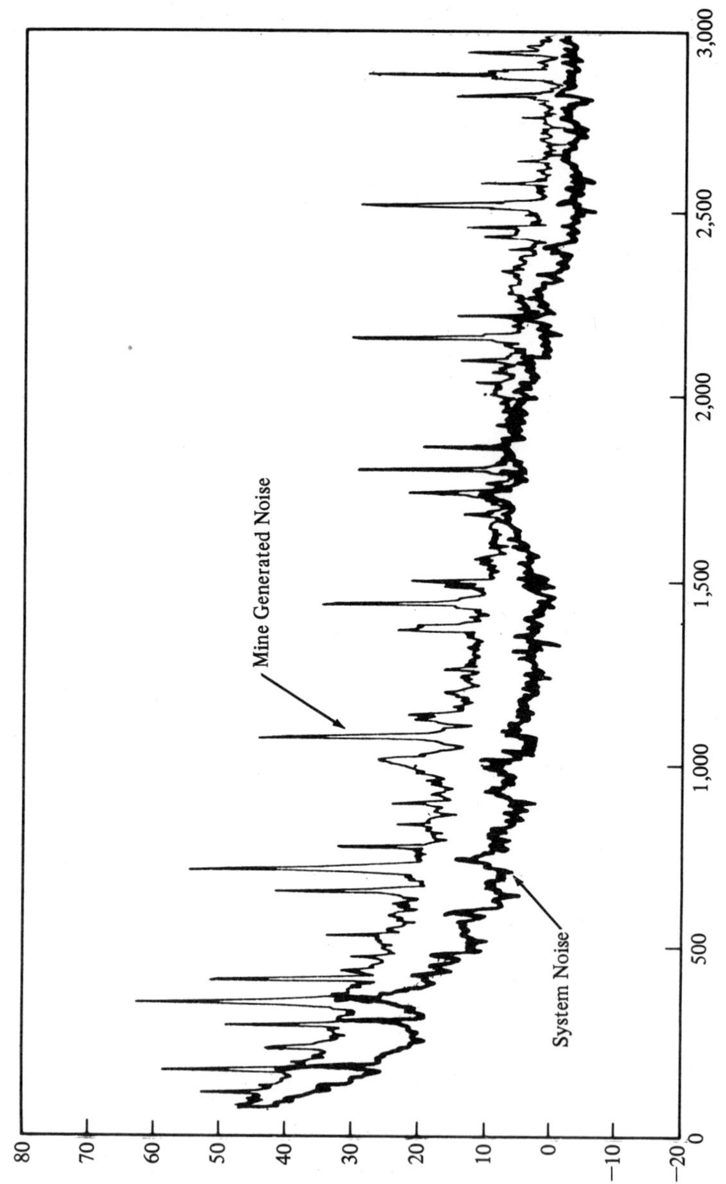

Fig. 4-20. Radiated noise spectra of an empty haulage train moving slowly past a magnetic-loop antenna at a distance of 3 m. Vertical component of the noise for the frequency range: 100 Hz to 3 Hz. (After Kanda et al., June 1974)

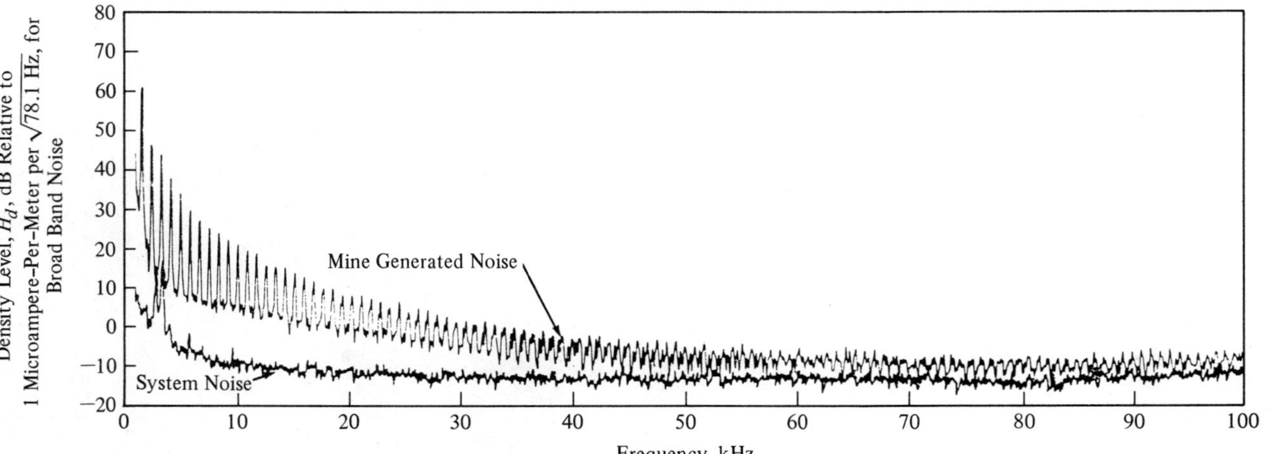

Fig. 4-21. Radiated noise spectra from a 6-yd^3 LHD machine, battery-charging alternator at a distance of 2 m. Vertical component of magnetic-field intensity. (After Adams et al., June 1974)

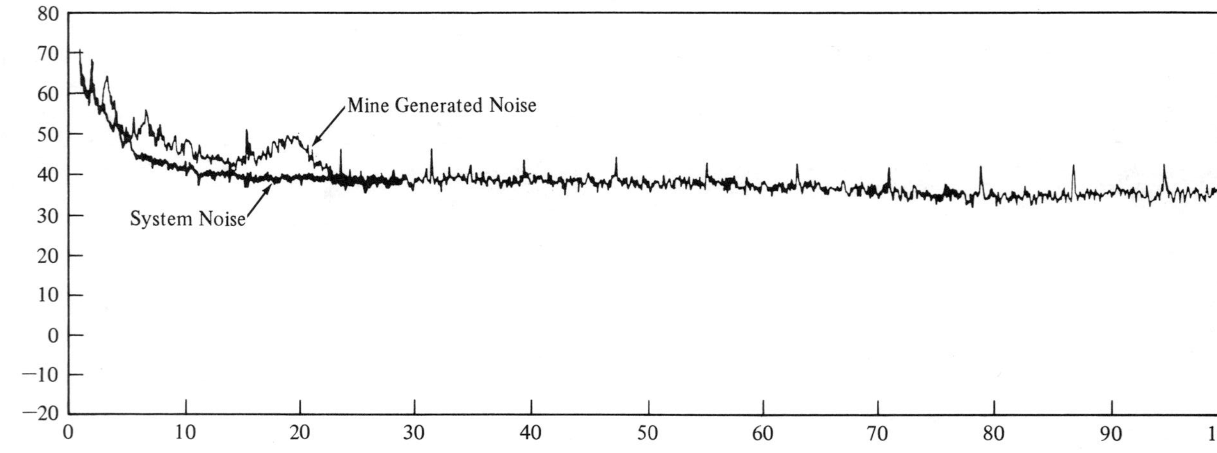
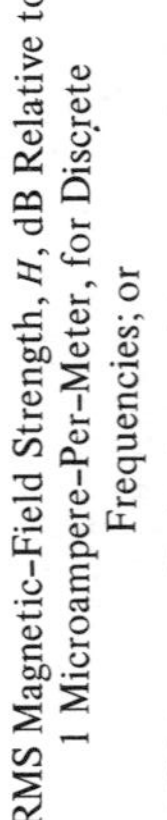

Fig. 4-22. Radiated noise spectra from a 1250-hp direct-current hoist motor. Vertical magnetic-field component. Observation distance: 2 m. (After Scott et al., Oct. 1974)

tensity emitted by a 6 cubic-yard capacity LHD machine is plotted in Fig. 4-21.[10] The vertical component of the measured magnetic field is shown at a distance of 2 m. Many single-frequency noise components are present in the noise spectra, with the fundamental and lowest falling at 425 Hz. Data of Fig. 4-21 were recorded for a constant engine speed. As expected, engine-speed changes proportionally alter the frequencies of the fundamental and harmonics of the noise. The source of the LHD noise components was determined to be the vehicle alternator used for starter battery charging. During typical vehicle usage, variation in engine rpm will cause the noise peaks to sweep the frequency band leaving few, if any, spectral intervals uncompromised by alternator radiation.

In many deep-shaft hard-rock ore mines, high-power electric motors are used to drive high-speed hoists employed for ore removal and the transport of personnel and equipment. Hoist motors of large horsepowers are usually set exterior to the mines and colocated with controls, gear boxes, and cable drums. Brush and commutator noise from direct-current motors and harmonics of the power-line frequency from induction motors produced in the vicinity of mine hoist complexes may be transmitted and reradiated within the mine via hoist cabling and communication circuits. Figure 4-22 presents the vertical magnetic-field intensity produced by a 1250-hp direct-current hoist motor at a distance of 2 m while lifting approximately 9 tons of rock at a rate of 1300 ft/min.[11] Brush and commutator noise are evident and exist in excess of instrumentation background levels up to a frequency of 22 kHz. The low-frequency portion of the noise spectrum from the 1250-hp motor has been enlarged in the presentation of Fig. 4-23, which extends downward from 3 to 0.1 kHz.[11] At the moment the data of Fig. 4-23 were recorded, the hoist was operating to raise an ore load at 1700 ft/min. Few, if any, harmonics of the 60-Hz power-line frequency are present in either spectrum. Most of the observed radiated noise is related to motor armature rotation.

MEDICAL EQUIPMENT

The largest sources of radiated noise among medical-related electrical equipment are radio-diathermy machines. Such instruments are func-

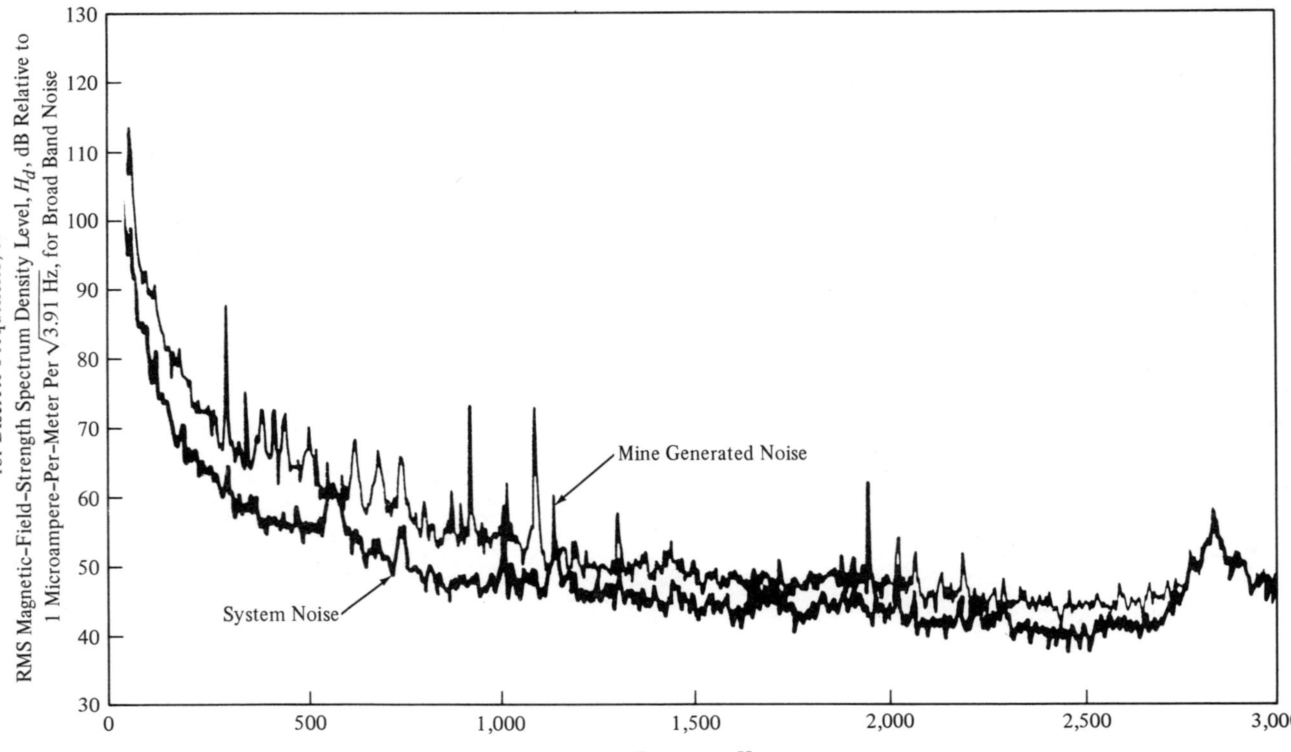

Fig. 4-23. Radiated noise spectra from a 1250-hp direct-current hoist motor. Spectral interval: 100 Hz to 3 kHz. Vertical magnetic-field component observed at a distance of 2 m. (After Scott et al., Oct. 1974)

tionally similar to the dielectric heating equipment used in industrial fabrication and material processing. Radio-frequency oscillators, either pulsed or continuous-wave, provide the power needed to induce body-tissue heating, which for diathermies currently available range from 50 to 1000 W. Typically, the fundamental radio frequency lies at one of the assignments: 13.56 or 27.12 MHz. Representative pulse-repetition rates of pulse-modulated diathermy oscillators fall between 80 and 2600 Hz. Pulse durations of 70 μsec may be used but these are not fixed by regulation nor by industrial standards. Radiation emitted by a diathermy oscillator manifests a line spectra, the fundamental frequency of which is intended to produce tissue heating. Oscillator design and circuit filtering are used to sup-

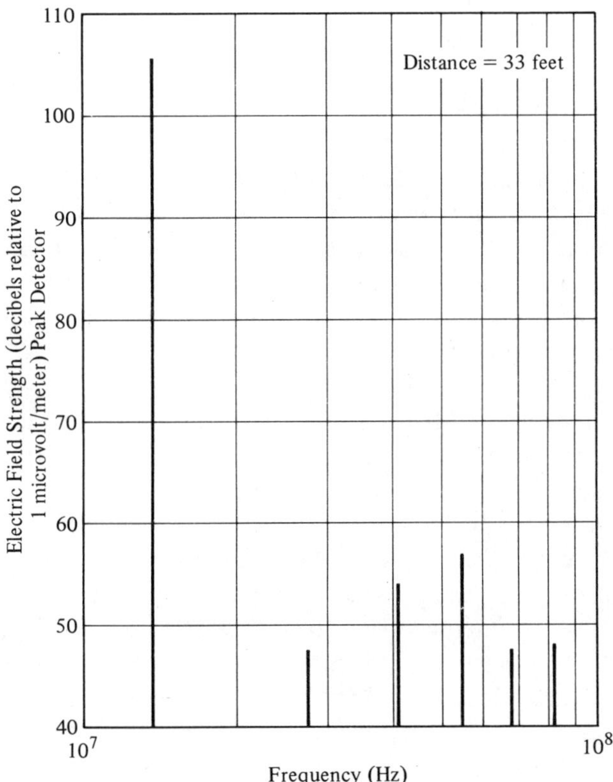

Fig. 4-24. Radiated electric-field intensity for a 1-kW medical diathermy machine. Vertical electric-field component observed at 33 ft. (After Martin and Tabor, 1972)

press oscillator harmonics as prescribed by licensing regulations.[1] Figure 4-24 depicts the radiated line spectrum of a rated 1-kW, pulsed diathermy machine observed at a distance of 33 ft.[5] The measured input power of the unit when examined was 750 W; the peak pulse power for which these data were obtained was observed to be 1.06 kW The vertical component of the electric field is shown in Fig. 4-24. Five harmonics of the design fundamental 13.56 MHz are evident above the instrumentation noise. Notice that all harmonics have been suppressed by at least 40 dB with respect to the fundamental frequency.

COMMERCIAL ELECTRICAL EQUIPMENT

In modern commerce, a multitude of electrically powered equipment exists that are known to incidentally generate radio-frequency radiation. Most of the radiation emitted by commercial equipment is of low intensity; consequently, few measurements have been performed or reported.

However, several types of electrical equipment employed in metropolitan commercial activities are known as radiators of substantial radio interference. Special attention has been accorded to the investigation of radiated signals from area illumination and advertising light sources (fluorescent and neon), radio-controlled door openers, and electrified trains and buses.

Area Illumination and Advertising Lights. Fluorescent lights in commercial and residential use are typically designed with tungsten-wire electrodes, separated by the length of a light column filled with a gas mixture containing argon and mercury. Radio emission is generated concurrently with the establishment of an alternating-current electric discharge produced by pulses of high-potential current from the electrodes. Pulsed current flow in the ionized gas column generates radio-frequency noise originating from the high-current-density region of the electrodes, and which may be augmented, in hot cathode tubes, by random and impulsive emissions from material defects occurring on one of the electrode heaters. Bright spot emissions, which may occur in hot cathode-tube designs, arise from heater-material defects that produce pulse emissions of very short duration and associated increases in the high-frequency radio-noise component.[12] Noise suppression in either hot or cold cathode-tube designs is achieved by the addition of electrode shunt capacitors and the use of radio-frequency shielding surrounding the lamp enclosure. A totally shielded enclosure is composed of an integral or bonded metal box covered on the light-emitting face by either a transparent conductively coated glass or an open metal grill electrically bonded to the box. Representative noise levels appearing in the radio spectrum are shown in Fig. 4-25 for three fluorescent lamps with partially shielded enclosures.[13] Peak detected measurements of the verti-

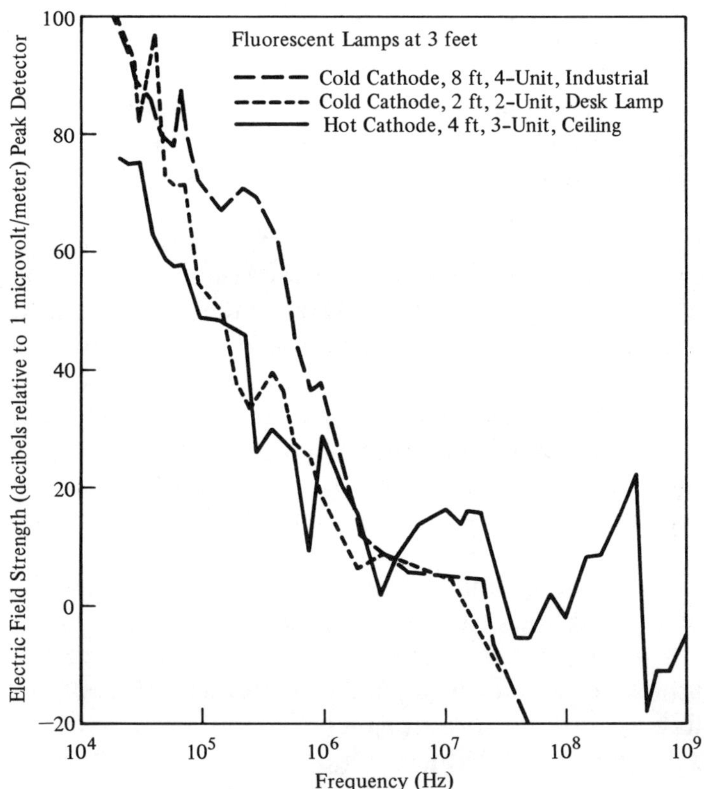

Fig. 4-25. Radiated electric-field intensity for three models of fluorescent lamps. Vertically polarized electric field. Peak detector. Observation distance: 3 ft. Length of fluorescent bulbs and number per fixture indicated. All lamps were partially shielded. (After Clark, 1961)

cal and horizontal electric-field components are shown for three models of fluorescent lamps, two designed with cold cathodes and one with a heated cathode. Each fixture was partially shielded, including bonding of the box seams, but not including shielding of the light-emitting surface. The point of observation for Fig. 4-25 was 3 ft from each lamp. Indicated in the legend are the lengths and numbers of fluorescent bulbs contained in each fixture. The additional radiation occurring for the hot cathode lamp above 10 MHz may possibly be attributable to the presence of bright spots on the heaters. Conventionally, impulsive noise emissions arising from fluorescent lamps produce a continuous spectrum upon which may be superimposed the power-line frequency. Some harmonics of the line frequency may be present depending upon the extent of the associated line filtering.

The very-low-frequency portion of the radiated noise spectrum of fluorescent fixtures taken with a magnetic-loop antenna in the interval between 100 Hz and 4 kHz is shown in Fig. 4-26.[10] The vertical component of the magnetic field is plotted for an observation point beneath the ceiling fixture at a distance of approximately 3 m. Superimposed upon a continuous noise background that exceeds instrumentation noise by 5 dB over most of the frequency interval are high-intensity harmonics of the 60-Hz primary power. Notice these harmonics achieve maximum intensities for the even integer values. The outer ordinate caption should be used to read the amplitude peaks of the harmonics and any other line spectral components present. Magnitudes of the continuous portions of the spectra are read from the inner caption, which provide the rms magnetic field strength in decibels relative to 1 μA/m for a detection bandwidth of 3.91 Hz.

Tubular, contoured, and sculptured lighting used in advertising and displays are fabricated from glass tubes and pipes, evacuated, and filled with neon gas at low pressure. To the electrodes inserted at the ends of the light column is applied a potential, either alternating or direct, in excess of 15 KV. Avalanche ionization of the gas column, initiated by the residual electron population, is produced and sustained by the applied electric field. Electron recombination and gas ionization occur simultaneously, causing the red glow associated with neon illumination.

Neon fixtures emit radio-frequency energy as a by-product of high-voltage transformer induction and rectification used to develop

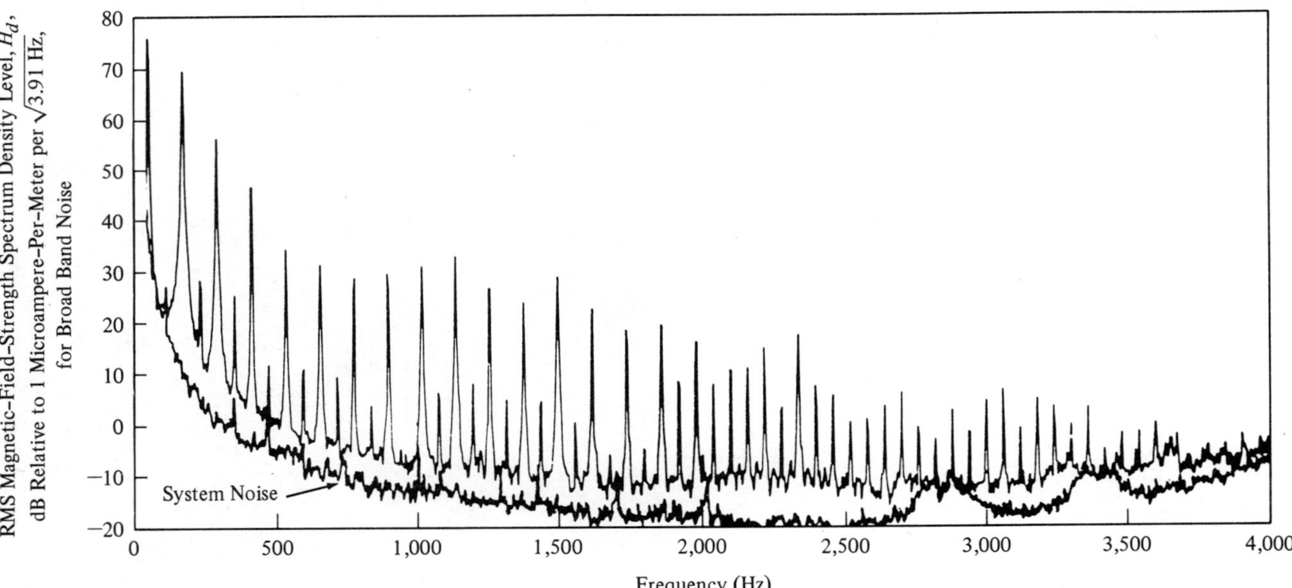

Fig. 4-26. Radiated noise field intensity for a fluorescent light fixture at a distance of 3 m. Vertical component of the magnetic field intensity. (After Adams et al., June 1974)

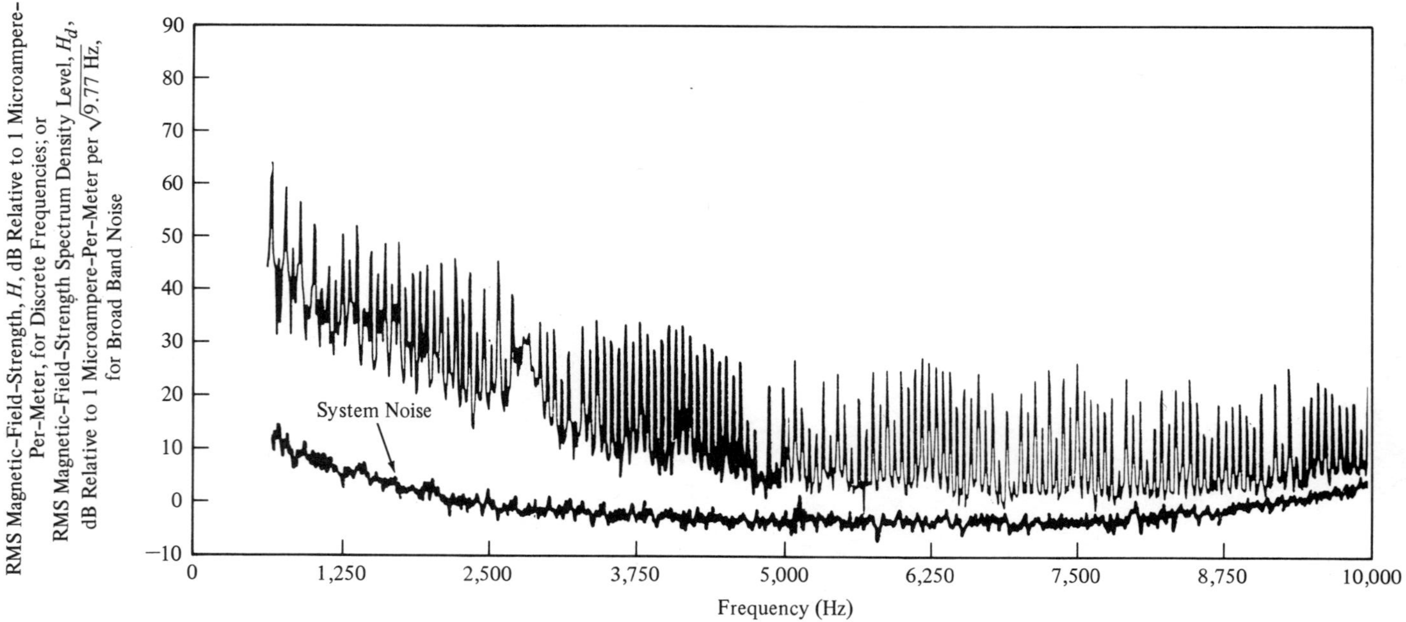

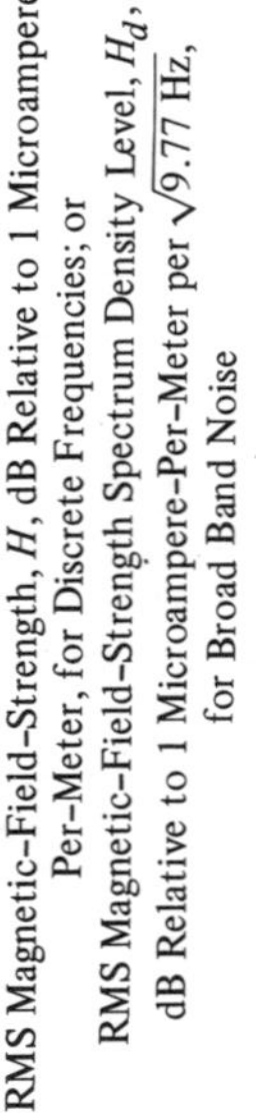

Fig. 4-27. Radiated noise-field intensity of a neon street-advertising display observed at a distance of 10 m. Vertical component of the radio-noise magnetic field intensity at a distance of 10 m. (After Scott et al., Oct. 1974)

the gas-column electrode potentials. It is possible that in addition to the fundamental and harmonics of the primary power frequency, a broad spectrum component of radio noise is produced by the ionized gas column. Figure 4-27 presents the measurement at 10 m of the vertical magnetic-field component of radio-frequency interference generated by a neon street-advertising display.[11] Figure 4-27 is predominantly a line spectrum produced by the odd harmonics of the 60-Hz power-line frequency. A broad-spectrum radio-interference component exists in the interharmonic intervals, primarily below 2.5 kHz, at a level equaling 20 dB above 1 μA/m per square root of the observation bandwidth, 9.77 Hz. Peaks of the line spectra, which at the low-frequency limit of 560 Hz in Fig. 4-27 attain a magnitude 63 dB above 1 μA/m, have been seen to increase with decreasing frequency until at 300 Hz they attain a maximum level of 90 dB above 1 μA/m.[11] For frequencies below 300 Hz, the neon radio-noise emissions decrease to 80 dB above 1 μA/m at 180 Hz.[11]

Measured but not shown were the horizontal magnetic-field components of the radio-frequency noise for the same range of frequencies, 560 Hz to 10 kHz. It was observed that one horizontal component was comparable in intensity to the vertical, while the other displayed an intensity at least 20 dB lower.[11]

RADIO-CONTROLLED DOOR OPENERS AND AERONAUTICAL BAND-FREQUENCY CONVERTERS

Remote control of building and garage doors may be affected using a very-high-frequency radio signal, which is detected by a sensitive receiver coupled via a switch to an electric-drive motor. The VHF- and UHF-band receivers presently in general use contain sufficient positive feedback to initiate input circuit oscillation upon reception of the correct frequency. The receiver, after breaking into oscillation when its excitation threshold is exceeded, is driven nonlinear; the current saturates, and the high-frequency oscillations are temporarily quenched until the charge stored in the feedback network is dissipated. This superregenerative action is reflected in the current variations present in the input circuit, which, because it is usually directly connected to the receiving antenna, gives rise to signal reradiation into the surrounding space, thus becoming a source of radio noise. Although these receivers are tuned to respond to a single VHF- or UHF-band frequency, nonlinear superregenerative oscillations produce a distributed-noise spectrum centered at the tuned frequency and extending several megahertz to either side. Measurements performed on three commercially available models of radio-controlled door openers have revealed the radiated-noise spectrum to be essentially continuous over a band of approximately 8 MHz centered at the tuned frequency. The observed radiated noise power was found to be proportional to receiver bandwidth, within an uncertainty of ±1 dB, for detector bandwidths lying between 1 and 300 kHz.[5] Results of average-radiated-noise-power measurements performed on three commercial door-opener models are plotted in Fig. 4-28 as (•) at the tuned frequency of each receiver. In some cases, the tuning frequency was seen to be offset from that of the remotely located transmitter 2 to 3 MHz. The number beneath each plotted point enclosed by ⟨ ⟩ indicates the distance in feet at which the data were observed.

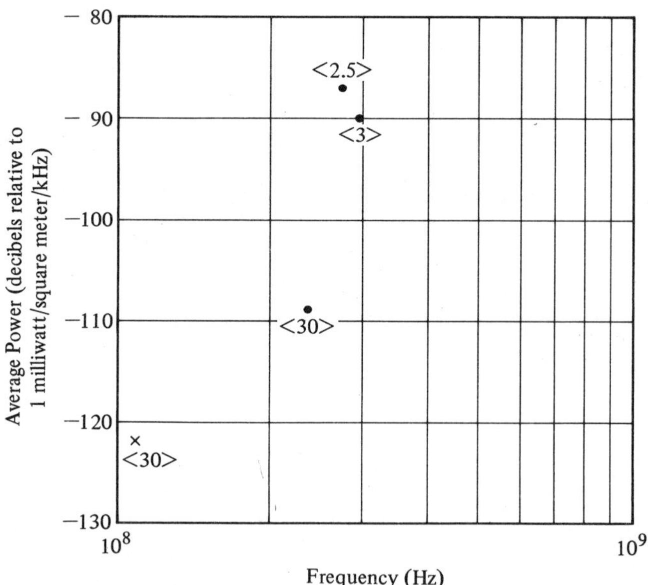

Fig. 4-28. Radiated average noise power produced by three models of radio-controlled door openers and one model of a VHF aeronautical-band receiver-converter. Measurement distance is shown within the brackets, in ft. (After Martin and Tabor, 1972)

VHF aeronautical-band-frequency converters, which are used to down-convert the VHF-channel carrier frequencies to the AM broadcast spectrum for convenient audio monitoring on a broadcast receiver, may be constructed using a superregenerative receiver similar to that employed in automatic door openers. As with self-quenched superregenerative door-opener receivers, the nonlinear oscillations occurring in VHF to AM converters are not inhibited from generating current components in the antenna circuit that radiate as noise into the surroundings. The radiated noise spectrum observed from two VHF to AM units was seen to be continuous within a minimum frequency interval of 10-MHz breadth, centered at the tuned frequency. The resulting spectral power density was found to be proportional to detector bandwidth for a range of bandwidths between 1 and 300 kHz, with an uncertainty less than ±1 dB.[5] The average noise power density for one model of a VHF to AM converter is plotted in Fig. 4-28 as x at a frequency of 109 MHz. The observation distance used was 30 ft and is marked within the brackets ⟨ ⟩.

ELECTRIC RAILROAD EQUIPMENT

Electric-powered trains contain main-drive motors; coupling circuitry to the external power bus; low-power subsystems including blower fans, control switches, door motors, and safety and warning devices, all of which may be sources of radio noise.[14] The noise spectrum generated by the low-powered auxiliary equipment lies below the high-frequency band, possessing a signal intensity much lower than that produced by the drive circuitry. Within the train-drive system, the principal radio-frequency noise source is the pantograph, which even at low speeds maintains sporadic contact with the external power bus. Gap-discharge breakdown occurs in the transitory airgaps appearing between the pantograph and power bus, producing an avalanche current flow that radiates short-duration pulses, the briefest of which are less than 1 μsec.[14] When a train is in movement, pulses are generated at rates of 10/sec or less. The spectrum generated by pantograph discharges extends downward from approximately 30 MHz as shown in Fig. 4-29 where the median values of many measurements of F_a are plotted.[15] These average noise-power data were measured at a location 50 ft from a railroad right-of-

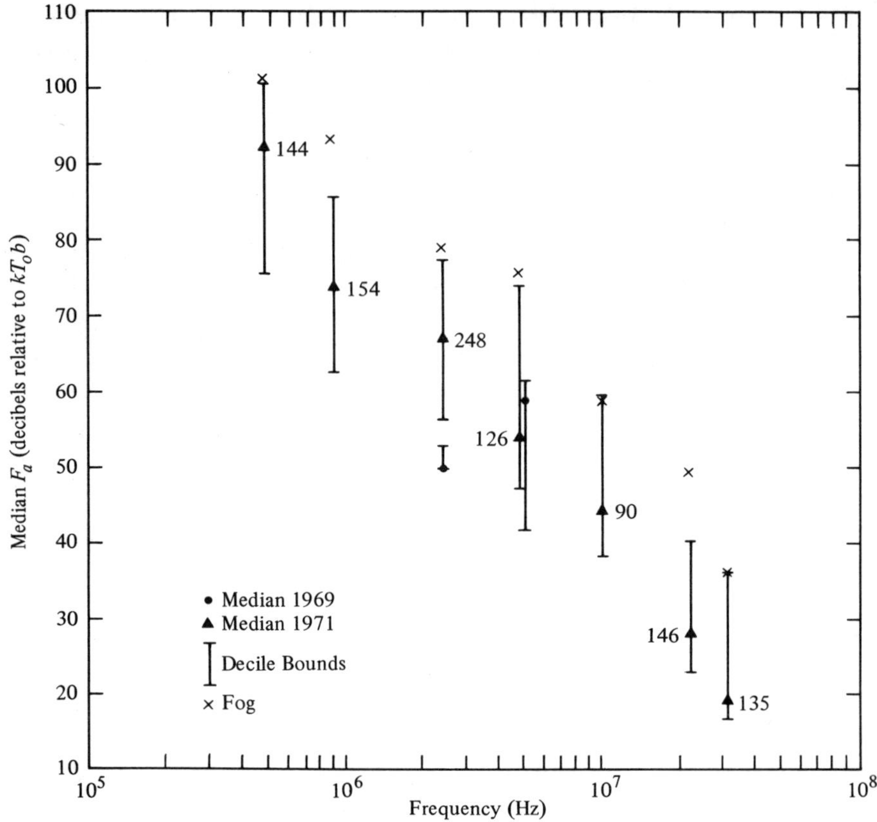

Fig. 4-29. Radiated noise power of an electric railroad measured at a distance of 50 ft. Median data obtained at one location separated by an interval of 2 yrs. Maximum-noise-power levels shown for foggy weather conditions (x). (After Hagn, 1972)

way on two occasions–in 1969 and in 1971. The entries beside each solid triangle indicate the number of data points; the vertical bars represent the interdecile ranges of the data sets, i.e., 10- to 90-percentile values. Marked by x are median values of F_a obtained while the local weather was foggy, the condition that yielded the largest radio-interference levels. The railway-noise spectrum is seen to be confined to and below the HF band. Notice that emitted noise power rises rapidly as the frequency decreases, displaying a decrement greater than −30 dB per decade frequency change.

References

1. Federal Communications Commission. Rules and Regulations. Volume II, Part 18.
2. Federal Communications Commission. Rules and Regulations. Volume II, Part 15.
3. Kernaghan, W. A., and Ford, R. R. Man-Made RFI Noise Tai Pei, Taiwan, Aug/Sept. 1970. PCA-EMA-70-47A. 1843rd Electronics Engineering Squadron, HQ Pacific Communications Area, May 28, 1971.
4. Pearce, S. F., and Bull, J. H. Interference from Industrial, Scientific and Medical Radio Frequency Equipment. Report No. 5033. Electrical Research Association, Leatherhead, Surrey, England, 1964.
5. Martin, H. and Tabor, F. Radio Frequency Emission Characteristics and Measurement Procedures of Incidental Radiation Devices and Industrial, Scientific and Medical Equipment. Electromagnetic Compatibility Analysis Center, Report FAA-RD-72-80, September 1972.
6. Garlan, H., and Whipple, L. G. Field Measurements of Electromagnetic Energy Radiated by RF Stabilized Arc Welders. FCC Technical Division Report No. T-6401, February 10, 1964.
7. Bensema, W. D., Kanda, M., and Adams, J. W. Electromagnetic Noise in Robena No. 4 Coal Mine. National Bureau of Standards Technical Note 654, April 1974.
8. Kanda, M., Adams, J. W., and Bensema, W. D. Electromagnetic Noise in McElroy Mine. National Bureau of Standards Report, NBSIR 74-389, June 1974.
9. Bensema, W. D., Kanda, M., and Adams, J. W. Electromagnetic Noise in Itmann Mine. National Bureau of Standards Report, NBSIR-74-390, June 1974.
10. Adams, J. W., Bensema, W. D., and Kanda, M. Electromagnetic Noise in Grace Mine. National Bureau of Standards Report, NBSIR-74-388, June 1974.
11. Scott, W. W., Adams, J. W., Bensema, W. D., and Dobroski, H. Electromagnetic Noise in Lucky Friday Mine. National Bureau of Standards Report, NBSIR 74-391, October 1974.
12. Radiation from fluorescent tubes. *Wireless World* 93–94 (March 1950).
13. Clark, D. B. Evaluation of Interference Suppression of Fluorescent Lamps. U.S. Naval Civil Engineering Laboratory, Technical Report 166, October 1961.
14. Amamiya, Y. A research on radio noise in medium and high frequency regions generated by electric cars. *Bulletin I.R.C.A.* 192–203 (1962).
15. Hagn, G. H. MF and HF Man-Made Radio Noise and Interference Survey-Brennerhaven, Germany. Stanford Research Institute Technical Memorandum No. 2, Project 1022-1, February 1972.

5
Theory of the Envelope Statistics of Man-Made Radio Noise

In the presentation of noise-emission signatures derived from experimental measurements performed upon the various types of man-made noise sources, the amplitude probability distribution (APD) of the noise-envelope voltage observed at the output of a linear receiving system has frequently been employed. The cumulative distribution, or level exceedance probability, of the instantaneous-output-envelope voltage from a linear network, and the cumulative distribution of instantaneous signal phase at the same circuit point, are together the first-order statistical functions of the random noise processes. Related to each distribution function are the density functions of instantaneous envelope amplitude and signal phase and the first, second, and higher moments of each, which collectively characterize the noise envelope and phase distributions. All play important roles in treating the effect of noise upon radio systems as, for example, in the determination of the rms value of the noise envelope, which is the square root of the second moment of the instantaneous envelope voltage. First-order statistical characterizations of noise signals provide an important but only an initial description of the noise phenomena. Remaining undescribed by first-order instantaneous envelope-amplitude and instantaneous signal-phase statistics are the important noise properties of time correlation, pulse-duration distribution, and pulse-spacing distribution. The complexity encountered in analytically describing non-Gaussian noise processes, which includes man-made incidental radio noise, in a formulation permitting both numerical evaluation and physical understanding of the interactions of the independent variables is sufficiently formidable to currently limit successes to the first-order statistics of instantaneous envelope voltage and signal phase.

By assuming that man-made incidental noise is reasonably well represented by a pattern of temporally and spatially, randomly occurring, independent noise impulses that are superimposed upon one another and upon a Gaussian distributed background, a Poisson-Gaussian model has been developed. This model yields an analytical representation of instantaneous envelope and signal phase. Significant physical understanding is maintained, and, to the extent it has been compared with experimental results, the model is creditably accurate.[1, 2]

The most general form of the analysis contains representations of the density of man-made noise sources within the area surrounding the observer and of the signal-attenuation behavior of a representative radio path lying between a noise source and the observer. Simplifications of the generalized source densities and propagation processes are introduced to yield analytically tractable results. Two principal subcases or classes of the results have been separately identified:

1. Class A. The frequency components of Class A noise emissions are constrained to a spectral width that is less than the bandwidth of the narrowest element of a linear receiving system. Consequently, no significant transient impulses are produced in the receiver by the noise signal.
2. Class B. The frequency components of Class B noise extend over a spectral range that is greater than the bandwidth of a linear receiving system, which may experience impulse excitation and manifest exponential signal buildup, signal decay, and damped oscillations.

Determination of the bandwidth-limiting element of a linear receiver in every case requires assessment of the frequency characteristics of the antenna, antenna coupler, and the radio-frequency and intermediate-frequency circuitry.

The arrangement of material used in the following sections for each noise class are drawn from Middleton (1974, 1976), commencing with a development of the characteristic function for the noise-envelope statistics. By a transformation of the characteristic function, the envelope-density function is formed and used to calculate both the cumulative distribution function (APD) and the moments of the noise, which include the average and rms envelope voltages.

DEVELOPMENT OF THE CHARACTERISTIC FUNCTION

Within the area surrounding an observation point, there is assumed to exist a domain within which man-made radio-noise sources exist. The boundaries of the domain may be described by a measure of radial distance from the observer that is equal to ct_T, where c is the velocity of light and t_T is transmission time of a noise source. The maximum time interval during which a receiver may accept a noise signal may be noted as $t_{T\max}$, which defines an upper limit to the radial distance between the observer and furthest noise source $ct_{T\max}$. The minimum time delay, t_{To}, and the associated lower range limit, ct_{To}, provide the lowest bound to the source domain, which may be zero.

Three cases of interest are shown in Fig. 5-1:

1. Case I: When the noise source distribution extends to a radial distance $ct_{T\Lambda}$ greater than the range represented by the upper time limit of the sampling window $ct_{T\max}$, some portion of the noise emission is not admitted to the receiver during the time $0 \leqslant t_T \leqslant t_{T\max}$.
2. Case II: When the sampling window extends from time zero to a time $t_{T\max}$, which permits the reception of noise pulses over a radial distance $ct_{T\max} > ct_{T\Lambda}$, where $ct_{T\Lambda}$ is the outer range limit of the noise-source-effective domain $ct_{T\Lambda}$.
3. Case III: In Case III, the lower time bound of the sampling window $ct_{To} \neq 0$, and the source distribution extends a radial distance beyond the upper observation limit $ct_{T\max}$. Thus, an annular area of the noise-source domain contributes to the observed noise envelope

$$ct_{To} \leqslant ct_T \leqslant ct_{T\max} \leqslant ct_{T\Lambda}.$$

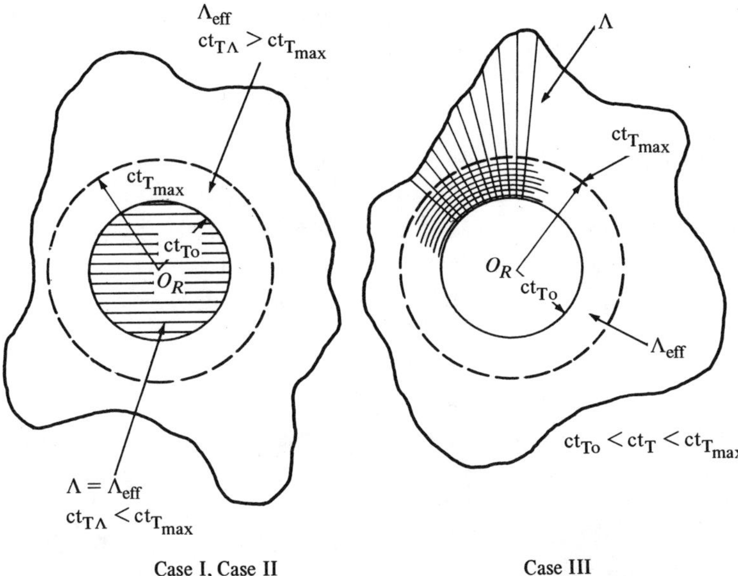

Fig. 5-1. Planar representation of the effective source domain for three cases of observation time duration.

The term *effective domain* symbolized by Λ_{eff} conveys the meaning that only those noise sources of sufficient amplitude to be distinguished above the thermal noise of the receiver comprise the domain of residence. In the following development, Case I is presumed to represent the most commonly encountered situation. Although Fig. 5-1 is drawn for a planar distribution of noise source, it may be extended to include a volume-noise-source distribution should it be necessary.[3] Normally, a planar distribution of noise emitters is adequate and accurate for man-made incidental-radio-noise analyses wherein noise-emitting equipment is dominantly confined to or within a few hundred feet of the earth surface.

The effective source domain, Λ_{eff}, may be subdivided into small spatial intervals, $\Delta\lambda_N$, within each of which N independent random-noise-emission events occur during an epoch $\hat{\epsilon}$. The noise emitted from the jth spatial interval possesses a waveform representable at the output of the linear receiver as a function of observer time, t, by $u(t,\ \hat{\epsilon},\ \theta)_n$. The receiver linearly superimposed the noise emissions from concurrent events to form the total observed noise waveform

$$X(t)_j = \sum_{n=0}^{N} U(t, \hat{\epsilon}, \underset{\sim}{\theta})_n . \tag{5-1}$$

The symbol $\underset{\sim}{\theta}$ represents any time-independent statistical parameter, e.g., mean noise-signal amplitude. Because it is assumed that each noise emission is independent and Poisson distributed, the conditional probability that exactly N events will occur within the jth spatial interval, $\Delta\lambda_j$, and epoch interval, $\Delta\hat{\epsilon}$, is given by

$$P_j(N|t', \Delta\hat{\epsilon}, \Delta\lambda_j) = \frac{1}{N!} [\rho(\Delta\lambda_j, \Delta\hat{\epsilon}) \cdot \Delta\lambda_j] \exp(\rho\Delta\lambda_j) \tag{5-2}$$

The point density of the process is represented by $\rho(\Delta\lambda_j; \Delta\hat{\epsilon})$ and is the average number of noise-emission events occurring per spatial interval $\Delta\lambda_j$ during time interval $\Delta\hat{\epsilon}$.

The joint distribution density, $W_j(X_1, t_1, \cdots, X_N t_N | t', \Delta\hat{\epsilon}, \Delta\lambda_j)$, that N noise waveforms will be received and observed at the receiver output at time t' from the spatial interval $\Delta\lambda_j$ is given by:

$$W_j(X_1, t_1, \cdots, X_N, t_N | t', \Delta\hat{\epsilon}, \Delta\lambda_j)$$

$$= \sum_{n=0}^{\infty} P_j(N|t', \Delta\hat{\epsilon}, \Delta\lambda_j) \cdot w_N(X_1, t_1, \cdots, X_N, t_N | N, t', \Delta\hat{\epsilon}, \Delta\lambda_j) \tag{5-3}$$

in which $w_N(X_1, t_1, \cdots, X_N, t_N | N, t', \Delta\hat{\epsilon}, \Delta\lambda_j)$ is the joint conditional probability density, given N events in time $\Delta\hat{\epsilon}$ and spatial interval $\Delta\lambda_j$. The characteristic function of $W_j(X_n t_N | t', \Delta\hat{\epsilon}, \Delta\lambda_j)$ represented by $F_j(i\xi' | t', \Delta\hat{\epsilon}, \Delta\lambda_j)_X$, where ξ is the transform variable and X represents the instantaneous amplitude of the waveform, is obtained as the Fourier transform of Equation 5-3 as:

$$F_j(i\xi | t', \Delta\hat{\epsilon}, \Delta\lambda_j)_X = \sum_{N=0}^{\infty} P_j(N|t', \Delta\hat{\epsilon}, \Delta\lambda_j)$$

$$\cdot F_N(i\xi_1, t_1, \cdots, i\xi_N, t_N | N, t', \Delta\hat{\epsilon}, \Delta\lambda_j). \tag{5-4}$$

Herein, $F_N(i\xi_1, t_1, \cdots, i\xi_N, t_N | N, t', \Delta\hat{\epsilon}, \Delta\lambda_j)$ is the characteristic function of

$$w_N(X_1, t_1, \cdots, X_N, t_N | N, t', \Delta\hat{\epsilon}, \Delta\lambda_j)$$

which is, by definition, the following expression, the Nth-order Fourier transform of the joint distribution function $W(\theta_N)$ of N set of random variables.

$$F_N(i\xi_1, t_1, \cdots, i\xi_N, t_N | N, t', \Delta\hat{\epsilon}, \Delta\lambda_j)$$
$$= \left\{\exp\left[i \sum_{n=0}^{N} \xi_n U(t, \hat{\epsilon}, \underset{\sim}{\theta})_n\right\}^N\right.$$
$$= \left[\int W(\theta_N) \exp\left\{i \sum_{n=0}^{N} \xi_n U(t, \hat{\epsilon}, \underset{\sim}{\theta})_n\right\} d\theta\right]. \qquad (5\text{-}5)$$

Equation 5-5 is applicable to N noise waveforms emitted within the spatial interval $\Delta\lambda_j$ during time interval $\Delta\hat{\epsilon}$.[3, 4] By substituting Equations 5-2 and 5-5 into Equation 5-4, the nth order characteristic function for the jth spatial interval is obtained:

$$F_j(i\xi|\Delta\hat{\epsilon}, \Delta\lambda_j) = \exp\left\{\rho(\Delta\lambda_j, \Delta\hat{\epsilon}) \cdot \Delta\lambda_j \left\langle \exp\left[i \sum_{n=0}^{N} \xi_n U(t, \hat{\epsilon}, \underset{\sim}{\theta})_n\right] - 1\right\rangle_{\underset{\sim}{\theta}}\right\} \qquad (5\text{-}6)$$

where $\langle\,\rangle_{\underset{\sim}{\theta}}$ indicates that the average has been taken over the range of the variables, $\underset{\sim}{\theta}$. The simplification of form obtained in Equation 5-6 has been accomplished by using the infinite series representation for the exponential, that is,

$$e^x = \sum_{n=0}^{\infty} \frac{x^n}{n!}$$

and by factoring the product $\rho(\Delta\lambda_j, \Delta\hat{\epsilon}) \cdot \Delta\lambda_j$ from the result. In the limit of

$$\begin{matrix} \Delta\epsilon \longrightarrow d\hat{\epsilon} \longrightarrow 0 \\ \Delta\lambda_j \longrightarrow d\lambda \longrightarrow 0 \end{matrix} \text{ as } j \longrightarrow \infty$$

$$F(i\xi) = \prod_j F_j(i\xi|\Delta\hat{\epsilon}, \Delta\lambda_j),$$

which may be obtained by integrating the exponent of Equation 5-6 over the effective domain Λ_{eff},

$$F(i\xi)_X = \exp\left\{\int_{\hat{\epsilon},\Lambda_{eff}} \rho(\lambda,\hat{\epsilon}) \cdot \left\langle \exp\left[i\sum_{n=1}^{N} \xi_n U(t,\hat{\epsilon},\underset{\sim}{\theta})_n\right] - 1\right\rangle_{\underset{\sim}{\theta}} d\lambda\, d\hat{\epsilon}\right\}. \quad (5\text{-}7)$$

The range of integration for the epoch time $\hat{\epsilon}$ is the observation period T and is determined by the receiver sampling interval. When N in Equation 5-7 is placed equal to 1, the first-order characteristic function is obtained,

$$F(i\xi)_X = \exp\left[\int_{\hat{\epsilon},\Lambda_{eff}} \rho(\lambda,\hat{\epsilon}) \left\langle \exp i\xi U(t,\hat{\epsilon},\underset{\sim}{\theta}) - 1\right\rangle_{\underset{\sim}{\theta}} d\lambda d\hat{\epsilon}\right]. \quad (5\text{-}8)$$

The observed noise waveform, $u(t, \hat{\epsilon}, \underset{\sim}{\theta})$, may be resolved into the orthogonal quadrature components as follows:

$$U_c = U(t,\hat{\epsilon},\underset{\sim}{\theta}) \cos\psi = B_o \cos(\phi_s' + \mu_d \hat{\epsilon}\,\omega_0)$$

$$U_s = U(t,\hat{\epsilon},\underset{\sim}{\theta}) \sin\psi = B_o \sin(\phi_s' + \mu_d \hat{\epsilon}\,\omega_0) \quad (5\text{-}9)$$

where B_o = the observed noise signal; a function of time and the number and location of sources but which may be assumed to be independent of the relative doppler velocity existing between the set of noise sources and the observer

ϕ_s' = a collective-phase variable composed of the separate phase contributions of (1) the noise sources referenced to an appropriate origin and (2) the phase shift introduced by the antenna and antenna coupler used in the observing equipment

μ_d = the relative doppler coefficient produced by motion of observer relative to the source domain and equal to $1 + \frac{2v}{c}$ where v-the relative source domain-observer velocity. Most often, $\frac{2v}{c} \ll 1$ and may be ignored.

The two dimensional form of the characteristic function may now be written as follows:

$$F_1(i\xi, i\eta) = \exp\left[\int_{\hat{\epsilon},\Lambda_{eff}} \rho(\lambda, \hat{\epsilon})\,\langle \exp(i\xi U_c + i\eta U_s) - 1\rangle\, d\lambda\, d\hat{\epsilon}\right]. \tag{5-10}$$

Orthogonal components U_c and U_s of the observed signal are represented above in a Cartesian coordinate system (ξ, η). A simplification of Equation 5-10 is achieved by transforming the (ξ, η) coordinate representation to polar notation where the radius vector r and polar angle ϕ are used.

$$\hat{F}_1(ir, \phi) = \exp\left\{\int_{\hat{\epsilon},\Lambda_{eff}} \rho(\lambda, \hat{\epsilon})\,\langle \exp[irB_o \cos(\phi_s' + \mu_d\hat{\epsilon}\,\omega_0 - \phi)] - 1\rangle_{\underset{\sim}{\theta}}\, d\hat{\epsilon}\, d\lambda\right\} \tag{5-11}$$

The probability density function of the observed signal envelope is derived from Equation 5-11 by means of the two-dimensional Fourier transform expressed in polar form:

$$W_1(E, \psi) = \frac{E}{(2\pi)^2}\int_0^\infty r\,dr \int_0^{2\pi} d\phi\,\hat{F}_1(ir, \phi)\, e^{-iEr\cos(\chi-\phi)},$$

$$E > 0;\ 0 \leqslant \psi \leqslant 2\pi, \tag{5-12}$$

using ψ to represent the total angular variable as presented in Equation 5-9. The polar form of the transformation is derived from the Cartesian representation by replacement of the unit vectors (ξ, η) by $(r\cos\phi, r\sin\phi)$ and multiplication of $F_1\,(ir\cos\phi, i\sin\phi)$ by the Jacobian $\left|\dfrac{\partial(u_c, u_s)}{\partial(E, \phi)}\right| = E$. The Jacobian multiplier is used to transform the joint probability density function from the variable pair (u, ψ) to (E, ψ), which were introduced in Equation 5-9 and in which E represents the instantaneous value of the observed-signal envelope.

The envelope APD or exceedance probability, $P_1(E > E_0)$, is required for both the experimental comparisons that will be presented later and for further development of the basic theory. In the definition of $P_1(E > E_0)$, the value of the envelope E_0 represents a selected level for computation of the likelihood that E_0 is exceeded; thus,

$$P_1(E > E_0) = \int_{E_0}^{\infty} W_1(E)\, dE$$

$$= 1 - \int_0^{E_0} W_1(E)\, dE. \tag{5-13}$$

Using Equation 5-12,

$$P_1(E > E_0) = 1 - \int_0^{E_0} \frac{E}{2\pi} \left[\int_0^{\infty} r\, J_0(rE)\, dr \right] dE \int_0^{2\pi} \hat{F}_1(ir, \phi)\, d\phi. \tag{5-14}$$

Rearranging yields

$$P_1(E > E_0) = 1 - \frac{1}{2\pi} \int_0^{\infty} r \left[\int_0^{E_0} E\, J_0(rE)\, dE \right] dr \int_0^{2\pi} \hat{F}_1(ir, \phi)\, d\phi$$

$$= 1 - \frac{E_0}{2\pi} \int_0^{\infty} \left[r\, J_1(rE_0) \int_0^{2\pi} \hat{F}_1(ir, \phi)\, d\phi \right] dr, \tag{5-15}$$

which has been obtained using the integral relationship:*

$$\int_0^{z} z\, J_0(z)\, dz = zJ_1(z).$$

In the development of Equations 5-11 and 5-15, representing the characteristic function and exceedance probability, respectively, only a Poisson distribution of radio-noise-source locations and emission times within a spatial domain have been assumed. No restrictions have, at this point in the development, been placed upon the source parameters, such as phase.

*Reference 5, Equation 11.3.20.

The inner exponential existing in the representation of the noise process characteristic function, Equation 5-11, may be simplified by accepting a restriction usually found to exist for most radio receiving equipment, namely, that the instantaneous bandwidth of the antenna and receivers, Δf_{AR}, is much less than the observation frequency, f_0. When this assumption is admissible, the inner exponential of Equation 5-11, which is representable by the general form[5] as:

$$\exp(ia\cos\phi) = J_0(a) + 2\sum_{k=1}^{\infty}(-1)^k J_{2k}(a)\cos(2k\phi) + i2\sum_{k=1}^{\infty}(-1)^k J_{2k+1}(a)\cos[(2k+1)\phi] \tag{5-16}$$

may be simplified by omitting all terms $m > 0$, which oscillate rapidly with time $\hat{\epsilon}$ and cancel with the three other parameters, i.e., envelope amplitude E, noise sources and receiving system phase ϕ_s', and process density $\rho(\lambda, \hat{\epsilon})$ that are time-independent or very slowly varying functions of time. Upon applying Equation 5-16 to Equation 5-11 and discarding all terms with $m > 0$, the characteristic function becomes:

$$\hat{F}_1(ir) = \exp\left[\int_{\epsilon,\Lambda_{eff}} \rho(\lambda, \hat{\epsilon})\,\langle J_0(rB_o) - 1\rangle_{\underset{\sim}{\theta}}\, d\hat{\epsilon}\, dt'\right], \tag{5-17}$$

which is independent of the phase term, ϕ. Thus, when Equation 5-17 is introduced into Equations 5-12 and 5-15, and integration over ϕ is performed,* there results:

$$W_1(E, \psi) = \frac{E}{2\pi}\int_0^{\infty} rJ_0(rE)\,\hat{F}_1(ir)\,dr \tag{5-18}$$

and

$$P(E > E_0) = 1 - E_0\int_0^{\infty} J_1(rE_0)\,\hat{F}_1(ir)\,dr. \tag{5-19}$$

*Using Reference 5, Equation 9.1.21.

The Poisson point process density $\rho(\lambda, \hat{\epsilon})$ appearing in Equation 5-17 was introduced by the use of Equation 5-2 to express the conditional probability of occurrence of N noise emissions in the jth spatial interval, $\Delta\lambda_j$, and may be represented in terms of the surface density $\sigma_\Lambda(\lambda)$ of the noise sources in domain Λ, i.e.,

$$\rho(\lambda, \hat{\epsilon}) = \sigma_\Lambda(\lambda) \left| J \begin{pmatrix} r\partial\phi & \partial r \\ \partial\phi & \partial\hat{\epsilon} \end{pmatrix} \right| \tag{5-20}$$

where J ( ) represents the Jacobian. By expressing the differential area element in polar coordinates with the radius vector r expressed as $c\hat{\epsilon}$, where c, the velocity of light, yields $J(\,) = c^2\hat{\epsilon}$, and the point process density becomes:

$$\rho(\lambda, \hat{\epsilon}) = \sigma_\Lambda(\lambda)\, c^2\hat{\epsilon}. \tag{5-21}$$

The form of the noise-source surface density $\sigma_\Lambda(\lambda)$ must be established independently and, in general, experimentally. To proceed with the analyses, a general form for $\sigma_\Lambda(\lambda)$ will be assumed, a function only of radial distance, i.e., time $\hat{\epsilon}$ as

$$\sigma_\Lambda(\lambda) = A_K\hat{\epsilon}^{-\mu}.$$

This allows the point process density to be written as:

$$\rho(\lambda, \hat{\epsilon}) = A_K\, c^2\hat{\epsilon}^{(1-\mu)} \tag{5-22}$$

The density function expressed by Equation 5-22 requires normalization to unity with the aid of the constant A_K adjusted so that

$$\int_0^{\hat{\epsilon}_{max}} \rho(\lambda, \hat{\epsilon})\, d\hat{\epsilon} = A_K \int_0^{\hat{\epsilon}_{max}} c^2\hat{\epsilon}^{(1-\mu)}\, d\hat{\epsilon} = 1,$$

yielding:

$$A_K = c^{-2}\,(2-\mu)\,\hat{\epsilon}_{max}^{(\mu-2)}. \tag{5-23}$$

Introducing the following definitions and symbols permits a simplified presentation of the characteristic function and exceedance probabilities to be developed:

$\nu_T(\hat{\epsilon})$ = the average number of emissions per source, per time interval $d\hat{\epsilon}$ in period T.

$$A_{\hat{\epsilon},T} = \int_T \nu_T(\hat{\epsilon})\, d\hat{\epsilon} = \text{the average number of noise emissions per noise source in period } T.$$

It is apparent that the probability of a noise source being located at some coordinate point in domain Λ is independent of the probability that it or any other noise source in the domain will emit at time $\hat{\epsilon}$ in period T. The left-hand side of Equation 5-21 may therefore be factored as

$$\rho(\lambda, \hat{\epsilon}) = \rho_\Lambda(\lambda)\, \nu_T(\hat{\epsilon}), \tag{5-24}$$

and where

$A_\Lambda = \int_\Lambda \rho_\Lambda(\lambda)\, d\lambda$ is the average number of noise-emitting sources in domain Λ

$A_T = \int_{\Lambda,T} \rho(\lambda, \hat{\epsilon})\, d\lambda d\hat{\epsilon} = A_\Lambda A_{\hat{\epsilon},T}$ is the average number of noise emissions in period T.

Using the preceding definitions for $\rho_\Lambda(\lambda)$, A_Λ, $\nu_T(\hat{\epsilon})$, and $A_{\hat{\epsilon},T}$, probability densities $W_1(\lambda)$ and $W_1(\hat{\epsilon})_T$ are definable as

$$W_1(\lambda) = \rho_\Lambda(\lambda)/A_\Lambda$$
$$W_1(\hat{\epsilon})_T = \nu_T(\hat{\epsilon})/A_{\hat{\epsilon},T}. \tag{5-25}$$

With the aid of the preceding definitions, it becomes possible to prepare the exponent of $\hat{F}_1(ir)$ (Equation 5-17) for a final simplification. Representing this quantity by $\hat{I}_T$ and using Equations 5-24 and 5-25 leads to

$$\hat{I}_T = A_\Lambda A_{\hat{\epsilon},T} \int_{\Lambda,\hat{\epsilon}} W_1(\lambda)\, W_1(\hat{\epsilon})\, \langle J_0(rB_0) - 1\rangle_{\underset{\sim}{\theta}}\, d\lambda d\hat{\epsilon}. \tag{5-26}$$

The noise waveform B_o appearing in Equation 5-26 will be recalled to be a function of time, number, and location of the radio-noise sources. At the output of the receiver IF amplifier, a noise waveform will occur whose duration T_s is a random variable. A stable distribution of sources will permit representing the ensemble of observed noise waveforms as having an average duration $\overline{T}_s$. In this manner, one may use z to indicate a random variable, that is unitless and related to event time as:

$$t + \hat{\epsilon} = \overline{T}_s \cdot z. \tag{5-27}$$

Transforming the variable of integration in Equation 5-26 by means of Equation 5-27 and performing the integration over λ provides the mean of the integral with respect to z and the following form for $\hat{I}_T$:

$$\hat{I}_T = T^{-1} A_\Lambda A_{\hat{\epsilon},T} \, \overline{T}_s \left\langle \int_0^{T_s/\overline{T}_s} T \, W_1(z) \, [J_0 \, (rB_o) - 1] \, dz \right\rangle_{\underset{\sim}{\theta}, \lambda, T_s}, \tag{5-28}$$

which has been written after multiplying and dividing the integrand by the observation period T. Noting the definitions introduced earlier, the first factors appearing on the right of Equation 5-28 yield a quantity $\overline{\nu}_T$ designated as the average number of noise emissions per second occurring during observation period T, i.e.,

$$T^{-1} A_\Lambda \, A_{\hat{\epsilon},T} = \overline{\nu}_T.$$

Furthermore, it is assumed that during the observation period T, the statistics of the noise sources remain stationary and that, during the observation period, the noise emissions are uniformly distributed in time, i.e., $W_1(z) = T^{-1}$, then

$$\hat{I}_T = \overline{\nu}_T \, \overline{T}_s \left\langle \int_0^{T_s/\overline{T}_s} [J_0 \, (rB_o) - 1] \, dz \right\rangle_{\underset{\sim}{\theta}, \lambda, T_s}. \tag{5-29}$$

In the limiting observation interval $T \longrightarrow \infty$

$$\overline{\nu}_T \longrightarrow \overline{\nu}_\infty$$

and the product

$$\overline{\nu}_\infty \overline{T}_s = A_\infty, \tag{5-30}$$

where A_∞ is designated the impulsive index, which measures the degree of temporal noise pulse overlap observed arising from a domain of sources. When A_∞ is small, few noise waveforms will be found to overlap at the output of the receiver, and the total noise pattern will be impulsive in form dominated by these few waveforms appearing singularly. When A_∞ assumes large values, many waveforms will concurrently contribute to the total observed noise pattern, which in the limit of large numbers will appear as thermal noise and representable by Gaussian statistics.

Rewriting Equation 5-17, using $\hat{I}_T$ from Equation 5-29 and A_∞ from Equation 5-30, provides the final form of the characteristic function that will be used in subsequent development of the statistical features of the man-made radio-noise envelope

$$\hat{F}_1(ir) = \exp \; A_\infty \left\{ \left\langle \int_0^{T_s/\overline{T}_s} [J_0(rB_0) - 1] \, dz \right\rangle_{\underset{\sim}{\theta}, \lambda, T_s} \right\}. \tag{5-31}$$

STATISTICAL DEVELOPMENTS FOR CLASS A AND CLASS B NOISE

Classification of the observed radio noise employed in this chapter depends upon identifying the features of the noise envelope produced by translation of the noise signals through a linear receiver, reflecting the manner in which the receiver modifies the noise-signal envelope. A designation given to radio noise when it is narrowband relative to the instantaneous bandwidth of the receiver and thereby productive of no significant transient responses is *Class A*. Noise that is broadband in comparison to the receiver instantaneous bandwidth and capable of inducing a transient response consisting of pronounced exponential modifications of the leading and trailing edges of the received signal envelope and/or damped oscillations is called *Class B*. Since it is possible for an intermediate noise *Class C* to exist embodying features of both A and B, this is also identified. If the source-emission duration T_S is considered to be fixed, and if the positive portion of the noise-source-emission waveform incident on the receiver is designated by $S(z)$, which exists for $0 \leqslant z \leqslant T_{in}$, then Fig. 5-2 portrays the receiver output responses as a function of the reduced time variable z. Shown in Fig. 5-2 is an exponential rise of the amplifier output, $U_{oA}(z)$, occurring at the advent of a pulse signal of duration T_{SA}, which is terminated at time $t = \overline{T}_{SA}$ by the onset of the transient decay, with decay coefficient $\tilde{\beta}$ persisting

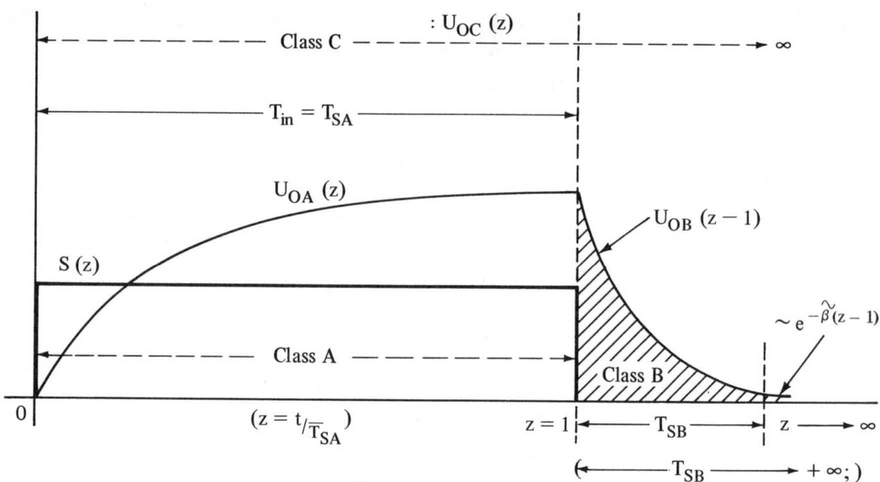

Fig. 5-2. Output response of an IF amplifier in a linear receiver to an input rectangular pulse of duration T_{SA} and magnitude $E_o(z)$. Class A and Class B positive envelope components are shown as $U_{oA}(z)$ and $U_{oB}(z - 1)$. The combined envelope component $U_{oC}(z)$ is shown and noted as Class C noise. (After Middleton, 1976)

indefinitely and representable by $U_{oB}(z - 1)$ for the positive portion of the output envelope. The subscripts A, B, and C appearing on U_o indicate the receiver responses associated with the three classes of noise. The $U_{oC}(z)$ response function equals the sum of $U_{oA}(z) + U_{oB}(z - 1)$. Furthermore, $\overline{T}_{SA}$ is the mean value of the observed source emission time for Class A noise, while T_{SB} denotes the observed duration for Class B. T_{SB} depends upon the response characteristics of the amplifier in addition to the temporal features of the noise sources waveform.

Observations of man-made radio noise frequently record a mixed distribution consisting of impulsive noise superimposed upon a background of thermal noise possessing a Rayleigh distributed envelope, which derives from a Gaussian distribution of the quadrature signal component amplitudes. The term most frequently used to designate the thermal background component is *Gaussian noise*. The Gaussian noise component appearing in man-made noise emissions may arise from either the man-made noise that is transmitted to the observation point by the same propagation process that governs the impulsive noise or from the radio receiver employed in the measurements. Noise from both Gaussian sources must be presumed to be present in any series of tests and thus requires representation in the analytical formulations. Each Gaussian source may be considered to independently arise and, furthermore, may be assumed to be respectively independent of the man-made impulsive noise. Statistical independence of the Gaussian and impulsive noise distributions allows the joint characteristic function to be written as the product of Equation 5-31 and the characteristic function for a Gaussian distributed bivariate $(i\xi, i\eta)$:

$$F_{1G}(i\xi, i\eta) = \exp\{-\tfrac{1}{2}(\xi^2 + \eta^2)\,\sigma_G^2\} = \exp\{-\tfrac{1}{2}r^2\sigma_G^2\} \quad (5\text{-}32)$$

where the Cartesian quadrature components of the observed signal are related to the polar angle and radius vector in the polar system by

$$\xi = r\cos\varphi;\ \eta = r\sin\varphi$$

and for which it has been assumed the angle variable φ is uniformly distributed $0 \leqslant \varphi \leqslant 2\pi$. The variances of the total Gaussian noise component is represented in Equation 5-32 by σ_G^2, which is equal to

$$\sigma_G^2 = \sigma_R^2 + \sigma_E^2 \quad (5\text{-}33)$$

in which subscripts R and E indicate the receiver and external Gaussian sources, respectively.

Class A Noise Statistics

Development of the Characteristic Function. The initial and principal step to be taken in the construction of the statistical formulations for Class A noise is the development of the characteristic function for the observed signal envelope. The Class A noise characteristic function, augmented by inclusion of the Gaussian background component, is constructed by combining Equations 5-31 and 5-32 to form:

$$\hat{F}_1(ir)_{A+G}$$

$$= e^{-1/2r^2\sigma_G^2}\exp\left\{A_{\infty,A}\left\langle\int_0^{T_S/\overline{T}_S}[J_o(rB_o) - 1]\,dz\right\rangle_{\underline{\theta},\lambda,T_S}\right\}. \quad (5\text{-}34)$$

The added subscript A appearing on A_∞ is used to relate this quantity and associated equation with Class A noise. The averaging, indicated in Equation 5-34 by the brackets $\langle\ \rangle_{\underset{\sim}{\theta},\lambda,T_S}$, is to be performed over the process variables $\underset{\sim}{\theta}$, source-emission time T_S, or the reduced variable z, and the noise-domain dimensions λ. The latter average for Class A interference leads to a particularly simple result and may be introduced at this time to further simplify the expressions to follow. The indicated average in Equation 5-34 over λ may be written, using the process density $\rho(\lambda, \hat{\epsilon})$ of Equation 5-22, as

$$\langle\ \rangle_{\hat{\epsilon}} = \int_0^\infty \left\langle \int_0^{T_S/\overline{T}_S} dz\ [J_o(rB_o) - 1] \right\rangle_{\underset{\sim}{\theta},T} \rho(\lambda, \hat{\epsilon})\, d\hat{\epsilon}$$

$$= \int_0^\infty \left\langle \int_0^{T_S/\overline{T}_S} dz\ [J_o(rB_o) - 1] \right\rangle_{\underset{\sim}{\theta},T} A_K c^{2(1-\mu)}\, d\hat{\epsilon}.$$

The upper limit for the integration over $\hat{\epsilon}$ may be set $\equiv \infty$ because for Class A noise the observed waveform, B_{oA}, for $z > 1$ is $\equiv 0$. Noting that the variable of integration for averaging $\hat{\epsilon}$ is a time measure derived from the representation of radial distance $R = C\hat{\epsilon}$ and is independent of reduced source time z, the order of integration may be interchanged to yield

$$\langle\ \rangle_{\hat{\epsilon}} = \int_0^\infty A_K c^2 \hat{\epsilon}^{(1-\mu)}\, d\hat{\epsilon} \left\langle \int_0^{T_S/\overline{T}_S} [J_o(rB_{oA}) - 1]\, dz \right\rangle_{\underset{\sim}{\theta},T_S}. \tag{5-35}$$

However, from Equation 5-23, the normalization condition yields unity for the first integration and permits the operational notation of Equation 5-34 to be simplified as:

$$\hat{F}_1(ir)_{A+G} = e^{-1/2 r^2 \sigma_G^2} \exp\left\{ A_{\infty,A} \left\langle \int_0^{T_S/\overline{T}_S} [J_o(rB_{oA}) - 1]\, dz \right\rangle_{\underset{\sim}{\theta},T_S} \right\}. \tag{5-36}$$

The averaging operation over T_S indicated in Equation 5-36 may be performed subject to a condition that the input signal duration observed by the receiver, T_{in}, is equal to the source emission duration. By virtue of this stipulation, the observation period is adjusted to permit the receiver to accumulate the full noise-emission waveform:

$$\left\langle \int_0^{z_o} [J_o(rB_{oA}) - 1]\, dz \right\rangle_{\underset{\sim}{\theta}, T_S} = -\langle z_o \rangle_{\underset{\sim}{\theta}} + \left\langle \int_0^{z_o} J_o(rB_{oA})\, dz \right\rangle_{\underset{\sim}{\theta}}$$

for $z_o = T_{in}/\overline{T}_s$. However,

$$\langle z_o \rangle = \langle T_{in} \rangle \langle T_s \rangle^{-1} = \langle T_s \rangle \langle T_s \rangle^{-1} = 1.$$

Consequently,

$$\left\langle \int_0^{z_o} [J_o(rB_{oA}) - 1]\, dz \right\rangle_{\underset{\sim}{\theta}, T_S} = -1 + \left\langle \int_0^{z_o} J_o(rB_{oA})\, dz \right\rangle_{\underset{\sim}{\theta}}. \tag{5-37}$$

It is appropriate to expand the Bessel function appearing in Equation 5-37 as an aid in constructing a form for the characteristic function, which possesses a greater convenience for computation of the envelope moments and yet contains the essential dependences of the complete form. The independence of the variables $\underset{\sim}{\theta}$ and z permit executing the average shown in Equation 5-37 over the integrand, i.e.,

$$\left\langle \int_0^{z_o} J_o(rB_{oA})\, dz \right\rangle_{\underset{\sim}{\theta}} = \int_0^{z_o} \langle J_o(rB_{oA}) \rangle_{\underset{\sim}{\theta}}\, dz \tag{5-38}$$

By rewriting the kernal of Equation 5-38 as

$$\begin{aligned} \langle J_o(rB_{oA}) \rangle_{\underset{\sim}{\theta}} &= \exp\left(-\tfrac{1}{4} r^2 \langle B_{oA}^2 \rangle_{\underset{\sim}{\theta}}\right) \langle J_o(rB_{oA}) \exp\left(\tfrac{1}{4} r^2 \langle B_{oA}^2 \rangle\right) \rangle_{\underset{\sim}{\theta}} \\ &= \exp\left(-\tfrac{1}{4} r^2 \langle B_{oA}^2 \rangle_{\underset{\sim}{\theta}}\right) \langle J_o(rB_{oA}) \rangle_{\underset{\sim}{\theta}} \exp\left(\tfrac{1}{4} r^2 \langle B_{oA}^2 \rangle_{\underset{\sim}{\theta}}\right) \end{aligned} \tag{5-39}$$

and employing the following representation,*

$$\begin{aligned} &\langle J_o(rB_{oA}) \rangle_{\underset{\sim}{\theta}} \exp\left(\frac{1}{4} r^2 \langle B_{oA}^2 \rangle_{\underset{\sim}{\theta}}\right) \\ &\quad = 1 + \sum_{\ell=2}^{\infty} \frac{\langle B_{oA}^2 \rangle_{\underset{\sim}{\theta}}\, r^{2\ell}}{2^{2\ell} \ell!}\; {}_1F_1\left(-\ell, 1, \frac{B_{oA}^2}{\langle B_{oA}^2 \rangle_{\underset{\sim}{\theta}}}\right), \end{aligned} \tag{5-40}$$

*Reference 6, Equation 13. 107b.

in which ${}_1F_1(\alpha, \beta, x)$ is the notation for the hypergeometric function.[5] Replacing $\langle J_o(rB_{oA})\rangle_{\underset{\sim}{\theta}}$ in Equation 5-38 with Equations 5-39 and 5-40 yields

$$\left\langle \int_0^{z_o} J_o(rB_{oA})\,dz \right\rangle_{\underset{\sim}{\theta}} = \exp\left(-\frac{1}{4}r^2 \langle B_{oA}^2\rangle_{\underset{\sim}{\theta}}\right)\left\{1 + \sum_{\ell=2}^{\infty} \frac{\langle B_{oA}^2\rangle_{\underset{\sim}{\theta}}\, r^{2\ell}}{2^{2\ell}\,\ell!} F_\ell\right\} \tag{5-41}$$

in which

$$F_\ell = \left\langle \int_0^{z_o} {}_1F_1\left(-\ell, 1, \frac{B_{oA}^2}{\langle B_{oA}^2\rangle_{\underset{\sim}{\theta}}}\right) dz \right\rangle_{\underset{\sim}{\theta}}.$$

The characteristic function, Equation 5-36, by substitution of Equations 5-37 and 5-41, becomes

$$\hat{F}_1(ir)_{A+G} = e^{-1/2 r^2 \sigma_G^2 - A_{\infty,A}}$$

$$\cdot \left\{\exp\left[A_{\infty,A} \exp\left(-\frac{1}{4}r^2\langle B_{oA}^2\rangle_{\underset{\sim}{\theta}}\right)\left\{1 + \sum_{\ell=2}^{\infty} \frac{\langle B_{oA}^2\rangle_{\underset{\sim}{\theta}}\, r^{2\ell}}{2^{2\ell}\,\ell!} F_\ell\right\}\right]\right\}. \tag{5-42}$$

By expanding the exponential factor on the right, the product of two infinite series is obtained. In the second infinite series, all terms beyond the first ($\ell = 0$) are negligible in virtue of each being multiplied by $\exp\,(-\frac{1}{4}r^2\langle B_o^2\rangle_{\underset{\sim}{\theta}})^\ell$ for integer ℓ, $0 \leqslant \ell < \infty$. The approximate form for $\hat{F}_1(ir)$ becomes:

$$\hat{F}_1(ir)_{A+G} = e^{-A_{\infty,A}} \sum \frac{A_{\infty,A}^m}{m!} \exp\left(-\frac{1}{2}\, m\, \langle B_{oA}^2\rangle_{\underset{\sim}{\theta}} - \sigma_G^2\right)$$

$$\cdot \frac{1}{2}r^2\left[1 + \frac{A_{\infty,A}\,\langle B_{oA}^2\rangle_{\underset{\sim}{\theta}}^2\, r^4\, \hat{C}_4}{4^3}\, e^{-1/4 r^2 \langle B_{oA}^2\rangle_{\underset{\sim}{\theta}}}\right.$$

$$\left. + \frac{A_{\infty,A}\langle B_{oA}^2\rangle_{\underset{\sim}{\theta}}^3\, r^6\, \hat{C}_6}{4^3(3!)^2}\, e^{-1/4 r^2 \langle B_{oA}^2\rangle_{\underset{\sim}{\theta}}} \cdots\right] \tag{5-43}$$

where $\hat{C}_{2\ell} = (-1)^\ell\, \ell!\, F_\ell$ has been used. In effect, the preceding approximation yields a representation for $\hat{F}_1(ir)_{A+G}$, which most accurately represents the impulsive noise for small values of $A_{\infty,A}$ and large envelope values, $E >> 0$. Terms with r to powers of 4 and greater appearing in the [ ] are corrections, which may be included when intermediate cases of $A_{\infty,A}$ and E are of concern. In the development of the characteristic function representing Class A noise, Equation 5-43 and specifically in the step represented by Equation 5-35, explicit dependence upon the noise-source distribution within the domain Λ_{eff} was shown to disappear. An implicit dependence remains, consolidated within the impulsive index, $A_{\infty,A}$, which contains the averages of the source emission rate, emission duration, and the number of sources. Class A noise-envelope statistics will not, therefore, permit determination of noise-source concentration on a local scale. Thus, with regard to the propagation law governing the transmission of a time-dependent noise signal, $G_o(t)$, originating at a source location to the point of observation where the waveform B_o is observed, i.e.,

$$B_o = G_o(t)/\hat{\epsilon}^{\gamma}, \tag{5-44}$$

only the average of $\langle B_{oA}^2 \rangle_{\underset{\sim}{\theta}}$ enters into $F_1(ir)_{A+G}$. In Equation 5-44, the distance variable is represented by the propagation time $\hat{\epsilon}$ with C^{γ} absorbed into $G_{oA}(t)$. As only $\langle B_{oA}^2 \rangle_{\underset{\sim}{\theta}}$ appears in the statistics for Class A noise, the propagation process is masked and unresolvable from the first-order envelope relationships.

Calculations of the Probability Distributions, Density Function, and Moments of Class A Noise

Probability Distributions. The cumulative probability, $P(E > E_o)_A$ given by Equation 5-19 and $\hat{F}_1(ir)_{A+G}$ from Equation 5-43 is now written by (1) dropping terms in r with powers of 4 and greater, (2) using several variable changes to simplify the notation, and (3) discarding the ∞ notation on $A_{\infty,A}$, as in the following:

$$\hat{F}_1(ia\chi)_{A+G} = e^{-A_A} \sum_{m=0}^{\infty} \frac{A_A^m}{m!} \exp\left[-\sigma_{mA}^2 \tfrac{1}{2} a_A^2 \chi^2\right] \tag{5-45}$$

The new symbols introduced in Equation 5-45 are:

$$\sigma_{mA}^2 = \frac{1}{2}\left(\frac{m}{A_A} + \Gamma'_A\right)(1 + \Gamma'_A)^{-1}; \Gamma'_A = \sigma_G^2/\Omega_{2A}$$

$$\Omega_{2A} = \frac{1}{2} A_{\infty,A} \langle B_{oA}^2 \rangle_{\underset{\sim}{\theta}}; r = a_A \chi; a_A^{-1} = [2\Omega_{2A}(1 + \Gamma'_A)]^{1/2}. \tag{5-46}$$

The introduction of the variable identities presented in Equation 5-46, in effect, normalize the noise envelope E and exceedance level E_o allowing a new variable pair $\mathcal{E}$ and $\mathcal{E}_o$ to be used, i.e.,

$$\mathcal{E} = E[2\Omega_{2A}(1+\Gamma'_A)]^{-1/2} \qquad \mathcal{E}_o = E_o[2\Omega_{2A}(1+\Gamma'_A)]^{-1/2}$$
$$= a_A E \qquad\qquad = a_A E_o \tag{5-47}$$

Introducing Equations 5-47 and 5-45 into Equation 5-19 yields, for $P(\mathcal{E} > \mathcal{E}_o)_A$,

$$P_1(\mathcal{E} > \mathcal{E}_o)_A$$
$$= 1 - \mathcal{E}_o e^{-A_A} \sum_{m=0}^{\infty} \frac{A_A^m}{m!} \int_0^{\infty} J_1(\mathcal{E}_o \chi) \exp\left[-\sigma_{mA}^2 \frac{a_A^2 \chi^2}{2}\right] d\lambda. \tag{5-48}$$

The Gaussian subscript has been dropped at this point for simplification of the notation. In all subsequent expressions, a Gaussian background component is implicitly included. Execution of the integration in Equation 5-48 is performed by means of the Bessel integral relationship:*

$$\int_0^{\infty} J_\nu(a_A z)\, z^{\mu-1}\, e^{-b^2 z^2}\, dz$$
$$= \frac{\Gamma\left(\frac{\nu+\mu}{2}\right)\left(\frac{a_A}{2b}\right)^{\nu}}{2b^{\mu}\Gamma(\nu+1)}\ {}_1F_1\left(\frac{\nu+\mu}{2}, \nu+1, -\frac{a_A^2}{4b^2}\right)$$
$$\operatorname{Re} a_A^2 > 0;\ \operatorname{Re}(\nu+\mu) > 0. \tag{5-49}$$

The notation ${}_1F_1(a, b, z)$ or $M(a, b, z)$ represents the confluent hypergeometric function.† A useful approximation to a special and commonly encountered case of this function is developed in Reference 6 and repeated below for reference:

$$ {}_1F_1(1, 2, -z) \cong (1 - e^{-x})/x^{-1}. \tag{5-50}$$

*Reference 5, 1965, Equation 11. 4. 28.

†Discussed and tabulated in Reference 5, Chapter 13, in association with its mathematical properties.

$P(\mathcal{E} > \mathcal{E}_o)_A$ is obtained by applying Equation 5-49 to Equation 5-48 and using the special case representation for ${}_1F_1(a, b, z)$ given in Equation 5-50; i.e.,

$$P_1(\mathcal{E} > \mathcal{E}_o)_A = 1 - e^{-A_A} \sum_{m=0}^{\infty} \frac{A_A^m}{m!} \frac{\mathcal{E}_o^2}{2\sigma_{mA}^2} \, {}_1F_1\left[1, 2, -\frac{\mathcal{E}_o^2}{2\sigma_{mA}^2}\right]$$

$$= e^{-A_A} \sum_{m=0}^{\infty} \frac{A_A^m}{m!} e^{-\mathcal{E}_o^2/(2\sigma_{mA}^2)} \qquad (5\text{-}51)$$

In obtaining the final simplified form of $P(\mathcal{E} > \mathcal{E}_o)_A$, use has been made of the identity:

$$e^{-A_A} \sum_{m=0}^{\infty} \frac{A_A^m}{m!} = 1.$$

Examination of the form of Equation 5-51 reveals that the exceedance probability contains a series of terms, each of which has the Rayleigh distribution form but with a variance and a scale factor $(A_A^m/m!)$ that are dependent upon the term index. Furthermore, the variance, σ_{mA}^2, as seen in Equation 5-46, increases with the index value.

Figure 5-3 plots the exceedance probability, $P(\mathcal{E} > \mathcal{E}_o)_A$ for values of $\mathcal{E}_o$ expressed in decibels relative to the rms value of $\mathcal{E}$ for several values of Γ'_A, $10^{-4} < \Gamma'_A < 10^{-1}$, using an impulsive index, $A_A = 0.1$.[2] The impulsive characteristic of $P(\mathcal{E} > \mathcal{E}_o)_A$ is seen by the sharp increase in slope in the rare-events portion of the plots. A scale of Fig. 5-3 is chosen to display Gaussian-distributed exceedance probabilities with a slope equal to $-\frac{1}{2}$; thus, on the right-hand portion of each curve, observe the transition of Equation 5-51 into a Rayleigh-distributed envelope function representative of the background Gaussian noise component upon which the large-magnitude, but less frequently occurring, component is superimposed.

Density Functions. Use of Equation 5-51 and the general definition for the density function of the normalized envelope $W_1(\mathcal{E})$ yields:

$$W_1(\mathcal{E}) = -\left.\frac{\partial P(\mathcal{E} > \mathcal{E}_o)}{\partial \mathcal{E}}\right|_{\mathcal{E}=\mathcal{E}_o}$$

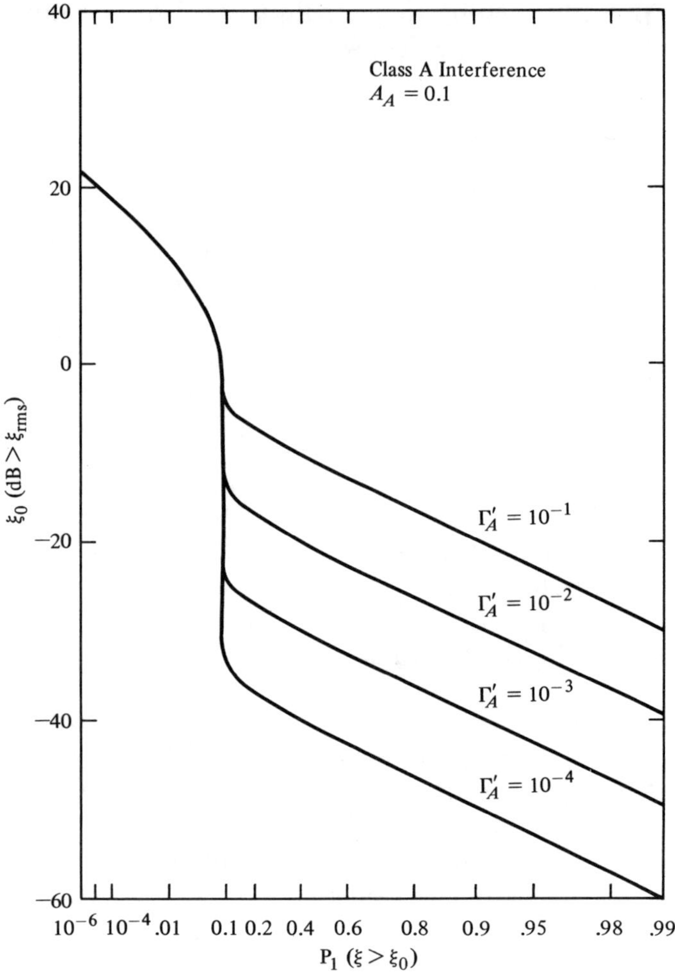

Fig. 5-3. Plot of exceedance probability, $P_1(\mathcal{E} > \mathcal{E}_o)_A$, for Class A man-made noise using Equation 5-51. (After Middleton, 1976)

$$W_1(\mathcal{E})_A = \mathcal{E} e^{-A_A} \sum_{m=0}^{\infty} \frac{A_A^m}{m!} \cdot \frac{\exp[-\mathcal{E}^2/(2\sigma_{mA}^2)]}{\sigma_{mA}^2}. \tag{5-52}$$

Envelope Moments. The envelope moments $\langle \mathcal{E}^\beta \rangle$ of the order β are developed using Equation 5-52 and the general definition as follows:

$$\langle \mathcal{E}^\beta \rangle = \int_0^\infty \mathcal{E}^\beta W_1(\mathcal{E})\, d\mathcal{E}. \tag{5-53}$$

Thus,

$$\langle \mathcal{E}^{\beta} \rangle_A = e^{-A_A}\, \Gamma\left(1 + \frac{1}{2}\beta\right) \sum_{m=0}^{\infty} \frac{A_A^m}{m!} \left(\frac{\dfrac{m}{A_A} + \Gamma'_A}{1 + \Gamma'_A} \right)^{1/2\beta} . \tag{5-54}$$

Typical values are:

$$\langle \mathcal{E} \rangle_A = \frac{1}{2}\sqrt{\pi}\, e^{-A_A} \sum_{m=0}^{\infty} \left(\frac{A_A^m}{m!} \cdot \left[\frac{\dfrac{m}{A_A} + \Gamma'_A}{1 + \Gamma'_A} \right]^{1/2} \right)$$

$$\langle \mathcal{E}^2 \rangle_A = e^{-A_A} \sum_{m=0}^{\infty} \left(\frac{\dfrac{m}{A_A} + \Gamma'_A}{1 + \Gamma'_A} \cdot \frac{A_A^m}{m!} \right). \tag{5-55}$$

That the second moment, $\langle \mathcal{E}^2 \rangle_A$, equals unity shows that the normalization of the characteristic function has been properly performed. In demonstrating the unity value of $\langle \mathcal{E}^2 \rangle_A$, it is convenient to use the relationship:

$$\sum_{m=0}^{\infty} \frac{m\, x^m}{m!} = x\, e^x .$$

Comparison of the Computed Envelope Distribution Function with Measurements for Class A Noise

In the preceding relationships for $P(\mathcal{E} > \mathcal{E}_o)_A$, $\langle \mathcal{E}^{\beta} \rangle_A$, and $W_1(\mathcal{E})_A$, three undetermined quantities occur, which must be evaluated from experimental data before the APD may be calculated. These are presented in Equation 5-46 and are identifiable as A_A, Γ'_A, and Ω_{2A}. For determination of this triplet, three independent relationships are needed, which may readily be the first and second moments, Equation 5-55, and the probability distribution for $\mathcal{E}$, $P_1(\mathcal{E} > \mathcal{E}_o)_A$, from Equation 5-51, expanded and specialized to small values of $\mathcal{E}$ where the Gaussian portion of the APD for impulsive noise is dominant; i.e.,

$$P_1(\mathcal{E} > \mathcal{E}_o)_A \cong 1 - \mathcal{E}_o^2 e^{-A_A} \sum_{m=0}^{\infty} \frac{A_A^m (1 + \Gamma'_A)}{m! \left(\dfrac{m}{A_A} + \Gamma'_A \right)} . \tag{5-56}$$

The approximation, Equation 5-56 results from replacing ${}_1F_1(1, 2, -\mathcal{E}^2/2\sigma_{mA}^2)$ by its equivalent form, Equation 5-50, followed by replacing the remaining exponential term by its power series expansion with all terms beyond the second disregarded. Contribution to the variance of the total Gaussian noise factor σ_G^2 arising from the receiver is separately measured by shorting the input stage.

In Fig. 5-4, a comparison is plotted of a measured APD for Class A man-made noise and an APD computed using the expressions and methodology discussed above for determining the necessary parameter triplet. These experimental results were obtained from a series of studies of the noise radiated by mining equipment, specifically from an ore-crushing machine.[7] Points extracted from an observed APD for the ore crusher are indicated by open circles, the solid interconnecting line presents the calculated APD using $A_A = 10^{-4}$ and $\Gamma'_A = 50$.[2] Inspection of Fig. 5-4 reveals that the distinctive features

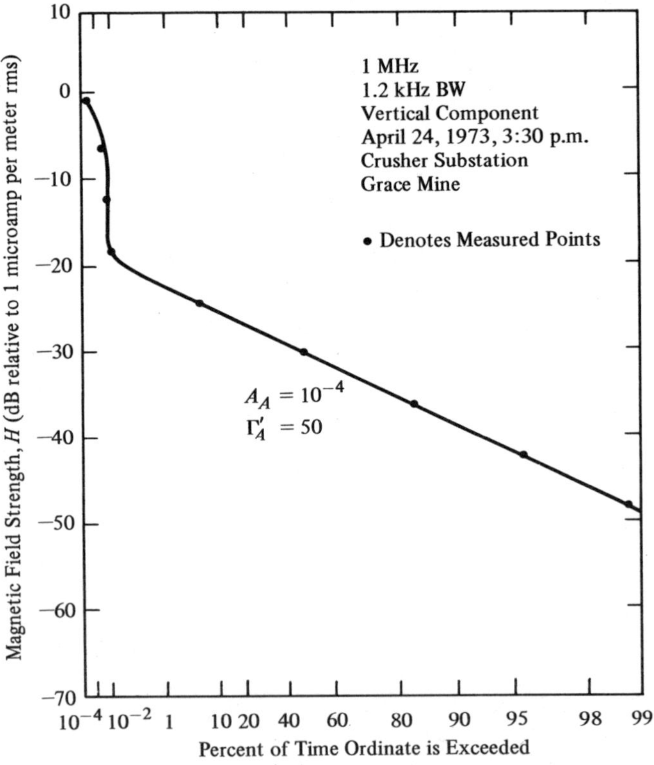

Fig. 5-4. Comparison of a measured and calculated APD for a sample of Class A man-made radio noise generated by mine equipment. (After Middleton, 1976)

of a Class A noise, APD, may be found in an experimental noise sample. Furthermore, it is seen that the calculated distribution function is in close agreement with the experimental results when the proper values of the parameter triplet are known.

Class B Noise Statistics

Development of the Characteristic Function. The characteristic function for Class B noise exists as the product of two independent quantities. The first, which is the simplest in form, arises from the thermal background whose quadrature components are Gaussian distributed, Equation 5-32. The second factor derives from the response of a receiver to the impulsive noise constituent during the reduced time interval $0 \leqslant z \leqslant 1$ of Fig. 5-2 and is representable by Equation 5-31 with the upper limit changed to ∞ and the subscript B added to B_o:

$$\hat{F}_1(ir)_{B+G} = e^{-1/2r^2\sigma_G^2} \exp\left\{A_{\infty,B}\left\langle\int_0^\infty [J_o(rB_{oB}) - 1]\,dz\right\rangle_{\underset{\sim}{\theta},\lambda}\right\}. \tag{5-57}$$

Notice that the indicated averages of the random variables $\underset{\sim}{\theta}$ and λ for the process and coordinate quantities, respectively, are taken. The presence of an infinite limit for the integral in z requires the use of the unaveraged integrand over domain z. Use of the integral relationship,

$$J_o(y) - 1 = -\int_0^y J_1(x),$$

permits rewriting Equation 5-57 as

$$\hat{F}_1(ir)_{B+G}$$

$$= e^{-1/2r^2\sigma_G^2} \exp\left\{-A_{\infty,B}\left\langle\int_0^\infty dz \int_0^{x=rB_{oB}} J_1(x)\,dx\right\rangle_{\underset{\sim}{\theta},\lambda}\right\}. \tag{5-58}$$

The upper limit on the inner integral for Class B noise is infinity in virtue of the receiver response to impulsive excitation and necessary inclusion of the transient decay of the receiver output in the total observed signal, Fig. 5-2. Consequently, the inner integral over x may be separated as follows:

$$\hat{F}_1(ir)_{B+G} = e^{-1/2r^2\sigma_G^2} \exp\left\{-A_{\infty,B}\left\langle \int_0^\infty dz \int_0^{x_o} J_1(x)\,dx \right\rangle_{\underset{\sim}{\theta},\chi} - A_{\infty,B}\left\langle \int_0^\infty dz \int_{x_o}^\infty J_1(x)\,dx \right\rangle_{\underset{\sim}{\theta},\chi} \right\}, \quad (5\text{-}59)$$

where

$$x = rB_{oB} = rG_o(t)/\hat{\epsilon}^\gamma. \quad (5\text{-}60)$$

Figure 5-5 depicts the range of variation of x of Equation 5-60 for Classes A and B interference as a function of $\hat{\epsilon}$. The time variable $\hat{\epsilon}$ relates the generated noise waveform $G_o(t)$ to the observed noise signal B_o by means of the propagation law which is expressed by Equation 5-44. Construction of the mean values of the integrals over the effective domain, Λ_{eff}, of the noise-source distribution, indicated in Equation 5-59 by the subscript χ, may be written in operational

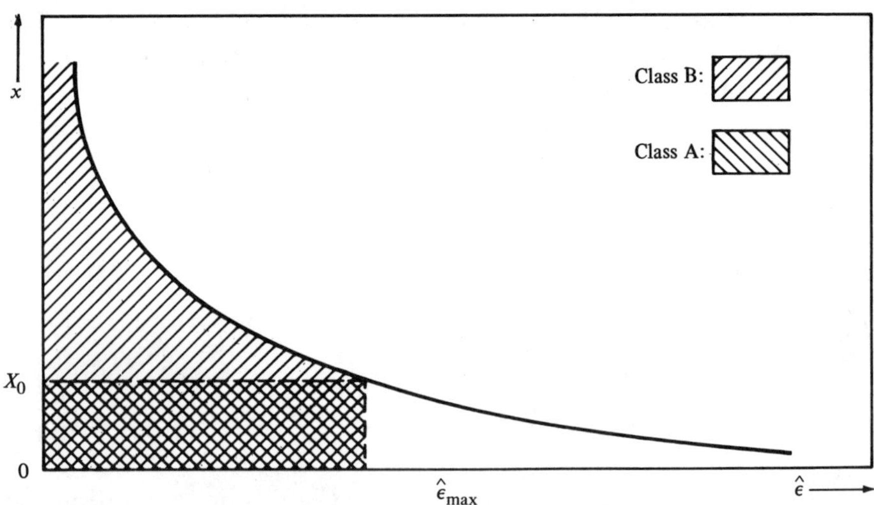

Fig. 5-5. Representation of the range for the integration variable $x = rG_o/\hat{\epsilon}^\gamma$ of the characteristic function.

form using the point density from Equation 5-22:

$$\hat{F}_1(ir)_{B+G} = e^{-1/2r^2\sigma_G^2} \exp\left[-A_{\infty,B} \int_0^{\hat{\epsilon}_{\max}} \cdot \left\langle \int_0^{\infty} dz \int_0^{x_O} J_1(x)\, dx \right\rangle_{\underset{\sim}{\theta}} \rho(\lambda, \hat{\epsilon})\, d\hat{\epsilon} - A_{\infty,B} \int_{\hat{\epsilon}}^{0} \cdot \left\langle \int_0^{\infty} dz \int_{x_O}^{\infty} J_1(x)\, dx \right\rangle_{\underset{\sim}{\theta}} \rho(\lambda, \hat{\epsilon})\, d\hat{\epsilon}\right]. \tag{5-61}$$

The order of integration indicated for both terms within the brackets may be interchanged because of the independence of the reduced time variable, z, and the derived distance variable, $\hat{\epsilon}$. Equation 5-61 may then be written with the representations of $\rho(\lambda, \hat{\epsilon})$ from Equation 5-22 and $\hat{\epsilon}$ from Equation 5-60 as

$$\hat{F}_1(ir)_{B+G} = e^{-1/2r^2\sigma_G^2} \exp\left[-A_{\infty,B} \int_0^{\hat{\epsilon}_{\max}} A_K c^2 \hat{\epsilon}^{(1-\mu)}\, d\hat{\epsilon} \cdot \left\langle \int_0^{\infty} dz \int_0^{x_O} J_1(x)\, dx \right\rangle_{\underset{\sim}{\theta}} - A_{\infty,B} \cdot \left\langle \int_0^{\infty} dz \int_{x_O}^{\infty} J_1(x)\, dx \cdot \int_{(rG_O/x)^{1/\gamma}}^{0} A_K c^2 \hat{\epsilon}^{(1-\mu)} d\hat{\epsilon} \right\rangle_{\underset{\sim}{\theta}}\right]. \tag{5-62}$$

Integration over $\hat{\epsilon}$ for the first term within the brackets yields unity in virtue of Equation 5-23. The second integral containing $\hat{\epsilon}$ is evaluated as

$$\int_{\hat{\epsilon}=(rG_O/x)^{1/\gamma}}^{0} A_K c^2 \hat{\epsilon}^{(1-\mu)}\, d\hat{\epsilon} = A_K c^2 \frac{\hat{\epsilon}^{(2-\mu)}}{2-\mu} = \left(\frac{\hat{\epsilon}}{\hat{\epsilon}_{\max}}\right)^{2-\mu} = \left(\frac{x_O}{x}\right)^{(2-\mu)/\gamma},$$

where A_K has been obtained from Equation 5-23 and the relationship between $\hat{\epsilon}$ and x given in Equation 5-50 has been used. Setting the exponent, $(2 - \mu)/\gamma = \alpha$, yields for Equation 5-62:

$$\hat{F}_1(ir)_{B+G} = e^{-1/2r^2\sigma_G^2} \exp\left[-A_{\infty,B}\left\langle \int_0^\infty dz \int_0^{x_o} J_1(x)\,dx \right\rangle_{\theta}\right.$$

$$\left. - A_{\infty,B}\left\langle x_o^\alpha \int_0^\infty dz \int_{x_o}^\infty J_1(x)\,x^{-\alpha}\,dx\right\rangle_{\theta}\right]$$

$$= e^{-1/2r^2\sigma_G^2} \exp\{-A_{\infty,B}\cdot(I_1 + I_\alpha)\} \qquad (5\text{-}63)$$

Convergence of the second integral over x in Equation 5-63 occurs for all values of $\alpha > -\frac{1}{2}$. The first Bessel integral, I_1, appearing in Equation 5-63 may be evaluated by expanding the integrand using the ascending series representation* and integrating the resulting terms individually to obtain

$$I_1 = \sum_{\ell=0}^{\infty} \frac{(-1)^\ell \langle x_o^{2\ell+2}\rangle_{z,\theta}}{\ell!\,(\ell+1)!\,2^{2\ell+1}(2\ell+2)}, \qquad (5\text{-}64)$$

in which the averaging operation indicated in the numerator in expanded form represents

$$\langle x_o^{2\ell+2}\rangle_{z,\theta} = \left\langle \int_0^\infty x_o^{2\ell+2}\,dx \right\rangle_{\theta} \qquad (5\text{-}65)$$

where $x_o = rG_o(t)\hat{\epsilon}^{-\gamma}$ from Equation 5-60. Representing the second integral over x in Equation 5-64 by I_α and using the following integral representation for $J_1(x)$,†

$$J_1(x) = \frac{1}{2\pi i}\int_{-i\infty}^{i\infty} \frac{\Gamma(-\tau)\left(\dfrac{x}{2}\right)^{1+2\tau}}{\Gamma(\tau+2)}\,d\tau$$

*Reference 5, Equation 1.1.10.
†Reference 5, Equation 9.1.26.

$$I_\alpha = \left\langle \frac{x_o^\alpha}{2\pi i} \int_{-i\infty}^{i\infty} \frac{\Gamma(-\tau)\, d\tau}{\Gamma(\tau+2)\, 2^{2\tau+1}} \int_{x_o}^{\infty} x^{2\tau+1-\alpha}\, dx \right\rangle_{z,\underline{\theta}}$$

$$= \left\langle \frac{x_o^\alpha}{2\pi i} \int_{-i\infty}^{i\infty} \frac{-\Gamma(-\tau)\, x_o^{2\tau+1-\alpha}\, d\tau}{\Gamma(\tau+2)\, 2^{2\tau+1}\,(2\tau+2-\alpha)} \right\rangle_{z,\underline{\theta}} .$$

The integrand has simple poles at $\tau = \frac{1}{2}(\alpha - 2)$ and at $\tau = 0, 1, 2, \cdots$ for the gamma function $\Gamma(-\tau)$. At the point $\tau = \frac{1}{2}(\alpha - 2)$, the residue of the function equals

$$\frac{1}{2\pi i} \cdot \frac{-\Gamma(1 - \frac{1}{2}\alpha)}{2^{\alpha-1}\,\Gamma(1 + \frac{1}{2}\alpha)} .$$

The residue at $\tau = \ell = 0, 1, 2, 3 \cdots$ equals $(-1)^\ell/\ell!$, permitting the second integral over x to be written as

$$I_\alpha = \frac{-\Gamma(1 - \frac{1}{2}\alpha)\,\langle x_o^\alpha\rangle}{2^{\alpha-1}\,\Gamma(1 + \frac{1}{2}\alpha)} - \sum_{\ell=0}^{\infty} \frac{(-1)^\ell\,\langle x_o^{2\ell+2}\rangle}{\ell!\,(\ell+1)!\,2^{2\ell+1}\,(2\ell+2-\alpha)} \tag{5-66}$$

where the notation $\langle x_o^\alpha\rangle$ and $\langle x_o^{2\ell+2}\rangle$ represent integration over the range $0 \leqq z \leqq \infty$ of the arguments as shown in Equation 5-65. Introducing the results obtained for the integrals I_1 and I_α given by Equations 5-64 and 5-66 into Equation 5-63 yields

$$\hat{F}_1(ir)_{B+G} = e^{-1/2 r^2 \sigma_G^2} \exp\left\{-A_{\infty,B}\left[\frac{(1+\frac{1}{2}\alpha)\,\langle x_o^\alpha\rangle}{2^{\alpha-1}\,\Gamma(1+\frac{1}{2}\alpha)}\right.\right.$$

$$\left.\left. + \sum_{\ell=0}^{\infty} \frac{(-1)^\ell\,\langle x_o^{2\ell+2}\rangle}{\ell!(\ell+1)!\,2^{2\ell+1}} \cdot \frac{4\ell+4-\alpha}{(2\ell+2)\,(2\ell+2-\alpha)}\right]\right\} \tag{5-67}$$

or, in simplified notation,

$$\hat{F}_1(ir)_{B+G} = \exp\left[-b_{1\alpha} A_{\infty,B}\, r^\alpha - (\sigma_G^2 + b_{2\alpha} A_{\infty,B})\,\tfrac{1}{2} r^2 \right.$$

$$\left. - \sum_{\ell=1}^{\infty} (-1)^\ell\, b_{\ell\alpha} A_{\infty,B}\, r^{2\ell+2}\right] \tag{5-68}$$

where

$$b_{1\alpha} = \frac{\Gamma(1 - \frac{1}{2}\alpha)}{2^{\alpha-1}\Gamma(1 + \frac{1}{2}\alpha)} \langle G_{o,B}^{\alpha} \rangle \hat{\epsilon}_{\max}^{-\alpha\gamma}$$

$$b_{2\alpha} = \left(\frac{4 - \alpha}{2 - \alpha}\right) \langle G_{o,B}^{2} \rangle \hat{\epsilon}_{\max}^{-2}$$

$$b_{\ell\alpha} = \frac{(4\ell + 4 - \alpha)\, 2^{-2\ell-1}}{\ell!(\ell + 1)!\,(2\ell + 2 - \alpha)\,(2\ell + 2)} \langle G_{o,B}^{2\ell+2} \rangle \hat{\epsilon}_{\max}^{-\gamma(2\ell+2)}. \tag{5-69}$$

Although exact, the characteristic function for Class B noise given by Equation 5-68 is cumbersome to employ, primarily because of the infinite series appearing in the third term of the exponent. Examination of the series,

$$\sum_{\ell=1}^{\infty} (-1)^{\ell}\, b_{\ell\alpha} A_{\infty,B}\, r^{2\ell+2},$$

reveals that, for all values of ℓ, $G_{o,B}$, and $A_{\infty B}$, the product $b_{\ell\alpha} A_{\infty,B} \leqslant 1$, and therefore the series is bounded above by

$$\sum_{\ell=1}^{\infty} (-1)^{\ell+1} r^{2\ell+2},$$

i.e.,

$$\sum_{\ell=1}^{\infty} (-1)^{\ell}\, b_{\ell\alpha} A_{\infty B} r^{2\ell+2} \leqslant \sum_{\ell=1}^{\infty} (-1)^{\ell+1} r^{2\ell+2} = r^4/(1 + r^2) \text{ for } r^4 < 1.$$

The range of the parameter r of importance for the representation of the low-event portion of the APD is $0 \leqq r < 1$ for impulsive interference processes; consequently, the contribution of the series term in the exponent of Equation 5-68 is smaller than either of the others, provided that $\alpha < 4$. This constraint on α is readily attained by observing that $\mu \geqq 0$ and $1 \leqslant \gamma \leqslant 2$, from which it is seen that $\alpha = (2 - \mu)/\gamma \leqslant 2$. It becomes possible now to approximate the function $\hat{F}_1(ir)_{B+G}$ of Equation 5-68 by the second term in the exponential, i.e.,

$$\hat{F}_1(ir)_{B+G} \cong \exp\{-(\sigma_G^2 + b_{2\alpha} A_{\infty,B})\, \tfrac{1}{2} r^2\}.$$

Noting that

$$-b_{2\alpha} A_{\infty,B} \tfrac{1}{2} r^2 \cong A_{\infty,B} (e^{-b_{2\alpha} 1/2 r^2} - 1),$$

the characteristic function becomes approximately for the low events or high envelope portion of the distribution,

$$_{II}\hat{F}_1(ir)_{B+G} \cong e^{-A_{\infty,B}} \exp \{A_{\infty,B}\, e^{-1/2 r^2 b_{2\alpha}} - \sigma_G^2 \tfrac{1}{2} r^2\} \quad (5\text{-}70)$$

The subscript II affixed to the characteristic function approximation is used to indicate that the expression is applicable to the range of envelope values $E > E_o \longrightarrow \infty$.

In the intermediate range of envelope values and lower, where the noise envelope approaches a Rayleigh distribution, typifying dominance of the thermal noise contribution over the impulsive portion, the first term appearing in the exponent of the characteristic function is retained in the approximate form, i.e.,

$$_{I}\hat{F}_1(ir)_{B+G} \cong \exp \{-b_{1\alpha} A_{\infty,B}\, r^{\alpha} - \Delta\sigma_G^2 \tfrac{1}{2} r^2\} \quad (5\text{-}71)$$

in which $\Delta\sigma_G^2 = G_b^2 + b_{2\alpha} A_{\infty,B}$.

The use of approximation Equations (5-70 and 5-71) necessitates that a fitting process be undertaken at an intermediate value of $P(E > E_o)$ to assure continuity of both the exceedance probability and at least its first derivative through the juncture interval. Constants appearing in the transition-point equations can only be evaluated from experimentally determined envelope distributions. A discussion of the steps involved and the constraints applicable to the intersection point has been postponed and placed in the section addressed to experimental and theoretical comparisons.

Calculation of the Probability Distributions, Density Functions, and Moments of Class B Noise

Cumulative Probability Distribution. In preparation for computing the cumulative probability distribution and density functions for the envelope of Class B noise, a new set of variables is introduced to facilitate the notation, as was done for Class A noise with the use of Equation 5-47, i.e.,

$$\mathcal{E}_o = E_o/[2\Omega_{2B}(1 + \Gamma_B')]^{1/2}$$

$$\mathcal{E} = E/[2\Omega_{2B}(1 + \Gamma_B')]^{1/2}$$

$$\Gamma'_B = \sigma_G^2/\Omega_{2B}$$
$$\Omega_{2B} = \tfrac{1}{2} A_{\infty,B} \langle B_{oB}^2 \rangle$$
$$a_B = [2\Omega_{2B}(1 - \Gamma'_B)]^{-1/2};\ r = a\lambda$$
$$\mathcal{E} = a_B E;\ \mathcal{E}_o = a_B E_o. \tag{5-72}$$

Rewriting Equations 5-71 and 5-70 using the preceding definitions and simplifying the subscript notation on the impulsive index (i.e., $A_{\infty,B} = A_B$) yields:

$$_{\mathrm{I}}\hat{F}_1(ia_B\lambda)_{B+G} \cong \exp[-b_{1\alpha}A_B a_B^\alpha \lambda^\alpha - \tfrac{1}{2} a_B^2 \lambda^2 \Delta \sigma_G^2] \tag{5-73}$$

$$_{\mathrm{II}}\hat{F}_1(ia_B\lambda)_{B+G} \cong e^{-A_B} \exp[A_B\, e^{-1/2 a_B^2 \lambda^2 b_{2\alpha}} - \tfrac{1}{2} a_B^2 \lambda^2 \sigma_G^2]. \tag{5-74}$$

Two cumulative distribution functions, which represent Class B interference separately within the low- and high-intensity portions of the noise-envelope range may be computed by employing the general expression of Equation 5-19 in which a change of variable from (E, E_o) to $(\mathcal{E}, \mathcal{E}_o)$, using the relationships of Equation 5-72, is made, followed by substitutions of the characteristic functions from Equations 5-73 and 5-74. The resulting exceedance probability for the low- and intermediate-amplitude segment of the noise-envelope range becomes

$$_{\mathrm{I}}P_1(\mathcal{E} > \mathcal{E}_o)_B \cong 1 - \mathcal{E}_o \sum_{n=0}^{\infty} (-1)^n \frac{(b_{1\alpha} A_B a_B^\alpha)^n}{n!} \cdot \int_0^\infty \lambda^{n\alpha} J_1(\lambda \mathcal{E}_o)\, r^{-1/2 a_B^2 \lambda^2 \Delta \sigma_G^2}\, d\lambda.$$

By employing the Bessel integral equation of Equation 5-49, the normalized exceedance probability is obtained as

$$_{\mathrm{I}}\hat{P}_1(\hat{\mathcal{E}} > \hat{\mathcal{E}}_o)_B \cong 1 - \Big| \hat{\mathcal{E}}_o^2 \cdot \sum_{n=0}^{\infty} \frac{(-1)^n \hat{A}_\alpha^n}{n!} \Gamma\left(1 + \frac{1}{2} n\alpha\right) {}_1F_1\left(1 + \frac{1}{2} n\alpha, 2, -\hat{\mathcal{E}}_o^2\right), \tag{5-75}$$

containing the new symbols:

$$\hat{\mathcal{E}} = \mathcal{E}N/2G_B\,;\hat{A}_\alpha = b_{1\alpha}A_B\;G_B^{-\alpha}\,a_B^\alpha$$
$$2G_B^2 = a_B^2\,\Delta\sigma_G^2\,;A_\alpha = 2^\alpha b_{1\alpha}a^\alpha A_B\,. \tag{5-76}$$

The Gaussian subscript has been dropped on ${}_I\hat{P}_1$ to simplify the notation. All subsequent relationships implicitly contain a contribution from the Gaussian-distributed background noise. The factor N appearing in the relations, Equation 5-76, is a scaling factor, which must be computed in order to unitize the second moment i.e., $\langle\mathcal{E}^2\rangle = 1$.

The presence in Equation 5-75 of the confluent hypergeometric function offers some manipulative problems when use is made of ${}_IP_1(\hat{\mathcal{E}} > \hat{\mathcal{E}}_o)_B$ to compute the exceedance probabilities. An asymptotic approximation to ${}_1F_1(1 + \frac{1}{2}\,n\alpha, 2, -\hat{\mathcal{E}}_o^2)$, accurate for values of $\hat{\mathcal{E}}_o^2 >> 1$ and applicable in the intermediate range of envelope amplitudes, may be used to expand Equation 5-75. Proceeding by employing the exponential approximation to ${}_1F_1(1, 2, -x)$ presented earlier in Equation 5-50 followed by the application of

$${}_1F_1(a, b, z) = e^z\,{}_1F_1(b - a, b, -z)^*$$

to Equation 5-75 yields:

$${}_I\hat{P}_1(\hat{\mathcal{E}} > \hat{\mathcal{E}}_o)_B \cong e^{-\hat{\mathcal{E}}_o^2}\left\{1 - \hat{\mathcal{E}}_o^2\right.$$
$$\left.\cdot\sum_{n=1}^{\infty}\frac{(-1)^n\hat{A}_\alpha^n}{n!}\,\Gamma\left(1 + \frac{1}{2}\,n\alpha\right)\,{}_1F_1\left(1 - \frac{1}{2}\,n\alpha, 2, \hat{\mathcal{E}}_o^2\right)\right\}\,. \tag{5-77}$$

The asymptotic approximation to the confluent hypergeometric function appearing in the preceding expression is representable by a converging series† as:

$${}_1F_1(\alpha, \beta, -x) \cong \frac{\Gamma(\beta)}{\Gamma(\beta - \alpha)}\,x^{-\alpha}\left[1 + \frac{\alpha(\alpha - \beta + 1)}{x}\right.$$
$$\left.+\frac{\alpha(\alpha + 1)\,(\alpha - \beta + 1)\,(\alpha - \beta + 2)}{2!\,x^2}\cdots\right]\quad\text{for } x^2 > 1,$$

*Reference 5, Equation 13.1.27.
†Reference 5, Equation 13.5.1.

which, when introduced into Equation 5-77, yields:

$$
{}_{\mathrm{I}}\hat{P}_1(\hat{\mathcal{E}} > \hat{\mathcal{E}}_o)_B \cong \sum_{n=1}^{\infty} (-1)^{n+1}
$$

$$
\cdot \frac{\hat{A}_\alpha^n}{n!} \frac{\Gamma(1 + \frac{1}{2} n\alpha)}{\Gamma(1 - \frac{1}{2} n\alpha)} \hat{\mathcal{E}}_o^{-n\alpha} \left[1 + \frac{(1 + \frac{1}{2} n\alpha)(n\alpha)}{2\hat{\mathcal{E}}_o^2} \cdots\right] \text{ for } \hat{\mathcal{E}}_o^2 >> 1.
$$

(5-78)

Figure 5-6 presents a plot of the exceedance probabilities for regime I computed from Equations 5-77 and 5-78 for several values of $\alpha = (2 - \mu)/\gamma$, the source density-propagation parameter, and the effective impulsive index for Class B noise, $A_\alpha = 1$. The high-intensity envelope range of the distribution function manifesting rare events or low occurrence rates is developed from Equation 5-74 by application of Equation 5-19 after introduction of the symbolism of Equation 5-72. The exceedance probability for regime II becomes

$$
{}_{\mathrm{II}}P_1(\mathcal{E} > \mathcal{E}_o)_B \cong 1 - \mathcal{E}_o e^{-A_B} \sum_{m=0}^{\infty} \frac{A_B^m}{m!} \int_0^{\infty} J_1(\mathcal{E}_o \lambda)\, e^{-1/2\lambda^2 \hat{\sigma}_{mB}^2}\, d\lambda
$$

(5-79)

in which $\hat{\sigma}_{mB}^2$ has been set to represent the following binomial:

$$
\hat{\sigma}_{mB}^2 = (mb_{2\alpha} + \sigma_G^2)\, a_B^2 = (\Gamma_B' + m/\hat{A}_B)/(2(1 + \Gamma_B'))
$$

and $\hat{A}_B = A_B(2 - \alpha)/(4 - \alpha)$. The integral over λ appearing in Equation 5-79 may now be treated by again employing Equations 5-49 and 5-50 to obtain the approximation to ${}_{\mathrm{II}}P_1(\mathcal{E} > \mathcal{E}_o)_B$ for $\mathcal{E} > \mathcal{E}_B$, i.e.,

$$
{}_{\mathrm{II}}P_1(\mathcal{E} > \mathcal{E}_o)_B \cong \frac{1}{4G_B^2} e^{-A_B} \sum_{m=0}^{\infty} \frac{A_B^m}{m!} \exp\left(-\frac{\mathcal{E}_o^2}{2\sigma_{mB}^2}\right),
$$

(5-80)

where the lower bound of the regime of application lies in the transition range of the APD somewhat below, with respect to values of $\mathcal{E}$, the large negative slope and yet clearly not within the Rayleigh distributed portion, which is characteristic of Gaussian background noise.

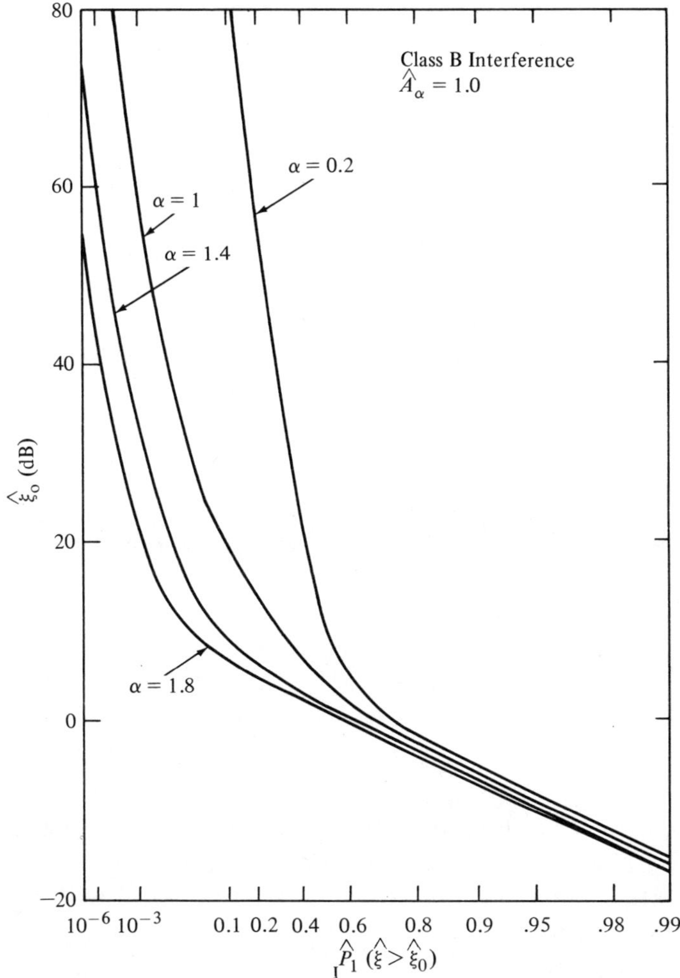

Fig. 5-6. ${}_{\mathrm{I}}P_1(\hat{\mathcal{E}} > \hat{\mathcal{E}}_o)_B$ as a function of source density-propagation parameter, α, for the effective impulsive index, $\hat{A} = 1$. (After Middleton, 1976)

Equations 5-75 and 5-78 for regime I and Equation 5-80 applicable within regime II, respectively, describe the low-amplitude Rayleigh and the rare-event high-intensity envelope regions of the cumulative distribution function for Class B noise. Each of the three expressions contains undetermined parameters, which require experimental data for evaluation, thereby permitting the respective approximations to be adjusted for continuity of both the functions and their first derivatives at the point of intersection. The point of intersection is chosen to lie within the Rayleigh-to-impulsive transition region at an envelope magnitude, $\mathcal{E}_B$, the inflection point of the composite approximation. Two additional constraints are imposed upon these three equations in the low-envelope-amplitude regime I, where it is required that the cumulative distribution function, accurate for the Rayleigh regime, and its first derivative both match the Rayleigh-regime extension of the regime II function. The rationale for these constraints is that both regime I and regime II representations must consistently describe the small-amplitude behavior of the distribution function. The preceding four constraints plus selection of the junction point (that is, the inflection point in the Rayleigh-to-impulsive transition region), result in the following four equations:

1. *Condition (a).* Equating regime I and II distribution functions at point $\mathcal{E}_B$, yields, using Equation 5-78 with Equations 5-76 and 5-80,

$$\text{(a)} \quad \frac{A_\alpha \Gamma(1+\frac{1}{2}\alpha)}{2^\alpha G_B^\alpha (1-\frac{1}{2}\alpha)} \left(\frac{\mathcal{E}_B N}{2G_B}\right)^{-\alpha} [1 + 0(\mathcal{E}_B N)]$$

$$= \frac{1}{4G_B^2} e^{-A_B} \sum_{m=0}^{\infty} \frac{A_B^m}{m!} \exp\left(-\frac{\mathcal{E}_B^2}{2\hat{\sigma}_{mB}^2}\right) \qquad (5\text{-}81)$$

2. *Condition (b).* Equating the first derivatives of ${}_{\mathrm{I}}P_1(\mathcal{E} > \mathcal{E}_o)_{B+C}$ and ${}_{\mathrm{II}}P_1(\mathcal{E} > \mathcal{E}_o)_{B+C}$ at the inflection point, $\mathcal{E}_B$ (i.e., $[\partial({}_{\mathrm{I}}P_1)/\partial\mathcal{E} = \partial({}_{\mathrm{II}}P_1)/\partial\mathcal{E}]_{\mathcal{E}=\mathcal{E}_o}$), while employing Equation 5-78 with Equations 5-76 and 5-80, leads to:

$$\text{(b)} \quad \frac{\alpha A_\alpha \Gamma(1+\frac{1}{2}\alpha)}{2^{\alpha-1} G_B^{\alpha-1} \Gamma(1-\frac{1}{2}\alpha)} \left(\frac{\mathcal{E}_B N}{2G_B}\right)^{-\alpha} N^{-1} [1 + 0(\mathcal{E}_B, N)]$$

$$= \frac{\mathcal{E}_B}{4G_B^2} e^{-A_B} \sum_{m=0}^{\infty} \frac{A_B^m}{m!} \exp\left(-\frac{\mathcal{E}_B^2}{2\hat{\sigma}_{mB}^2}\right) \qquad (5\text{-}82)$$

3. *Condition (c).* Stipulating in the Rayleigh region ${}_{\mathrm{I}}P_1(\mathcal{E} > \mathcal{E}_o)_{B+G}$ is identically equal to the extension of ${}_{\mathrm{II}}P_1(\mathcal{E} > \mathcal{E}_o)_{B+G}$, results in the following, when Equation 5-75 with Equations 5-76 and 5-80 are used:

$$\text{(c)}\quad \frac{1}{4G_B^2}\left[1+\mathcal{E}_o^2N^2\sum_{n=0}^{\infty}\frac{(-1)^n}{n!}\left(\frac{A_\alpha}{2^\alpha G_B^\alpha}\right)^n\Gamma\left(1+\frac{1}{2}n\alpha\right)\right]$$

$$=1+\frac{\mathcal{E}_o^2}{4G_B^2}\,e^{-A_B}\sum_{m=0}^{\infty}\frac{A_B^m}{m!}\,(2\hat{\sigma}_{mB}^2)^{-1}\qquad(5\text{-}83)$$

4. *Condition (d).* Requiring continuity of the first derivatives of ${}_{\text{I}}P_1(\mathcal{E}>\mathcal{E}_o)_{B+G}$ and ${}_{\text{II}}P_1(\mathcal{E}>\mathcal{E}_o)_{B+G}$ within the small signal envelope Rayleigh region, i.e.,

$$\frac{\partial({}_{\text{I}}P_1)}{\partial\,\mathcal{E}_o}=\frac{\partial({}_{\text{II}}P_1)}{\partial\,\mathcal{E}_o},$$

and utilizing Equation 5-75 with Equations 5-76 and 5-80 yields:

$$\text{(d)}\quad \sum_{n=0}^{\infty}\frac{(-1)^n}{n!}\left(\frac{A_\alpha}{2^\alpha G_B^\alpha}\right)^n\Gamma\left(1+\frac{1}{2}n\alpha\right)=e^{-A_B}\sum_{m=0}^{\infty}\frac{A_B^m}{m!}\,(2\hat{\sigma}_{mB}^2)^{-1}.\qquad(5\text{-}84)$$

Figure 5-7 presents a generalized diagram of the resulting composite cumulative distribution function formed by applying the preceding relationships in the transition region at point $\mathcal{E}_B$ and in the Rayleigh region of small envelope values.

Probability Density Function. Two probability density functions must be employed to represent Class B noise. Each is used separately in one of the two regimes of $\mathcal{E}$, ${}_{\text{I}}W_1(\mathcal{E})_B$ for small and intermediate values of the envelope and ${}_{\text{II}}W_1(\mathcal{E})_B$ for the upper range with $\mathcal{E}>\mathcal{E}_B$. Determination of the adjustable parameters appearing in the cumulative distribution functions by exploiting the relationships of Equations 5-81 through 5-84 and the proper selection of the inflection point, $\mathcal{E}_B$, produces a composite density function, $W_1(\mathcal{E})_B$, applicable throughout the range of $0\leqslant\mathcal{E}\leqslant\infty$.

Development of ${}_{\text{I}}W_1(\mathcal{E})_B$ proceeds from the general definition of the density function:

$$W_1(\mathcal{E})=-\left.\frac{\partial P_1(\mathcal{E}>\mathcal{E}_o)}{\partial\mathcal{E}_o}\right|_{\mathcal{E}_o\to\mathcal{E}}.\qquad(5\text{-}85)$$

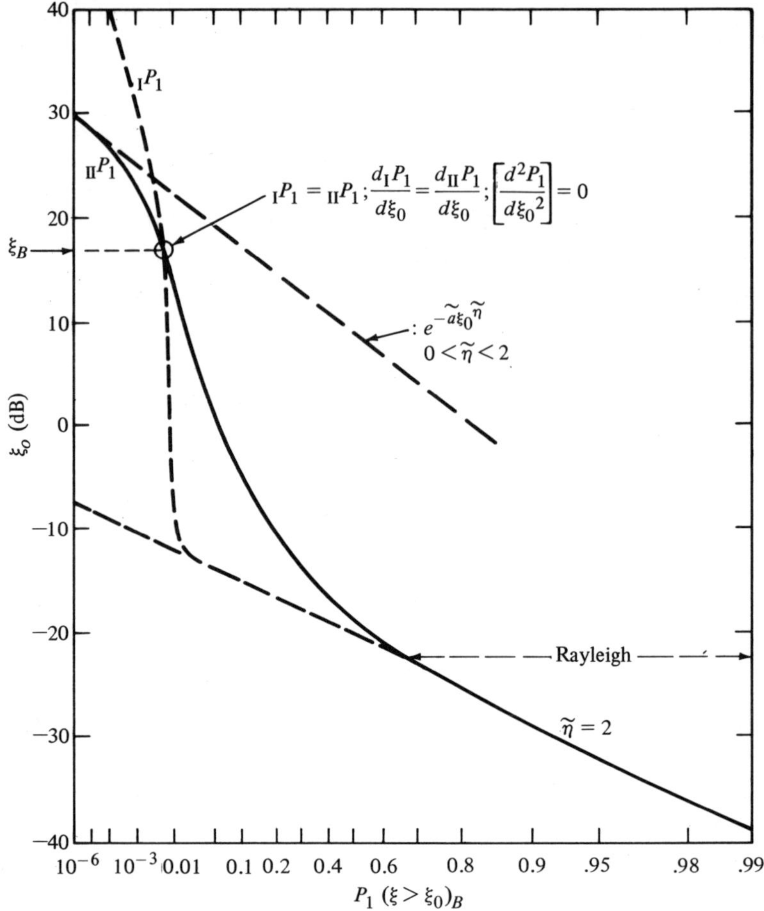

Fig. 5-7. Representation of a composite cumulative distribution function construction for Class B noise. (After Middleton, 1976)

Employing Equation 5-19 for $P_1(\mathcal{E} > \mathcal{E}_o)$ with the substitution of Equations 5-72 yields for $W_1(\mathcal{E})$:

$$W_1(\mathcal{E}) = \int_0^\infty J_1(\lambda \mathcal{E}_o)\, \hat{F}_1(ia\chi)\, d\chi$$

$$+\ \mathcal{E}_o \int_0^\infty \frac{\partial}{\partial \mathcal{E}_o}\ J_1(\lambda \mathcal{E}_o)\, \hat{F}_1(ia\chi)\, d\chi.$$

Using the Bessel recursion relationship, $J_1'(x) = J_o(x) - x^{-1}J_1(x)$, with $x = \chi\mathcal{E}_o$ in the preceeding expression for $W_1(\mathcal{E})$ permits simplification to:

$$W_1(\mathcal{E}) = \mathcal{E}\int_0^{\infty} \chi J_o(\chi\mathcal{E})\,\hat{F}_1(ia\chi)\,d\chi. \tag{5-86}$$

Employing Equation 5-73 for the characteristic function applicable to regime I, the Rayleigh distributed portion of the function $P_1(\mathcal{E} > \mathcal{E}_o)_{B+G}$, and the notation simplifications presented in Equation 5-76 results in:

$$\begin{aligned} {}_{\mathrm{I}}\hat{F}_1(ia_B\chi)_B &= \exp\left\{-\hat{A}_\alpha G_B^\alpha \chi^\alpha - \frac{1}{2}a_B^2\chi^2\Delta\sigma_G^2\right\} \\ &\cong \left(1 - \hat{A}_\alpha G_B^\alpha\chi^\alpha + \frac{\hat{A}_\alpha^2 G_B^{2\alpha}\chi^{2\alpha}}{2!}\cdots\right)\cdot \exp(-\chi^2 G_B^2). \end{aligned} \tag{5-87}$$

Substituting the preceding expansion, ${}_{\mathrm{I}}\hat{F}_1(ia_B\chi)_B$, into Equation 5-86 produces a series of Bessel exponential integrals each of which may be evaluated by means of the integral equation, Equation 5-49. The first three terms in ${}_{\mathrm{I}}W_1(\mathcal{E})_{B+G}$ corresponding to those shown in Equation 5-87, which result from the integrations, are given for reference:

$$\begin{aligned} {}_{\mathrm{I}}W_1(\mathcal{E})_B = \frac{\mathcal{E}}{2G_B^2}\,{}_1F_1(1,1,-\hat{\mathcal{E}}^2) - \hat{A}_\alpha\frac{\Gamma(1+\frac{1}{2}\alpha)}{2G_B^2}\,{}_1F_1\left(1+\frac{1}{2}\alpha, 1, -\hat{\mathcal{E}}^2\right) \\ + \frac{\hat{A}_\alpha^2\,\Gamma(1+\alpha)}{2G_B^2\,2!}\,{}_1F_1(1+\alpha, 1, -\hat{\mathcal{E}}^2)\cdots. \end{aligned}$$

Summation of all terms appearing in the density function for regime I yields:

$${}_{\mathrm{I}}W_1(\mathcal{E})_B = \hat{\mathcal{E}}G_B^{-1}\sum_{n=0}^{\infty}\frac{(-1)^n}{n!}\hat{A}_\alpha^n\,\Gamma\left(1+\frac{1}{2}n\alpha\right)\,{}_1F_1\left(1+\frac{1}{2}n\alpha, 1, -\hat{\mathcal{E}}^2\right). \tag{5-88}$$

In the rare-events portion of the envelope-distribution function where the impulsive characteristic of Class B noise is strongly evident and where the envelope amplitude $\mathcal{E}$ equals and exceeds the inflection point of the cumulative distribution function, $\mathcal{E}_B$; the envelope probability distribution function, may be developed from Equation 5-80 using the general definition, Equation 5-85, and becomes

$$_{\mathrm{II}}W_1(\mathcal{E})_B = \frac{\mathcal{E}}{4G_B^2} e^{-A_B} \sum_{m=0}^{\infty} \frac{A_B^m}{m!\,\hat{\sigma}_{mB}^2} \exp\left(-\frac{\mathcal{E}^2}{2\hat{\sigma}_{mB}^2}\right) \tag{5-89}$$

Envelope Moments. Computation of the moments for Class B noise proceeds using the general moments equation, Equation 5-53, applied to both regimes of the distribution function, which by superposition of the independent portions of the composite density function may be written as:

$$W_1(\mathcal{E})_B = {_{\mathrm{I}}W_1(\mathcal{E})_B} + {_{\mathrm{II}}W_1(\mathcal{E})_B}. \tag{5-90}$$

By replacing the terms of Equation 5-90 with their functional representations, Equations 5-88 and 5-89, and substituting the results in Equation 5-53, the moment equation of order β becomes

$$\begin{aligned} \langle \mathcal{E}^\beta \rangle_B &= \int_0^{\mathcal{E}} \mathcal{E}^\beta\, {_{\mathrm{I}}W_1(\mathcal{E})_B}\, d\mathcal{E} + \int_{\mathcal{E}}^{\infty} \mathcal{E}^\beta\, {_{\mathrm{II}}W_1(\mathcal{E})_B}\, d\mathcal{E} \\ &= G_B^{-1} \sum_{n=0}^{\infty} \frac{(-1)^n}{n!} \hat{A}_\alpha^n\, \Gamma\left(1 + \frac{1}{2} n\alpha\right) \\ &\cdot \int_0^{\mathcal{E}_B} \hat{\mathcal{E}}\, \mathcal{E}^\beta\, {_1F_1}\left(1 + \frac{1}{2} n\alpha, 1, -\hat{\mathcal{E}}^2\right) d\mathcal{E} + 4G_B^{-1} e^{-A_B} \\ &\cdot \sum_{m=0}^{\infty} \frac{A_B^m}{m!} \int_{\mathcal{E}_B}^{\infty} \frac{\mathcal{E}^{\beta+1}}{\hat{\sigma}_{mB}^2} \exp\left(\frac{-\mathcal{E}^2}{2\hat{\sigma}_{mB}^2}\right) d\mathcal{E}. \end{aligned} \tag{5-91}$$

The integral containing the confluent hypergeometric function is evaluated using a series representation* for ${_1F_1}(a, b, z)$, i.e.,

$${_1F_1}(a, b, z) = \sum_{n=0}^{\infty} \frac{(a)_n}{(b)_n} \cdot \frac{z^n}{n!}, \tag{5-92}$$

*From Reference 5, Equation 13.1.2.

in which

$$(a)_n = a(a+1)(a+2)\cdots(a+n-1); (a)_o = 1.$$

The use of Equation 5-92 in the first term of $\langle \mathcal{E}^\beta \rangle$ yields:

$$_{\text{I}}\langle \mathcal{E}^\beta \rangle_B = \frac{N}{2G_B^2} \sum_{n=0}^{\infty} \frac{(-1)^n}{n!} \hat{A}_\alpha^n \, \Gamma\left(1 + \frac{1}{2} n\alpha\right)$$

$$\cdot \sum_{k=0}^{\infty} \frac{(-1)^k \, \mathcal{E}^{(\beta+2k+2)} (1 + \frac{1}{2} n\alpha)_k}{(k!)^2 \, (\beta + 2k + 2)} \left(\frac{N}{2G_B}\right)^{2k+1} \quad (5\text{-}93)$$

In the second term of Equation 5-91, the integral

$$\mathcal{J} = \hat{\sigma}_{mB}^{-2} \int_{\mathcal{E}_B}^{\infty} \mathcal{E}^{\beta+1} \exp\left(-\frac{\mathcal{E}^2}{2\hat{\sigma}_{mB}^2}\right) d\mathcal{E}$$

may be reduced using the change of variable, $y = \mathcal{E}^2/(2\hat{\sigma}_{mB}^2)$, to form

$$\mathcal{J} = (2\hat{\sigma}_{mB})^{\beta/2} \int_y^{\infty} e^{-y} \, y^{\beta/2} \, dy,$$

which yields a solution in terms of the incomplete gamma function, I_c, by initially applying the following integral solution:*

$$\int_x^{\infty} e^{-t} \, t^{a-1} \, dt = \Gamma(a) - \gamma(a, x)$$

in which $\gamma(a, x) = \Gamma(x) \cdot I_c = \int_0^x t^{a-1} \, e^{-t} \, dt$. $\mathcal{J}$ then becomes:

$$\mathcal{J} = (2\hat{\sigma}_{mB}^2)^{\beta/2} \, \Gamma\left(1 + \frac{1}{2}\beta\right) \left[1 - I_c\left(\frac{\mathcal{E}^2}{2\hat{\sigma}_{mB}^2}, 1 + \frac{1}{2}\beta\right)\right]$$

yielding for the second term of Equation 5-91:

$$_{\text{II}}\langle \mathcal{E}^\beta \rangle_B = \frac{1}{4G_B^2} e^{-A_B} \sum_{m=0}^{\infty} \frac{A_B^m}{m!} (2\hat{\sigma}_{mB}^2)^{\beta/2}$$

$$\cdot \Gamma\left(1 + \frac{1}{2}\beta\right) \left\{1 - I_c\left(\frac{\mathcal{E}_B^2}{2\hat{\sigma}_{mB}^2}; 1 + \frac{1}{2}\beta\right)\right\}. \quad (5\text{-}94)$$

*From Reference 5, Equation 6.5.3.

The complete form for Class B moments of order β may now be written as

$$\langle \mathcal{E}^{\beta} \rangle_B = {}_{\mathrm{I}}\langle \mathcal{E}^{\beta} \rangle_B + {}_{\mathrm{II}}\langle \mathcal{E}^{\beta} \rangle_B, \tag{5-95}$$

where the expressions for the regimes I and II moments are given by Equations 5-93 and 5-94, respectively.

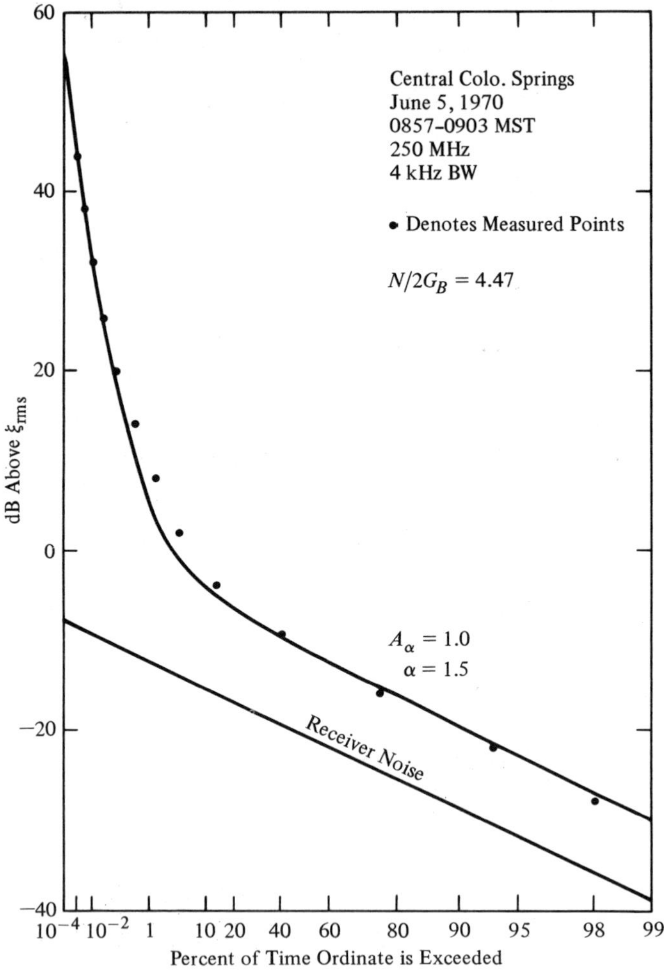

Fig. 5-8. Comparison of a measured and calculated APD for Class B man-made noise. Data obtained for automotive ignition radiation. (After Middleton, 1976)

Comparison of the Computed Envelope Distribution Function with Measurements for Class B Noise

In the preceding relationships for $P_1(\mathcal{E} > \mathcal{E}_o)_B$, $W_1(\mathcal{E})_B$, and $\langle \mathcal{E}^\beta \rangle_B$, six undetermined parameters exist whose values must be fixed from experimental data sources. The Class B noise sextuplet (A_B, Γ_B', Ω_{2B}, A_α, α, N) contains in the first three of this set the counterpart parameters arising for Class A noise. The latter three are parameters unique to Class B, as noted earlier in Equation 5-76 and by the definition $\alpha = (2 - \mu)/\gamma$. An additional quantity $\mathcal{E}_B$, which is positioned experimentally from the APD at the inflection point in the transition region between the Rayleigh and impulsive portions of the cumulative distribution, must be specified to complete the set. However, $\mathcal{E}_B$ is not independent of the first five members of the set, but rather is coupled to each through Equations 5-72 and 5-80 to Equation 5-83. The variable N, introduced to normalize $\langle \mathcal{E}^2 \rangle_B = 1$, is independent of $\mathcal{E}_B$.

Figure 5-8 provides an example of a computed APD for Class B noise using the preceding equations and procedures and a measured APD for automotive ignition interference. The open circles represent measurements, while the solid line is the computed functional form. The lower line with a slope of $-\frac{1}{2}$ arises from the thermal noise of the receiver. The parameters used for the computation are noted in the figure: $N/2G_B = 4.47$, $A_\alpha = 1$, and $\alpha = 1.5$. The envelope distribution is referenced to the rms value and is expressed in decibels.

Glossary of Principal Symbols

The following list comprises the principal mathematical symbols used in this chapter.

A_K	Normalization constant
$A_A; A_B; A_\infty; A_{\infty,A}; A_{\infty,B}$	Impulsive indices for Class A and Class B noise
A_α	Effective impulsive index
$A_{\hat{\epsilon},T}$	Average number of noise emissions per source in period T
A_T	Average number of noise emissions in period T

A_Λ	Average number of noise sources in domain Λ
$(a)_n$	$a(a+1)(a+2)\cdots(a+n-1)$
a, a_A, a_B	Normalization factors
APD	Amplitude probability distribution
α	Source density-propagation parameter
$B_o, \hat{B}_{oA}, \hat{B}_{oB}$	Observed waveform envelope
$b_{1\alpha}, b_{2\alpha}, b_{\ell\alpha}$	Weighted moments of the observed envelope
β	The order of an envelope moment
c	Velocity of light in a vacuum
E, E_o	Instantaneous observed envelope
$\mathcal{E}, \hat{\mathcal{E}}$	Normalized instantaneous observed envelope
$\mathcal{E}_o, \hat{\mathcal{E}}_o$	Envelope threshold
$\mathcal{E}_B$	Inflection point for Class B, APD
$\langle \mathcal{E}^\beta \rangle, {}_{\text{I}}\langle \mathcal{E}^\beta \rangle, {}_{\text{II}}\langle \mathcal{E}^\beta \rangle$	β moments of $\mathcal{E}$
$\hat{\epsilon}$	Impulse-time epoch
$\Delta\hat{\epsilon}$	Impulse-time epoch increment
F_j, F_N	Fourier transforms
$F_1, \hat{F}_1$	Characteristic functions
${}_1F_1(a, b, x)$	Confluent hypergeometric function
f	Frequency
η	Fourier transform variable
$G_o(t), G_{oA}(t), G_{oB}, G_o$	Noise-source waveform
$g(t_T)$	Geometrical factor of the observed waveform
Γ'_A, Γ'_B	Ratio of Gaussian component noise intensity to the impulsive component intensity
$\Gamma(x)$	Gamma function
γ	Range exponent of the propagation law
$I_1, I_\alpha, \mathcal{I}$	Integrals
$\hat{I}_T, \hat{I}_\infty$	Exponent of the characteristic function
I_c	Incomplete gamma function
$J_\nu(x)$	Bessel function of x, order ν, first kind
$\lvert J(\ )\rvert$	Absolute value of the Jacobian
$\Lambda, \Lambda_{\text{eff}}$	Noise-source domain

$\Delta\lambda, \Delta\lambda_N$	Noise-source spatial interval
λ	Noise-source spatial variable
μ	Exponent of the range dependence
μ_d	Normalized Doppler frequency
N	Second-moment normalization constant
ν_T	Noise source pulse-emission rate
$\overline{\nu_T}$	Average value of ν_T
$\overline{\nu_\infty}$	Limiting value of $\overline{\nu_T}$
ξ	Fourier transform variable
Ω_{2A}, Ω_{2B}	Mean intensity of non-Gaussian noise
v	Relative velocity between the observer and the noise source
ω	Angular frequency
ω_o	Carrier angular frequency
$P_1, {}_{\mathrm{I}}P_1, {}_{\mathrm{II}}P_1$	APD or exceedance probability of the noise envelope
P_j	Poisson density function
ψ, ϕ, ϕ_S'	Phase of narrowband observed waveform
ϕ_T	Phase of the noise source
R	Radial distance
r	Characteristic function variable
ρ	Poisson point density
$\rho_\Lambda(\lambda)$	Point density of sources in domain Λ
$S(z)$	Receiver input signal
$\sigma^2, \sigma_G^2, \hat{\sigma}^2, \hat{\sigma}_{mA}^2, \hat{\sigma}_{mB}^2, \Delta\sigma_G^2$	Variances
$\sigma_\Lambda(\lambda)$	Noise source surface density
T, T_{in}	Observation period
$T_s, \overline{T}_{SA}, \overline{T}_{SB}$	Noise emission durations
$t, t_n, t', t_{To}, t_T, t_{T\Lambda}$	Time variables
$\theta, \underset{\sim}{\theta}, \theta_N$	Random parameters of the observed waveform, time independent
Π_j	Infinite product
$U, U_{oA}, U_{oB}, U_{oC}, U_{NB}$	Basic observed noise waveform
U_c, U_s	Quadrature components of the observed noise waveform
$W_1, w_1, {}_{\mathrm{I}}W_1, {}_{\mathrm{II}}W_1$	Probability density functions
$X(t)$	Instantaneous observed waveform amplitude

x_o	Characteristic function variable
χ	Characteristic function variable
z, z_o	Normalized time

References

1. Middleton, D. First-Order Probability Models of the Instantaneous Amplitude—Part 1. U.S. Dept. of Commerce, Office of Telecommunications, OT Report 74-36, April 1974.
2. Middleton, D. Statistical-Physical Models of Man-Made and Natural Radio Noise—Part 2, First Order Probability Models of the Envelope and Phase. U.S. Dept. of Commerce, Office of Telecommunications, O.T. Report 76-86, April 1976.
3. Middleton, D. A statistical theory of reverberation and similar first-order scattered fields, part I. Waveforms and the general process. *IEEE Trans. Information Theory* **IT-13** (3): 372–392 (July 1967).
4. Furutsu, K., and Ishida, T. "On the theory of amplitude distribution of impulsive random noise. *J. Applied Physics* **32** (7): 1206–1221 (July 1961).
5. Abromowitz, M., and Stegun, I. A. Handbook of Mathematical Functions with Formulas, Graphs, and Mathematical Tables. U.S. Dept. of Commerce, National Bureau of Standards Applied Mathematics Series No. 55, December 1965. U.S. Government Printing Office, Washington, D.C.
6. Middleton, D. *An Introduction to Statistical Communication Theory*. New York: McGraw-Hill Book Co., Inc., 1960.
7. Adams, J. W., Bensema, W. D., and Kanda, M. Electromagnetic Noise, Grace Mine. National Bureau of Standards Report NBSIR-74-388, 1974.

6
Composite Metropolitan-Area, Surface Man-Made Noise

Examination of radio-frequency radiation intensity, spectral distribution, waveform statistics, range, and temporal variations of the major categories of incidental and restricted radiation devices is necessary to the regulation and suppression of radio-noise fields that can adversely affect the performance of sensitive wireless receiving equipment and specialized electronic products. The abundance and extensive distribution of incidental and restricted radiation sources in all urban areas necessitate enlargement of a man-made radio study beyond the treatment of individual source types to include an assessment of the geographical dependence, on a metropolitan scale, of the radio-noise fields.

The great abundance of incidental noise sources in most metropolitan areas contributes significantly to the extension of their radiations well beyond the urban core. Geographically varying mixes of the dominant incidental-noise emitters randomly produce appreciable location-dependent changes in the total incidental-noise level. In addition, the distinctive and sharply contrasting spectra of the major classes of incidental sources create a total incidental-noise spectral pattern possessing a complicated magnitude variation with frequency.

For fixed terminal communication systems or radio networks with few optional sites, system-performance analyses are able to establish the incidental-noise environment at each fixed location by examination of the type and number of the major incidental-noise sources in its vicinity. Success of any fixed-site noise-prediction method is predicated upon the existence of experimental data, which for the principal incidental-noise sources yield either average power or median electric-field intensity as a function of time, frequency, and surface distance from the source.

If a wireless receiving or transceiving system possesses a large number of surface terminals, if it possesses substantial metropolitan-area surface mobility, or if it is airborne, then an alternate analytical means must be sought to estimate the incidental noise levels affecting the receivers. Parameterization of incidental-noise level with respect to range, measured from the center of a metropolitan area either along[1] or above[2] the surface, has been the method employed for treating this class of communication problems.[3] Success of the approach relies upon proper establishment of the analytical dependence of either average incidental-noise power or electric-field intensity upon both frequency and distance from a collective geographical reference.

The two aforementioned classes of urban wireless-system-configurations are sufficiently distinct to benefit from separately constructed compilations of consolidated metropolitan-area, man-made radio-noise data. One compilation is based upon the subdivision of an area into activity classes; the other is based upon the range of an observer from the urban center. These are available for use and are the subject of this chapter.

The phrase *consolidated radio noise data* is presently used to designate all emissions from incidental and restricted radiation devices that are observable at one location, at one time, and in one spectral interval. The term *typical metropolitan area* will be employed to identify urbanizations comparable in population, geographical size, and degree of industrialization.

Subdivision of typical metropolitan areas into activity classes implies the reasonable possibility of finding therein geographical zones primarily devoted to a single human function. Such individualization of human urban functions is possible if the activity definitions are sufficiently broad, as is the case when *business*, *residential*, and *rural* designators are used. Attached to the use of these class terms are certain qualifications, namely, that the predominant activities are designated and that localized, intense noise sources (e.g., super highways in rural areas) are excluded.

ACTIVITY CLASSIFICATION OF COMPOSITE MAN-MADE METROPOLITAN-AREA RADIO NOISE

Assembly of man-made radio-noise data obtained from urban regions of comparable population, geographic area, and degree of industrialization into the general categories—business, residential, and rural—yields a valuable set of descriptors of the radio-noise environment if care is taken to exclude from each any large, localized noise anomalies. Restricting the principal activity of each category to the indicated functions admits of some localized noise-type variation—for example, the presence of small shopping centers in residential areas, without thereby significantly increasing the noise-measurement error. However, heavily traveled highways, large shopping centers, industrial parks, and factories must be excluded from zones designated as residential or rural. Likewise, parks and educational and institutional lands may be anomalous in business areas. More specifically, residential areas are those locations of single-or multiple-family dwellings with densities of two families or more per acre while rural areas are locations that possess a dwelling density of one or fewer per 5 acres, with an associated land use that is dominantly agricultural.[4]

The experimental data used to construct the metropolitan-activity-class radio-noise levels and associated statistical parameters were assembled from approximately 300 hr of recordings of the average noise power and various other envelope parameters, including V_d.[4] These data represent the noise envelopes for 10 frequencies lying at approximately 1-octave spacings within the interval 250 kHz to 250 MHz.

Contributing to the compilation of the business-area noise data were measurements performed in Washington, D.C.; Denver, Colorado; San Antonio, Texas; Cheyenne, Wyoming; Boulder, Colorado; and Colorado Springs, Colorado. Residential class information was derived from studies performed in all of these cities except Washington, D.C. Thirty-one rural regions dispersed within the states of Colorado, Wyoming, Virginia, and Washington yielded data for the rural class.[4]

For recording the noise data, short, vertical monopole antennas were used exclusively, thereby providing no azimuthal differentiation of the noise radiations. The radio-noise receivers employed provided instantaneous dynamic ranges of 85 dB and either 4 or 10 kHz bandwidths. Each was used to drive a magnetic tape-recording system possessing a 70-dB dynamic range for determination of average power and a 20-dB range for construction of the parameter V_d.[4]

In the lower portion of the frequency range investigated (specifically, below the ionospheric cutoff frequency), the possibility of incidental-noise-data contamination by atmospheric noise signals existed in the test area. The degree of contamination of the man-made noise ambient was reduced by performing most measurements during the daytime, usually in the morning, when atmospheric noise level was low.[5] During other observation periods, the data were audibly edited to eliminate the segments containing appreciable atmospheric noise.

Composite measurements of F_a as a function of frequency are shown in Figs. 6-1, 6-2, and 6-3 for business, residential, and rural areas, respectively. The upper and lower hourly decile values are shown by the range bars.[4] Fitted to each set of median values is a

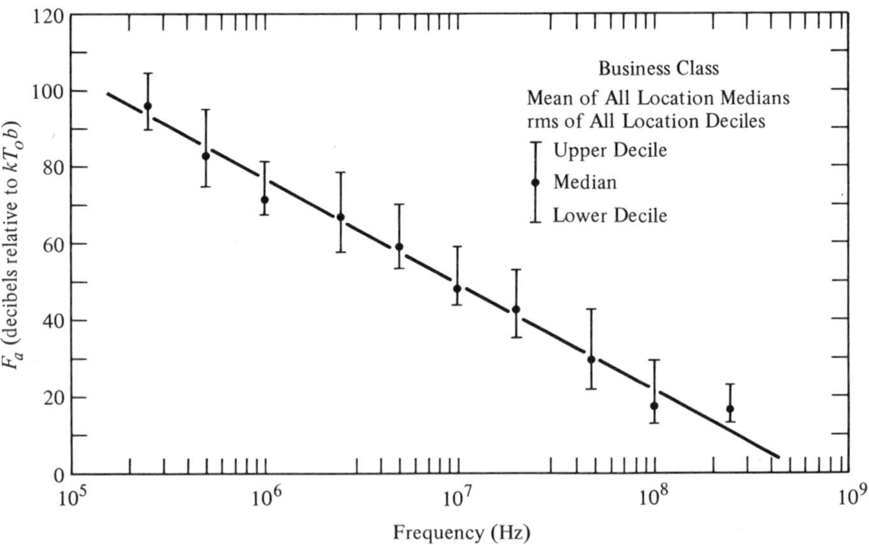

Fig. 6-1. Median values of F_a versus frequency for a business area. (After Spaulding and Disney, 1974)

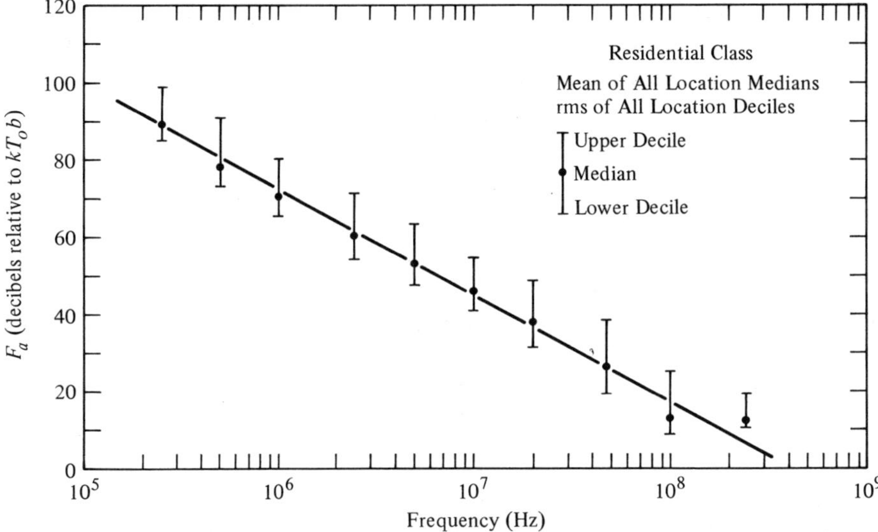

Fig. 6-2. Median values of F_a versus frequency for a residential area. (After Spaulding and Disney, 1974)

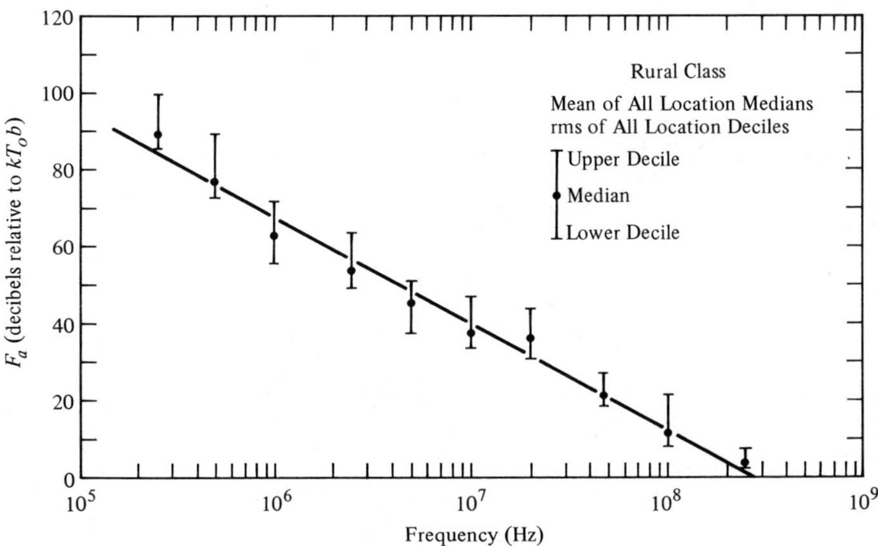

Fig. 6-3. Median values of F_a versus frequency for a rural area. (After Spaulding and Disney, 1974)

mean-squares regression line, which for the three activity classes has a slope of −27.7 dB per decade of frequency change. Examination of each plot reveals that the median value of F_a at 250 MHz lies above the regression line. The most pronounced separation occurs in the business and residential activity classes. The cause of the upward displacement is imperfectly understood and may in some part be due to internal noise of the receiving equipment or to a change in character of the composite noise distribution as the frequency of observation increased. The noise figures of the receivers used were 5 dB or less and comparable to the level of man-made noise in the high VHF band.

The mean-squares regression lines from Figs. 6-1 through 6-3 have been transferred to Fig. 6-4 and coplotted with values of F_a for both galactic and man-made noise in very quiet rural areas.[5] The associated standard errors of estimate obtained from Figs. 6-1 through 6-3 are: business class, 7 dB; residential class, 5 dB; rural class, 6.5 dB. A standard error of estimate is defined as the root variance of the median values of F_a about the mean-squares regression line within the indicated frequency interval.

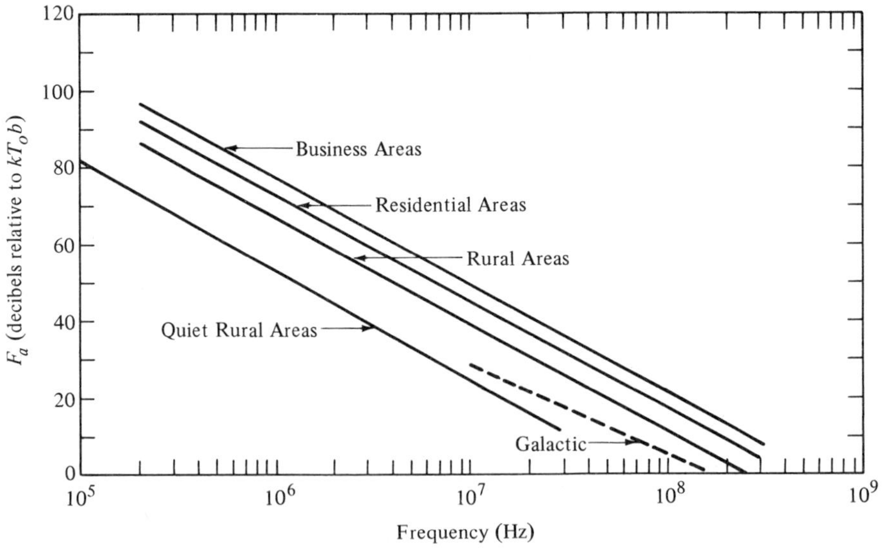

Fig. 6-4. Median values of F_a for three metropolitan-area activity classes versus frequency compared with data from a very quiet rural area and galactic noise. (After Spaulding and Disney, 1974)

Performance analyses of radio systems and receiving equipment routinely require a computation of service probability: the statistical liklihood that a specified or acceptable grade of reception will be attained. Typically, the determination of service probability is predicated upon an understanding of the randomness existing in the level of man-made noise to which a radio system or receiver is exposed. For this purpose, it is possible to associate with all median values of F_a in each activity class three statistical measures of variability—namely, the standard deviation σ_l of median F_a and the upper and lower rms decile levels, D_u and D_l, expressed in decibels. Collectively, this set of parameters provide a measure of the expected location variability of median F_a, which is applicable in any metropolitan area.

Figure 6-5 depicts the values of σ_l for each activity class and observation frequency within the interval 250 kHz and 250 MHz. No systematic variation of σ_l with frequency is evident in Fig. 6-5 for any activity class. It is probable that the irregular changes in σ_l observed in the data reflect concurrently the influences of both the

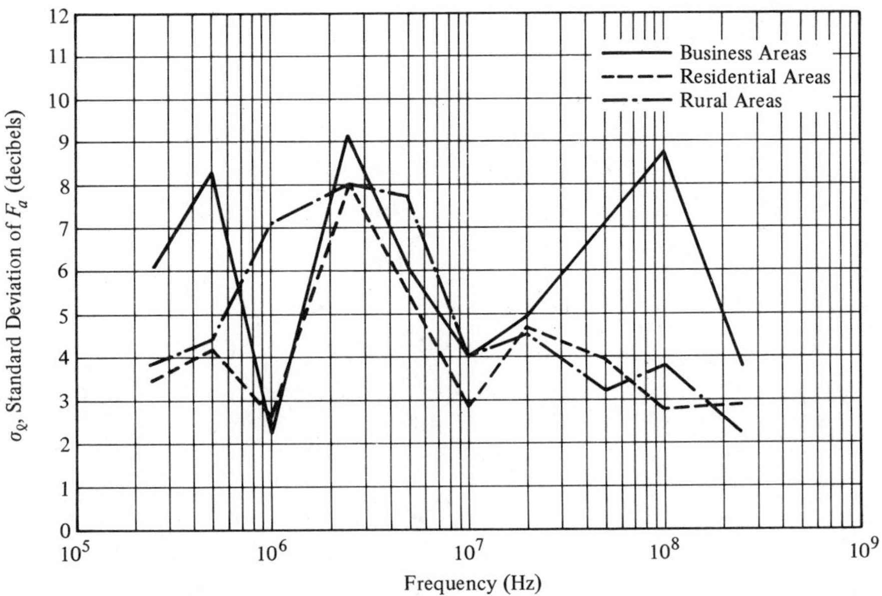

Fig. 6-5. Standard deviation of F_a versus frequency for each activity class. (After Spaulding and Disney, 1974)

comparatively small set of metropolitan locations studied and the uncertainty encountered in identifying a metropolitan location with one of the three classes.

The temporal variation of F_a observed in a Boulder, Colorado, residential area during a 1-hr test period is provided in Fig. 6-6. The cumulative distribution of 10-sec averages of F_a is shown for a 20-MHz observation frequency. The distribution is comprised of 360 samples of 50-sec duration, incremented in 10 sec steps. Note that the data suggest the existence of a mixed distribution by virtue of the discontinuity appearing in the slope. From comparable measurements of F_a performed at other frequencies between 0.25 and 250 MHz and at locations representative of each activity class, the upper and lower decile values, D_u and D_l, for 1-hr data samples have been determined. These results are presented in Fig. 6-7.[4]

The parameter V_d provides a measure of noise impulsiveness or,

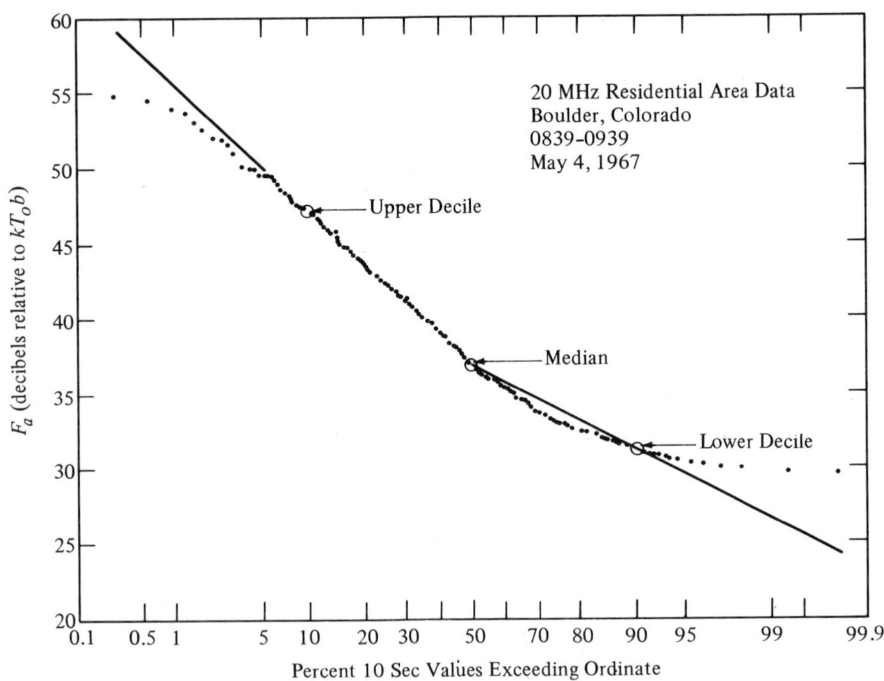

Fig. 6-6. Cumulative distribution of 10-sec averages of F_a at 20 MHz for a 1-hour sample of residential area noise. (After Spaulding and Disney, 1974)

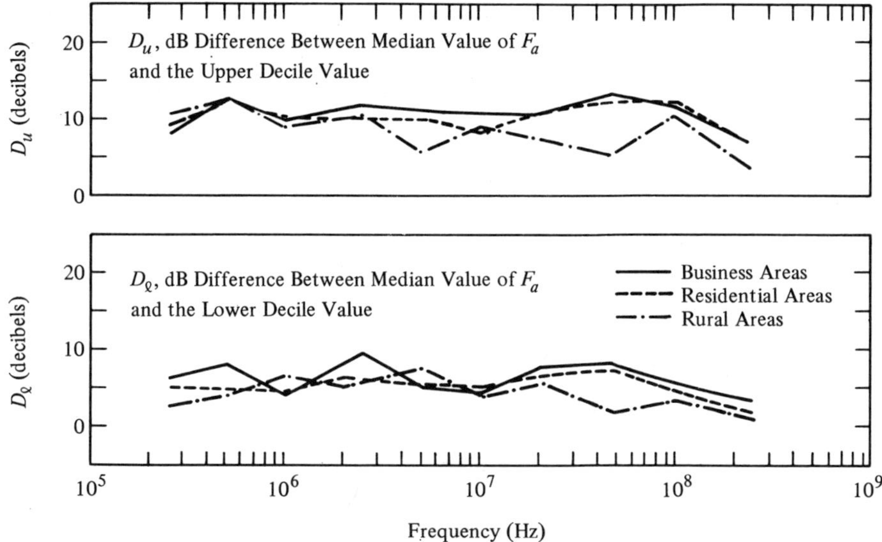

Fig. 6-7. Variations of the upper decile, D_u, and lower decile, D_l, values of the cumulative distribution of F_a with frequency as obtained for 1-hour observation periods of each activity class. Running averages of 50-sec duration and 10-sec step increments. (After Spaulding and Disney, 1974)

equivalently, a measure of the deviation of the noise-sample envelope distribution from a Rayleigh distribution. With increasing values of V_d above 1.05 dB, which is the limit attained when purely Gaussian noise is observed, the impulsive character of the observed interference increases. Noise-envelope filtering, however, injects a first-order effect upon the measured V_d. As the detection filter bandwidth decreases, pulse overlapping increases and the output envelope distribution approaches that of thermal noise, i.e., a Rayleigh distribution. Thus, identification of the detection bandwidth used in any study to measure V_d is essential. For a fixed detection bandwith, V_d varies with observation frequency and, to the extent independence exists, with activity class. Figs. 6-8 and 6-9 present mean values of V_d for the medians of the three activity classes as a function of frequency using two values of detection bandwidth: 4 and 10 kHz. The lower solid line on each figure represents a value of V_d of 1.05 dB for thermal noise. In Fig. 6-10, the cumulative distribution of V_d obtained at 20 MHz for a single residential location during a period of 1 hr is

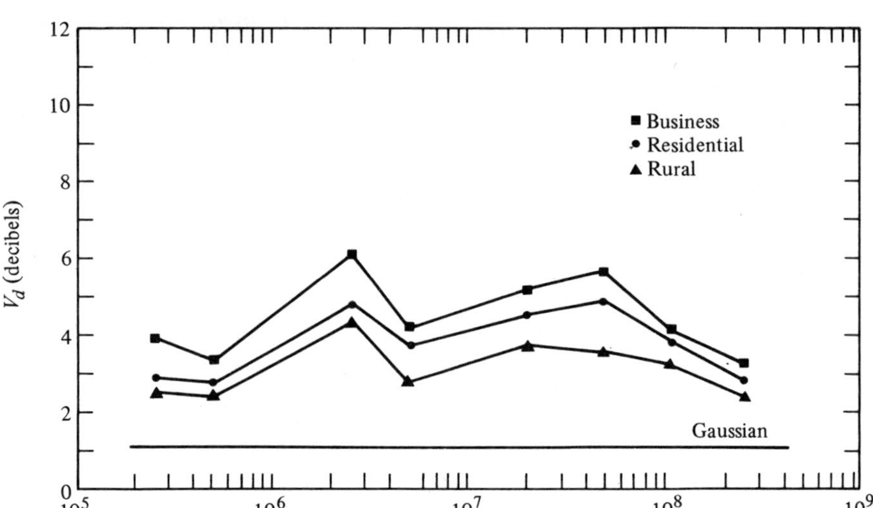

Fig. 6-8. Variation of V_d with frequency for three activity classes. A 4-kHz detection bandwidth. Gaussian noise-reference level is shown. (After Spaulding and Disney, 1974)

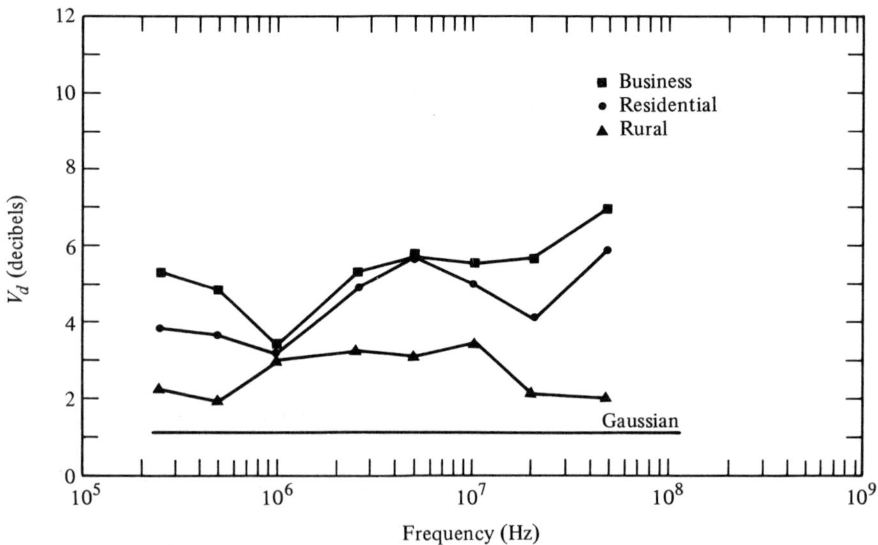

Fig. 6-9. Variation of V_d with frequency and activity class. A 10-kHz detection bandwidth. Gaussian noise-reference level is shown. (After Spaulding and Disney, 1974)

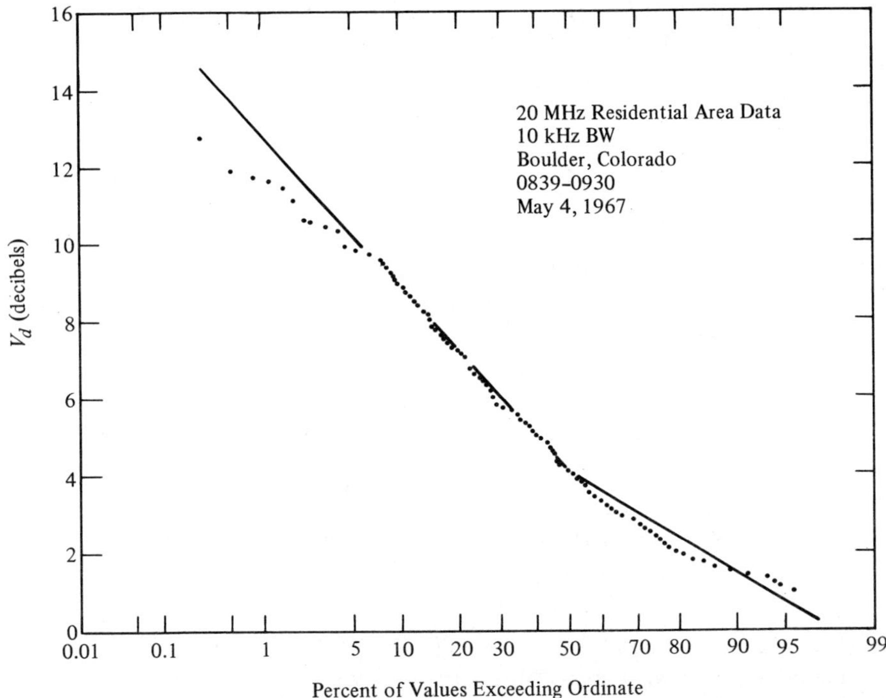

Fig. 6-10. Cumulative distribution of V_d for a residential location observed at 20 MHz. One-hour data recording with 50-sec integration period and 10-sec running average increments. (After Spaulding and Disney, 1974)

presented.[4] Running averages of 50-sec with step increments of 10 sec were used to record these results. Correlations of V_d variations with frequency and three activity classes are evident in Fig. 6-8 and 6-9 and suggest that the selected activity classifications are not totally independent for the metropolitan areas represented. Although the mean values of V_d at 20 MHz for the residential class obtained using a 10-kHz detection bandwidth is seen from Fig. 6-9 to be approximately 4 dB, the distribution of V_d for the 360, 50-sec samples shown in Fig. 6-10 discloses that 10 percent of the V_d samples exhibit envelope characteristics closely approaching thermal noise, i.e., $V_d \leqq 1.5$ dB. Thus, a single location during one hourly period may manifest a great variability in this fundamental characteristic of load man-made noise.

RANGE COMPUTATION OF COMPOSITE MAN-MADE METROPOLITAN-AREA RADIO NOISE

Consolidation of Experimental Data Sources

Initial efforts at consolidating man-made incidental noise measurements with respect to frequency and surface distance relative to the center of an urban area were forced to contend with limited and diverse sets of experimental data.[1,6] Sparseness in the early available data required commingling of results reported for average, quasi-peak, and peak detectors. This commingling necessitated the incorporation of detection-bandwidth and electric-field intensity-to-power transformation factors to arrive at an approximate dependence of available power upon frequency and urban-center separation distance. The complexities and uncertainties inherent in such transformations contributed to the residual errors of the composite data, much of which was removable by the advent of a larger data base.

As the available experimental data increased through efforts of several investigatory groups, it became possible to prepare composite surface-noise plots of both average incidental noise power and incidental quasi-peak electric-field intensity (or quasi-peak power) for the spectral interval 20 MHz to 1 GHz and within a maximum surface range of 35 miles from a city center.[7] Further enlargement of the available experimental data covering the frequency range 100 kHz to 1 GHz permitted a more exhaustive screening for accuracy. Concurrently, a more detailed subdivision of all data into subcategories of frequency and urban-center separation distance became possible.[8] In all instances, the primary information sources were screened to prevent the inclusion of measurements from any locale that was dominated by a single class of noise emitters. It has been the intent to compile incidental-noise data that represented a typical metropolitan area at any value of frequency and center separation range.

Separation of composite metropolitan-area man-made radio-noise data into two groups—one acquired with average power detectors and the other with quasi-peak detectors—has become possible. Each group spans a large segment of the 100-kHz to 1-GHz frequency interval, and various portions of the urban-center displacement range 0 to 23 miles. The data derived using either detector type were grouped by frequency interval and urban-center range. No *a priori* definitions were applied to constrain the subsets of either parameter, beyond that of reasonable utility in the results (e.g., unduly small or large intervals in either parameter were rejected). Approximately 1-octave frequency intervals were set as the preferable minima. Smaller partitions were rejected as inconvenient to exploit in subsequent studies, while larger segments threatened to mask small-scale frequency variability during least-squares-regression line construction. With respect to urban-center range compilation increments, point values are untenable because of the appreciable location variability existing in the data. However, were large distance increments in excess of 10 miles employed in the inner metropolitan zone, those range variations having correlation lengths that are small compared to the transition zone widths existing between heavy industrial and business areas or between the latter and residential areas would be suppressed. In addition, radio-noise data obtained at one location, for a single frequency, and by one experimenter could not be included in the composite representations because of their potentiality for distorting the slope of the signal-level-versus-frequency curve, i.e., the frequency decrement through the occurrence of differing equipment calibration or data-reduction accuracies or inherent noise-level location variability. The potential impact upon the frequency decrement of a linear function may be seen by examining Fig. 6-11. Fig. 6-11(a), two equal-length sets of data with a common

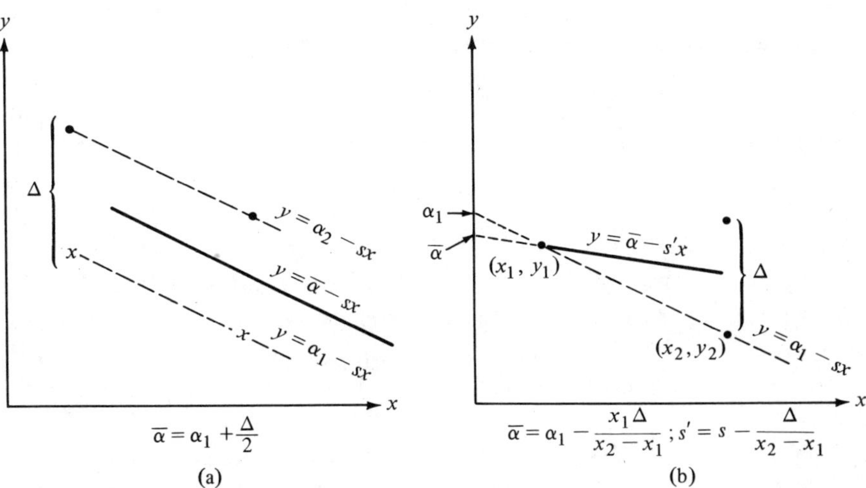

Fig. 6-11. Diagrams of the slope distortions introduced into linear functions by single point displacements.

slope of $-s$ but differing intercepts may be averaged to yield a mean line with an average intercept, $\overline{\alpha}$, and an unchanged slope of $-s$. If, however, all but one of the points of one of the data sets are omitted (i.e., only the upper right-hand point of the upper curve in Fig. 6-11(a) remains in Fig. 6-11), then the average line obtained from the remaining points differs in slope and intercept from the original mean line and from the component lines. This is seen by examining Fig. 6-11(b). Prevention of a slope distortion is possible if the component data sets are of approximately equal extent in the independent variable.

Consolidation of the available incidental man-made noise data by the preceding criteria yields four frequency groupings, labeled I through IV in Figs. 6-12 through 6-19. A further subdivision of each frequency interval in groups I, II, and IV, by distance from an urban center produces a total of eight composite plots encompassing the band of 250 kHz to 1 GHz and the range interval of 0 to 23 miles. Similarly, seven available sets of quasi-peak electric-field-strength data have yielded the two groupings labeled X and XI in Figs. 6-20 and 6-21. These data encompass 15 kHz to 200 MHz and extend outward to 7 miles from an urban center. The legends of Figs. 6-12 to 6-21 identify the original investigator and the locations of measurement. Details of the experiments and data-reduction steps employed are available in Reference 8.

Eleven sets of average incidental-noise power and seven sets of quasi-peak electric-field-strength data were grouped within eight frequency intervals during the construction of Figs. 6-12 through 6-21 in a manner that permitted the use of a maximum number of individual sets in each interval. To accomplish this, several of the larger data sets were segmented, and some portions were grouped with less extensive sets. No segregation was performed based upon antenna polarization. Vertically polarized, linear, constant-gain antennas were employed in the majority of the primary studies. Typically, the dependence of noise level upon antenna polarization for surface and near-surface observation points is small, less than 2 dB, with the differential favoring either polarization, depending upon local conditions.

The data in Figs. 6-12 through 6-21 are for weekday business hours and represent for each detector type the results of separate investigations performed at similarly disposed locations in several urban areas.

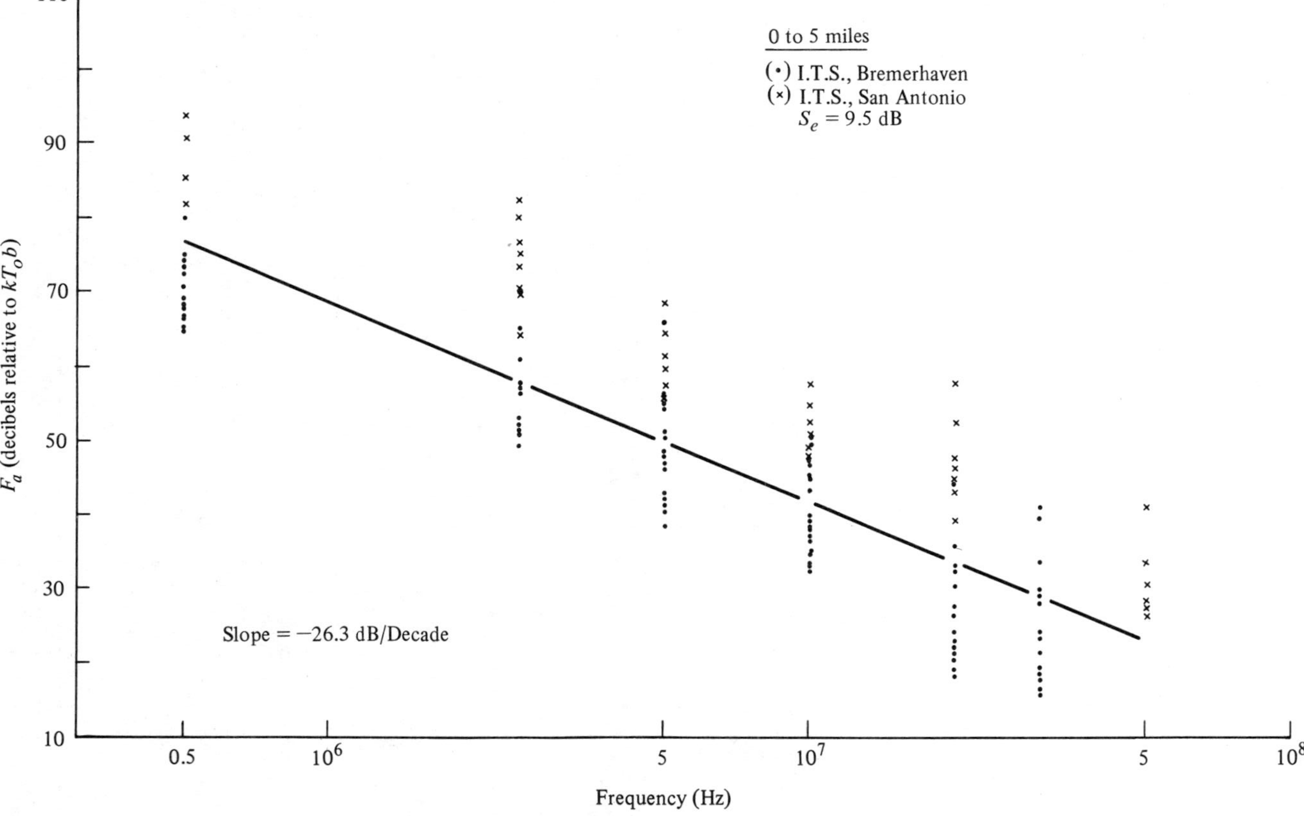

Fig. 6-12. F_a versus frequency for data Group IA.

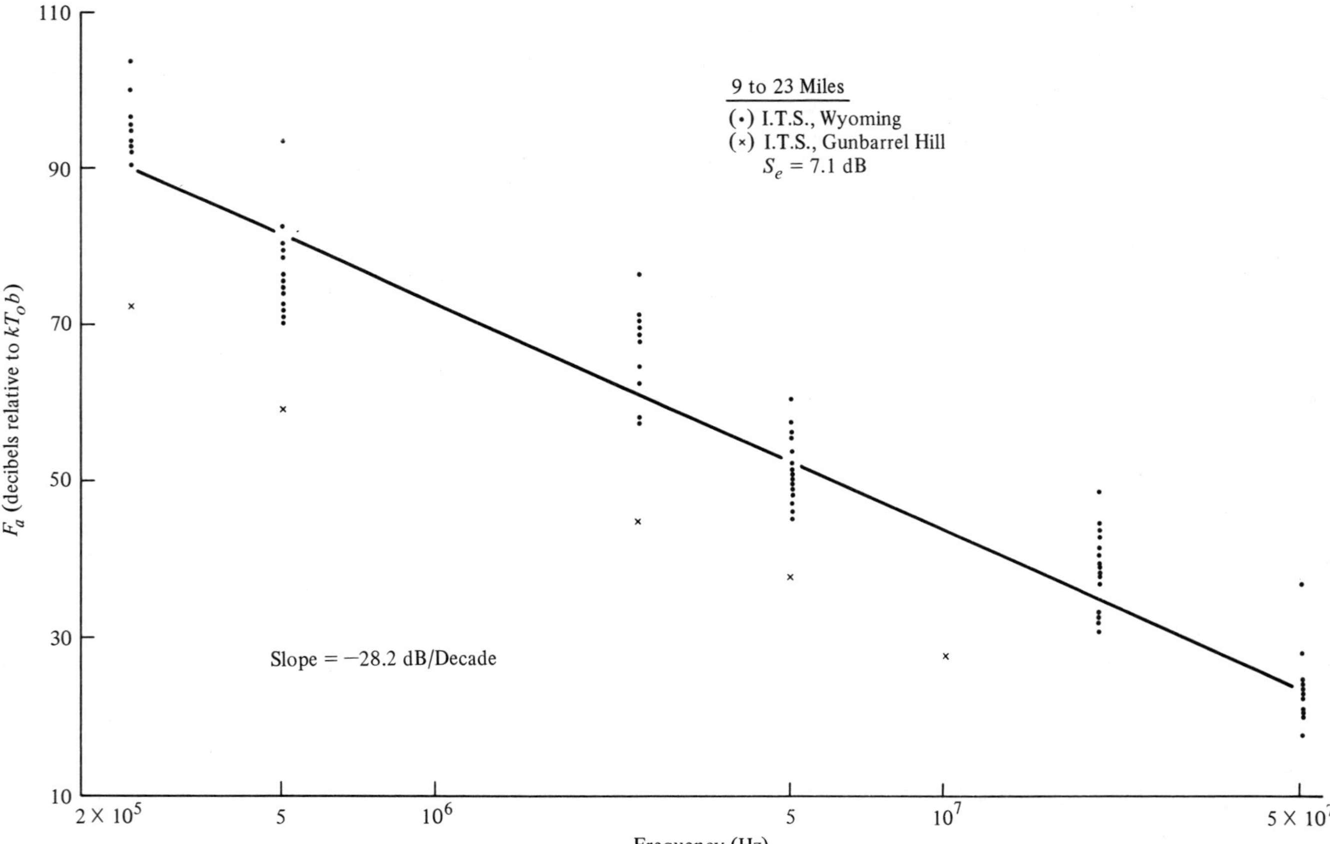

Fig. 6-13. F_a versus frequency for data Group IB.

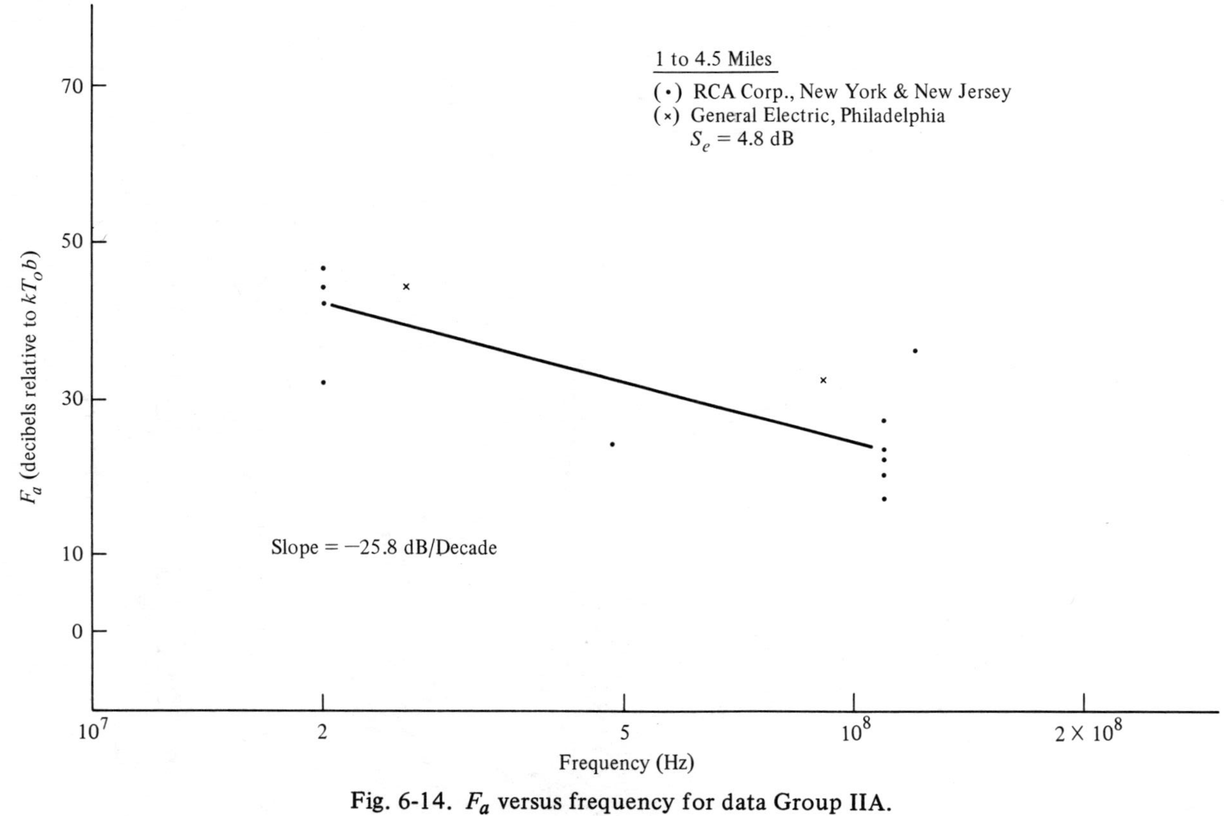

Fig. 6-14. F_a versus frequency for data Group IIA.

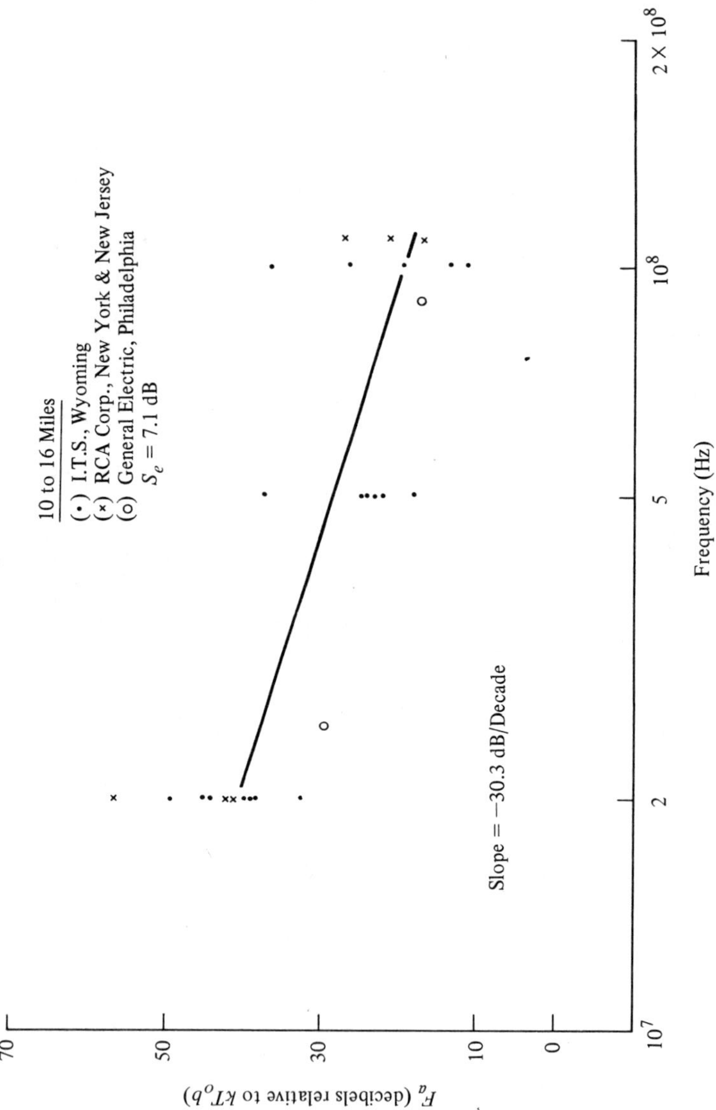

Fig. 6-15. F_a versus frequency for data Group IIB.

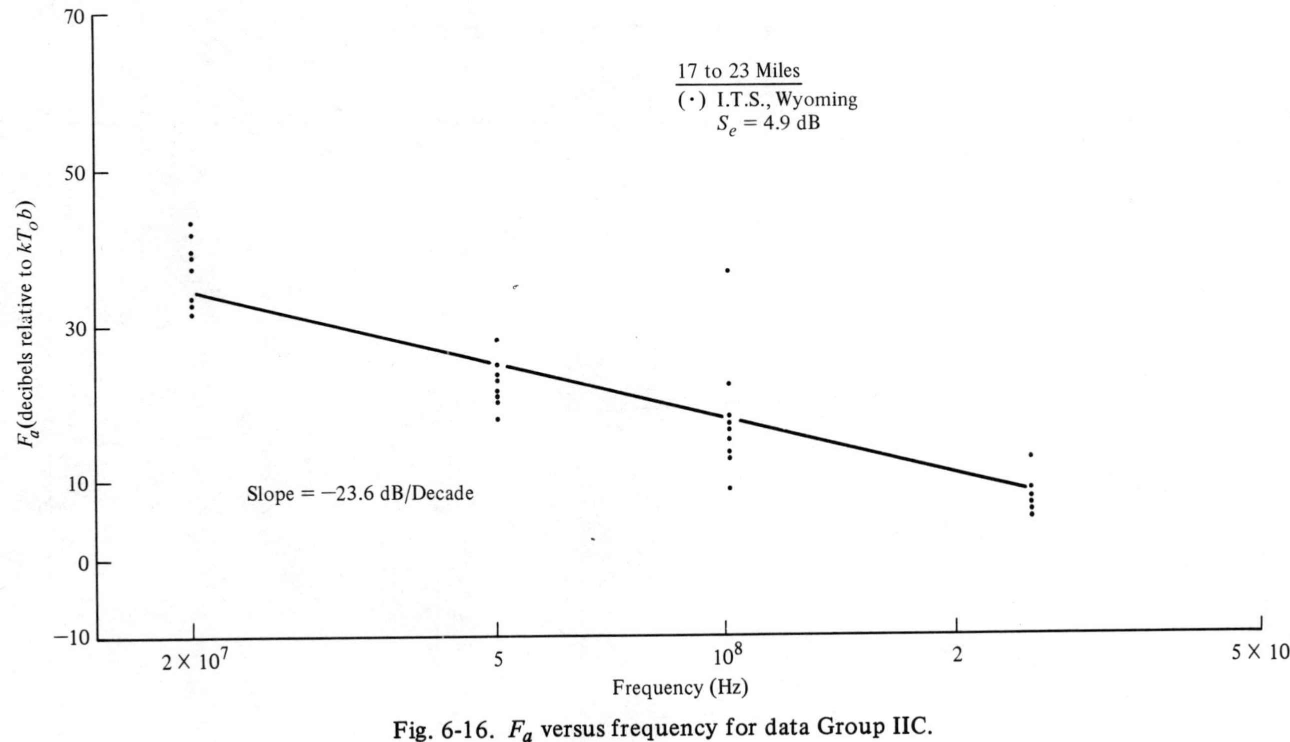

Fig. 6-16. F_a versus frequency for data Group IIC.

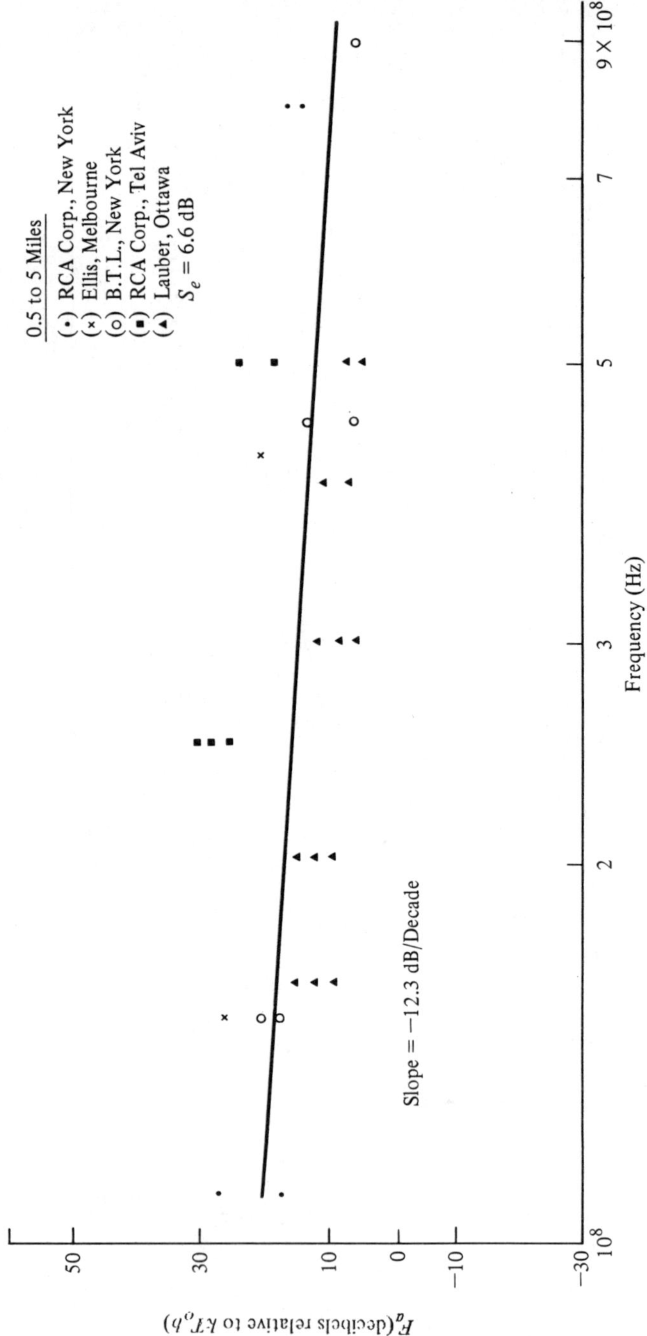

Fig. 6-17. F_a versus frequency for data Group III.

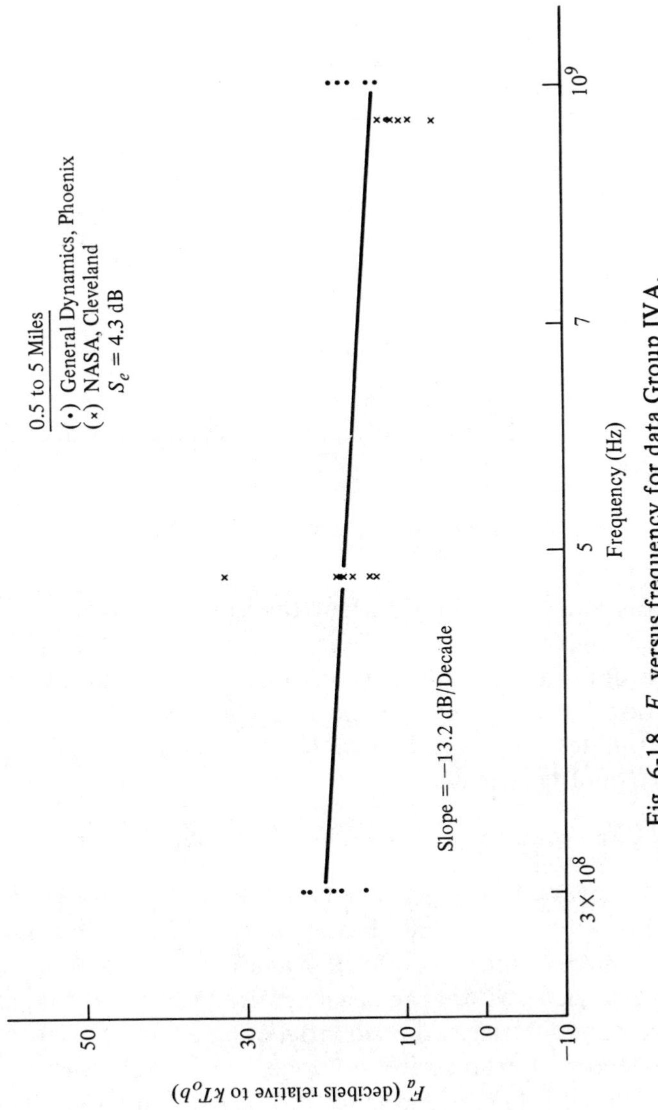

Fig. 6-18. F_a versus frequency for data Group IV A.

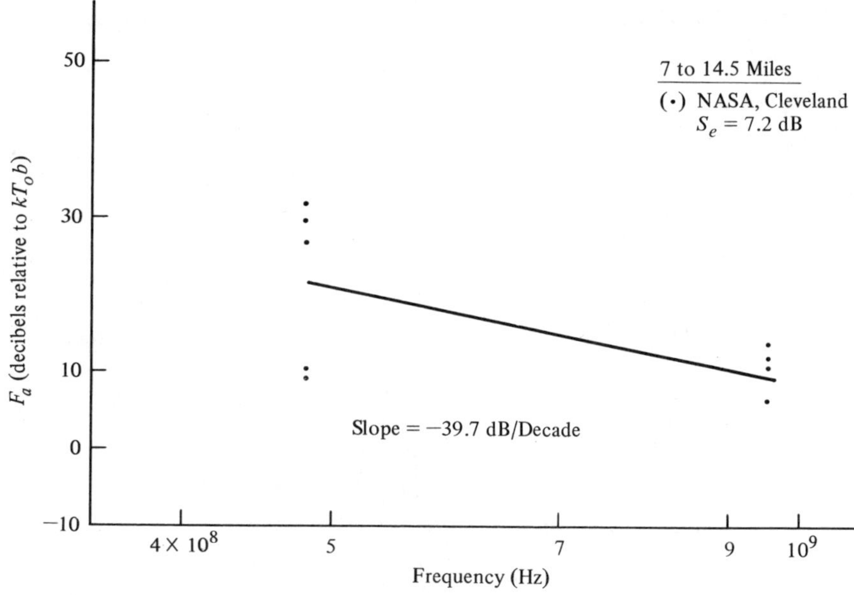

Fig. 6-19. F_a versus frequency for data Group IVB.

The diversity existing among the metropolitan areas from which these data were derived enlarges the variability of the resulting consolidations, yet simultaneously yields parametric representations of incidental-noise level for distance and frequency, which may be used as reliable estimates on noise factor, F_a, or electric-field strength for similar metropolitan areas.

Statistical Examination of Composite Radio-Noise Data

Because the level of incidental man-made radio noise is dependent upon the intensity of the individual emitters, the surface density of stationary sources, and the rate of passage of any mobile source in the observer's vicinity, the aggregate noise level would be expected to decrease as industrial and residential congestion of a metropolitan region decreases. Certain circumstances can mask these expected variations—namely, (1) the presence of localized, high-intensity emitters (for example, suburban and rural transformer substations, remote manufacturing sites, heavily traveled highways) or (2) long-distance transmission of noise in the lower portion of the spectrum

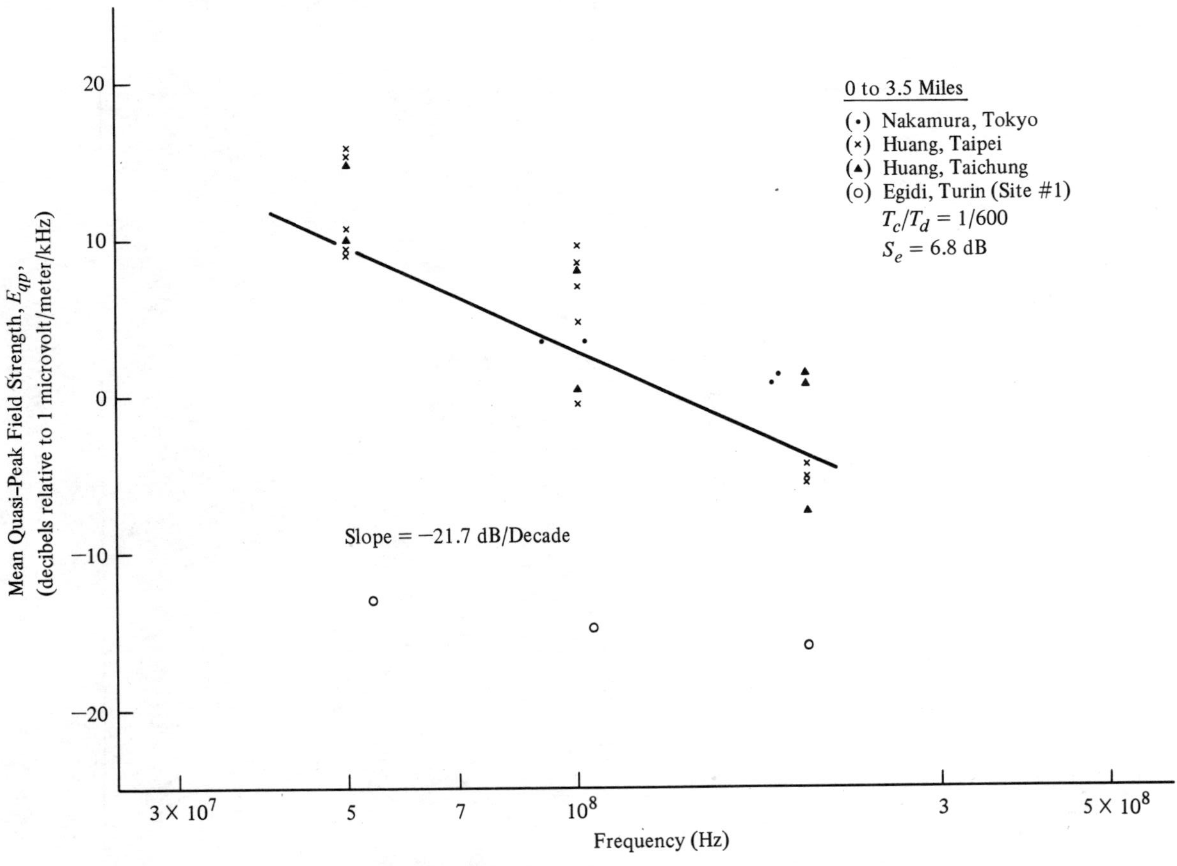

Fig. 6-20. Mean E_{qp} versus frequency for data Group X.

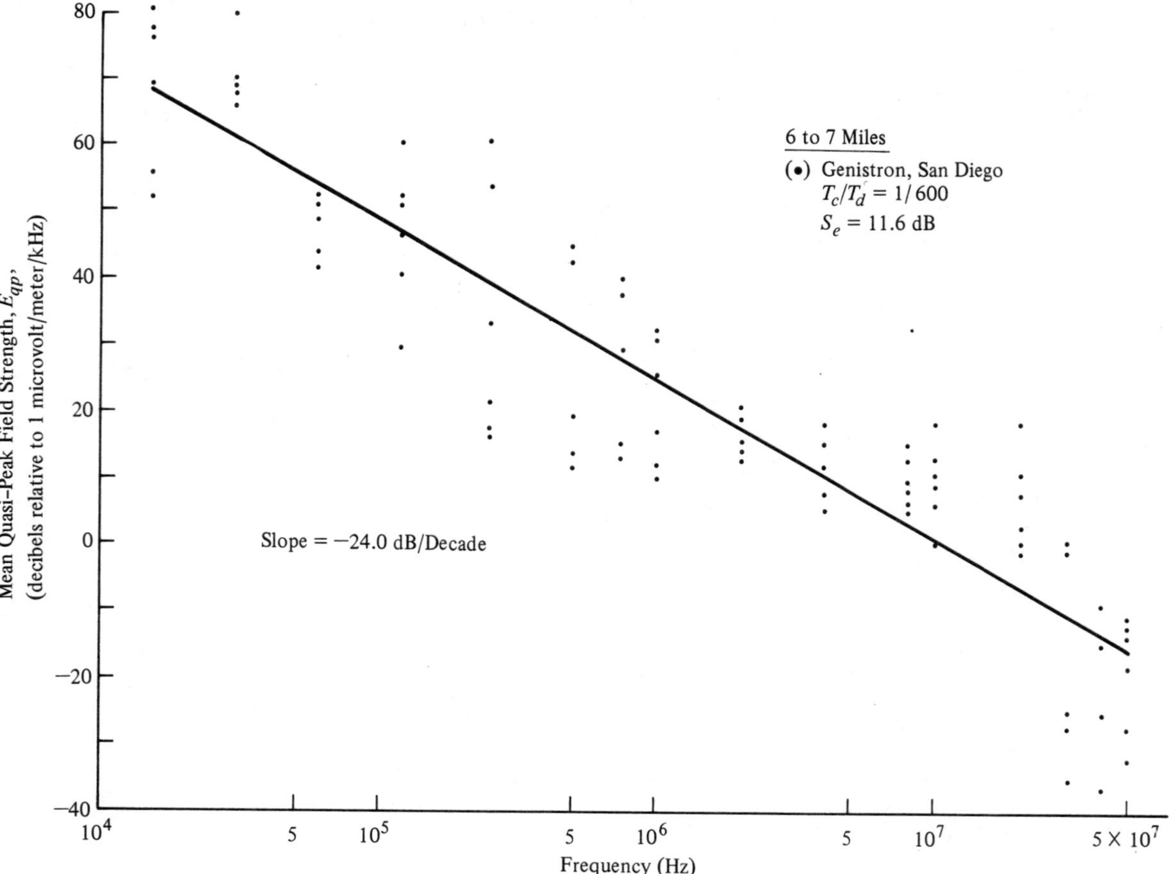

Fig. 6-21. Mean E_{qp} versus frequency for data Group XI.

(that is, the HF band and below, where low loss-propagation modes dominate).

To study the urban-center distance variations of surface incidental man-made noise, the Spearman Rank correlation coefficient has been computed for several groups of both average power and quasi-peak field-intensity data. Four sets each of average power and quasi-peak field-strength data were found to be sufficiently extensive to reliably permit these computations. The results are presented in Tables 6-1 and 6-2, along with identification of the investigators and the distance and frequencies for which the data were recorded. Note that a negative correlation coefficient indicates that the noise level decreases with increasing distance. Examination of Tables 6-1 reveals that the majority of the computed correlation coefficients are negative, 86% of those for the average power parameter, F_a, and 91% for the quasi-peak electric-field-strength data, F_{qp}. In some instances, the correlation is weak. Note too that the greatest positive values of the correlation coefficient occur at the lower frequencies where low radio-path attenuation occurs, making it possible for localized sources to conceal decreases in incidental-noise levels with increasing range from an urban center. The evident correlation between decreasing noise power

Table 6-1. Rank correlation coefficient for F_a.

	Distance Interval (miles)			
	0.5-3	2-5	0.5-14	12-23
Frequency (MHz)	I.T.S. Bremerhaven	I.T.S. San Antonio	NASA Cleveland	I.T.S. Wyoming
0.25	–	–	–	-0.13
0.5	-0.63	0.70	–	0.59
2.5	-0.31	0.10	–	-0.18
5	-0.32	-0.05	–	-0.45
10	-0.67	-0.65	–	–
20	-0.91	-0.21	–	-0.22
30	-0.75	–	–	–
50	–	-0.70	–	-0.19
102	–	–	–	-0.28
250	–	–	–	-0.37
480	–	–	-0.33	–
950	–	–	-0.17	–

Table 6-2. Rank correlation coefficient for E_{qp}.

	Distance Interval (miles)			
	0.8–1.9	0–6.1	0.6–3.1	0.3–12.4
Frequency (MHz)	Egidi Turin (1/600)[a]	Huang Chungli (1/600)	Huang Taipei (1/600)	Huang Taichung (1/600)
0.25	—	—	—	—
0.5	—	—	—	—
2.5	—	—	—	—
5	—	—	—	—
20	—	—	—	—
50	—	-0.70	-1.0	-0.93
54	-0.80	—	—	—
101	—	-0.70	-0.40	-0.98
110	-0.80	—	—	—
200	-1.00	—	-0.20	—
250	—	—	—	—
400	0.25	—	—	—

[a] (T_c/T_d).

or field intensity and increasing urban-center separation distance supports the concept of establishing range groupings of composite noise data.

Shown in plots for both F_a and E_{qp} are least-squares regression lines and the standard errors of estimate, S_e (Figs. 6-12 through 6-21).

Assessments of Composite Incidental-Noise Data. In the UHF, VHF, and upper HF bands, the results in Figs. 6-12 through 6-19 display values of F_a that decrease with distance from an urban center. This dependence of F_a upon distance varies among the plots. In the middle and lower HF band, the composite plots show increases of F_a with distance. A conservative interpretation of this behavior is that composite HF incidental-noise power is approximately independent of distance for ranges between 0 and 23 miles in the absence of high-intensity localized noise sources.

The frequency decrement manifests a nonuniform dependence upon observation frequency. Below approximately 100 MHz, the decrement is markedly negative, equaling -20 to -30 dB per decade. This slope moderates in the interval of 100 to 800 MHz, reducing to

-10 to -15 dB per decade. For frequencies above 600 MHz, Fig. 6-19 presents evidence of a return to a high negative decrement, which persists to 1 GHz, where the data terminate. Such return to a high-negative slope is consistent with the presence of a maximum in the UHF-band emission spectrum of automotive ignition interference, which is the dominant incidental-noise source.

The frequency decrement for quasi-peak electric-field-strength data in the interval below 200 MHz is observed to be greater than -20 dB per decade; appreciably larger than the frequency decrement for the rms electric-field strength observed for average noise power. This point becomes evident when it is recalled that average power is proportional to the rms electric-field strength squared, multiplied by the effective aperture of the observing antenna. Since the aperture area of constant-gain antennas (e.g., tuned monopoles and dipoles such as were used for acquiring the F_a data) is proportional to the inverse square of the frequency, the frequency decrement of the rms electric-field strength becomes equal to the decrement for F_a plus 20 dB. As an example, the frequency decrement for E_{rms} in the data of Fig. 6-12 equals -6.3 dB per decade, noticeably smaller than the slope of Figs. 6-20 or 6-21.

It is generally true that for all sources of incidental radio noise, the quasi-peak electric-field strength exceeds the rms electric-field intensity. Combining this fact with the preceding observations on the magnitude and sign of the quasi-peak and rms electric-field decrements permits the conclusion that for composite man-made noise, the quasi-peak electric-field strength approaches equality with the rms electric-field intensity, from above, as the observation frequency increases.

All measurements of E_{qp} presented in Figs. 6-20 and 6-21 were recorded for receivers and detectors meeting ANSI standards, (see Table 3-1); consequently these instruments possessed charge-to-discharge time-constant ratios, T_c/T_d, equaling 1/600. Although metropolitan-area, quasi-peak electric-field-strength measurements have been performed with receivers and detectors designed to CISPR standards, the available information is presently insufficient to allow construction of composite noise representations of acceptable reliability.

Special Representation of Quasi-Peak Detected Composite Metropolitan-Area Radio-Noise Data

The noise power available at the output terminals of a matched receiving antenna, p_a, is proportional to the square of the rms electric field, E, multiplied the effective aperture of the antenna, A_m :

$$p_a = E^2 A_m / Z_0 \tag{6-1}$$

where Z_0 = the impedance of free space. Equation 6-1 is valid whenever the observation point lies within the radiation field of signal E. The output voltage, V, of a receiver attached to the antenna is proportional to the incident electric-field strength, the constant of proportionality (ζ) containing the aperture efficiency of the antenna, the transmission-line losses, and the receiver gain, i.e.,

$$V = \zeta E. \tag{6-2}$$

Inserting the receiver output voltage into an attached detector produces a detected voltage, which for noise-voltage waveforms may be a function of the waveform characteristics such as average pulse spacing for impulsive noise, as well as the detector bandwidth and integrator time constant. If two independent and identical antennas and linear receivers are exposed to a common man-made impulsive noise field, but are used to feed a quasi-peak and rms voltage detector, respectively, the ratio of detected outputs is representable by

$$(V_{qp} / V_{rms}) = \mathfrak{F}^{1/2} \tag{6-3}$$

where $\mathfrak{F} > 1$. The factor $\mathfrak{F}$ is a unitless function of the noise bandwidth of the antenna-receiver-detector assembly, the average period of the noise impluses and the ratio T_c/T_d of the detector.[9]

Using Equations 6-1 through 6-3, one may write for p_a:

$$\begin{aligned} p_a &= \zeta^2 V^2{}_{\text{rms}} A_m / Z_o = \zeta^2 V^2{}_{qp} A_m / (\mathfrak{F} Z_o) \\ &= E^2{}_{qp} A_m / (\mathfrak{F} Z_o) = p_{qp} / \mathfrak{F}. \end{aligned} \tag{6-4}$$

In Equation 6-4, p_{qp} has been set equal to $E^2{}_{qp} A_m / Z_o$ and labeled *quasi-peak power*, as it possesses the units of power. The units used to represent p_{qp} in the subsequent applications will be decibels relative to 1 mW/kHz bandwidth (dBm/kHz) to avoid identification with p_a and the defination of F_a, both of which reference the available or average power of the observed noise signal. Specific application will be made of p_{qp} in Chapter 7 in an analysis of composite man-made radio noise observed above metropolitan areas.

Parametric Representation of Composite Man-Made Incidental-Noise Data. The construction of a parametric dependence of composite metropolitan-area incidental noise upon urban-center separation distance is required in preparation for the examination of airborne man-made noise in Chapter 7 and also to provide a foundation for studies of land-mobile-receiving systems and related networks. Assembling the available metropolitan area measurements of incidental radio noise in the manner described in the preceding section was intended to minimize distortion of the frequency decrement by the presence of limited groups of data. This method of compilation does not yield a set of consolidated plots from which a range representation of the data may be extracted. The inherent limitation is caused by irregularities that arise in the range partitions and frequency groupings.

It is preferable to proceed with the development of the range variation of composite noise, referenced to a metropolitan center, using a data-assembly method that yields the most ordered arrangement of separate data sources. Such an approach consists of placing the available sets of data in three concentric range intervals centered about an urban complex. The outer boundaries of each interval are designated by a radial distance measured from the metropolitan center and are fixed by the ratios, 1:3:7, with the innermost or urban-zone limit set at 5 miles. The boundaries of the three zones and their designations are

Zone	*Limit (miles)*
Urban	0 to 5
Suburban I	5 to 15
Suburban II	15 to 35

Notice that the width of each zone enlarges in the ratio 1:2:4. Within each zone, the frequency-range limits are sets by the extent of the available data. For this reason, the frequency limits vary somewhat among the zones.

Figures 6-22 through 6-24 present composite plots of average noise power as a function of frequency for each noise zone. Most of the primary data sources employed in the preceding analyses (Figs. 6-12 through 6-19) have been included, the principal exception being those studies that emphasized the band below 20 MHz. Two ordinate scales are provided, noise power, P_a, in decibels relative to 1 mW/kHz bandwidth, and F_a, in decibels. Associated with each data entry is a code identifying the original investigator and the location of the measurement.[7] To each set of data, a least-squares-regression line has been fitted, and the frequency decrement has been indicated. The legend contains the computed standard error of estimate, ${}_aS_e$. Notice that the urban zone data are the most extensive within the frequency interval 20 MHz to 1 GHz. The lower frequency limit on all plots is 20 MHz; however, the outermost zones are unrepresented above 500 MHz. Notice further that each zone displays a frequency decrement lying between -12 and -18 dB per decade, which is comparable to the decrements of Figs. 6-17 and 6-18.

Values of quasi-peak power, P_{qp}, expressed in decibels relative to 1 mW/kHz, are plotted for composite metropolitan-area noise in Figs. 6-25 and 6-26. The primary data sources used are indicated by the notation associated with each point. The quasi-peak data notation key, the details on the individual studies, and the manner of consolidation are available in Reference 10. Metropolitan-area, quasi-peak incidental-noise data are presently too sparse at ranges greater than 15 miles to permit consolidation and construction of a suburban II curve. Notice for the inner two zones that the frequency decrements are respectively -25.4 and -20 dB per decade of frequency change. Clearly, a much larger negative slope is represented by the P_{qp} data than by the average power composite curves of Figs. 6-22 and 6-23 for the rms electric-field intensity.

The circled dots of Figs. 6-22 to 6-26 indicated the average measured values at the respective frequencies. When only one entry occurs at a particular frequency, the point is circled and labeled with the source and location abbreviation. All averages shown were computed in the ordinate scale units and thus represent the geometrical means of the data when expressed in W/kHz.

At any frequency common throughout either the average-power or the quasi-peak composite metropolitan-area data presented in Figs.

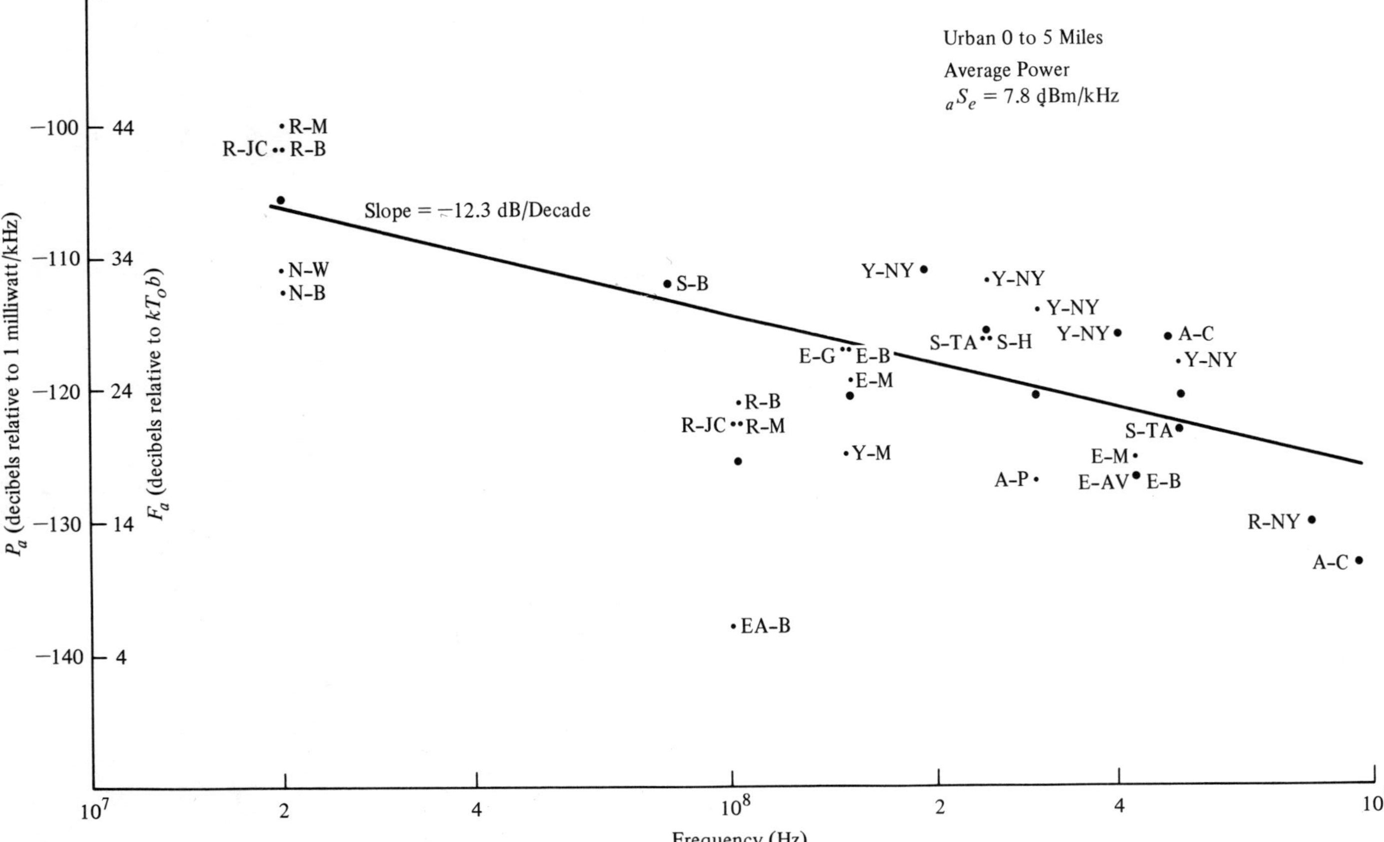

Fig. 6-22. Average composite man-made noise power for the urban zone, 0 to 5 miles.

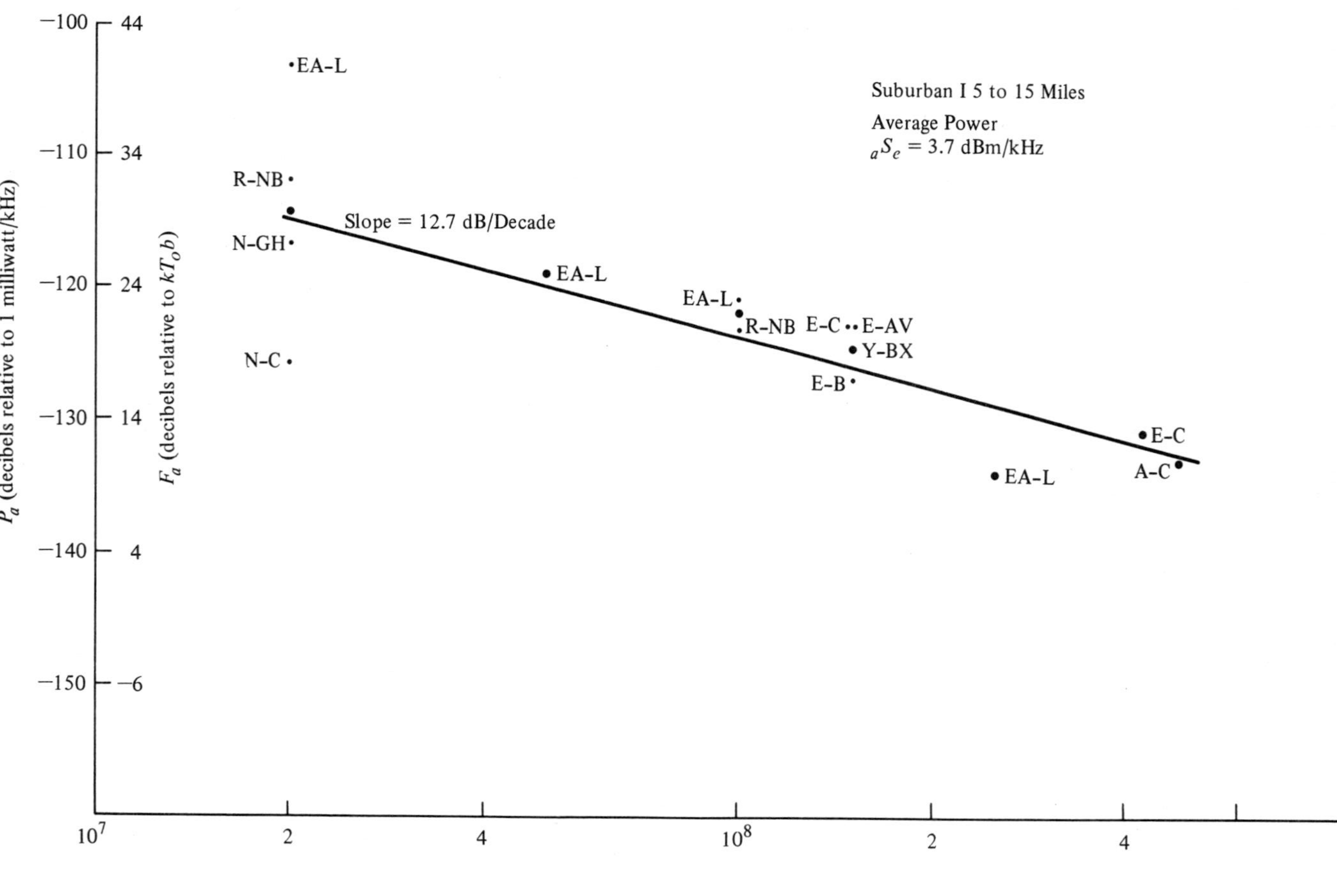

Fig. 6-23. Average composite man-made noise power for the suburban I zone, 5 to 15 miles.

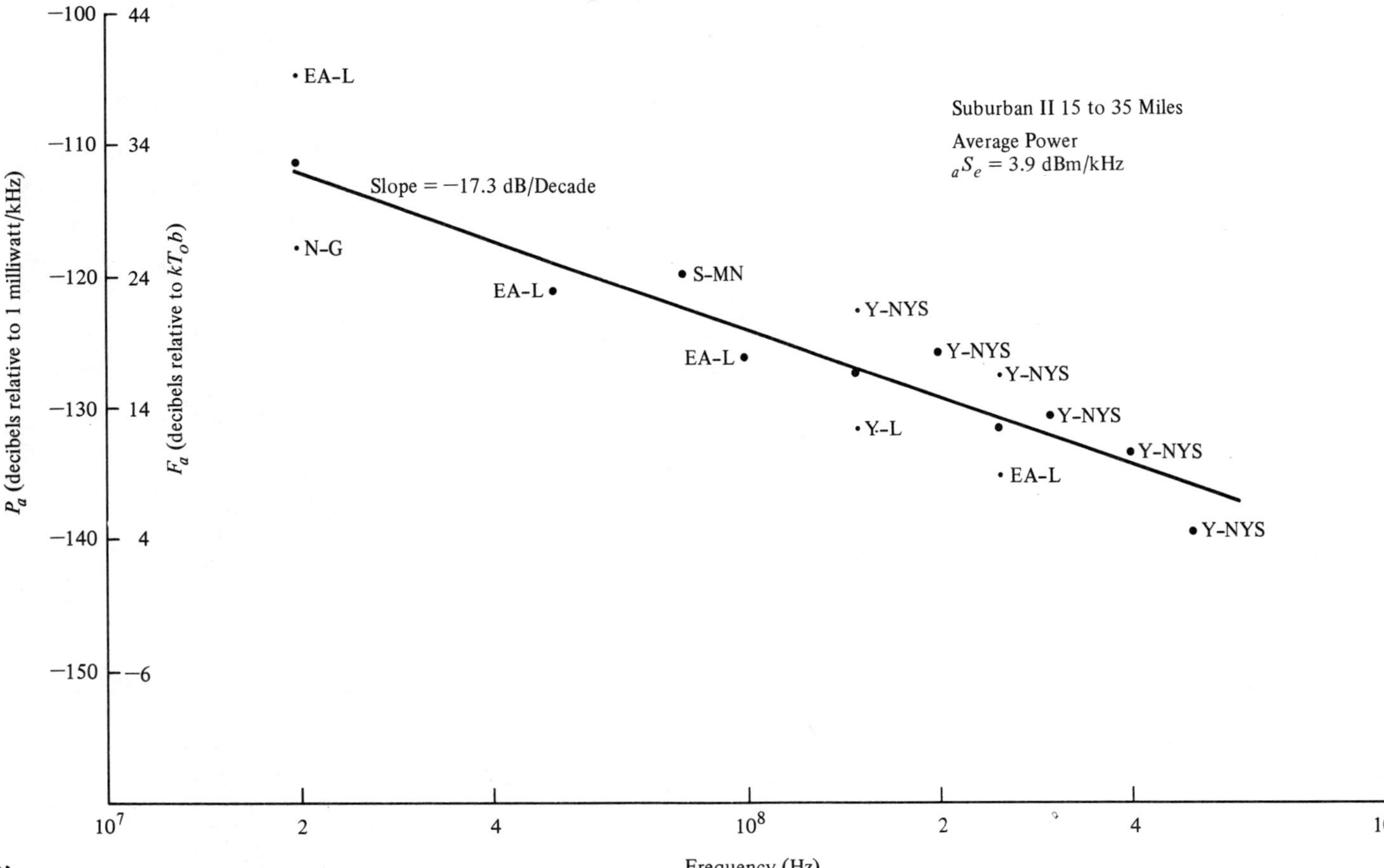

Fig. 6-24. Average composite man-made noise power for the suburban II zone, 15 to 35 miles.

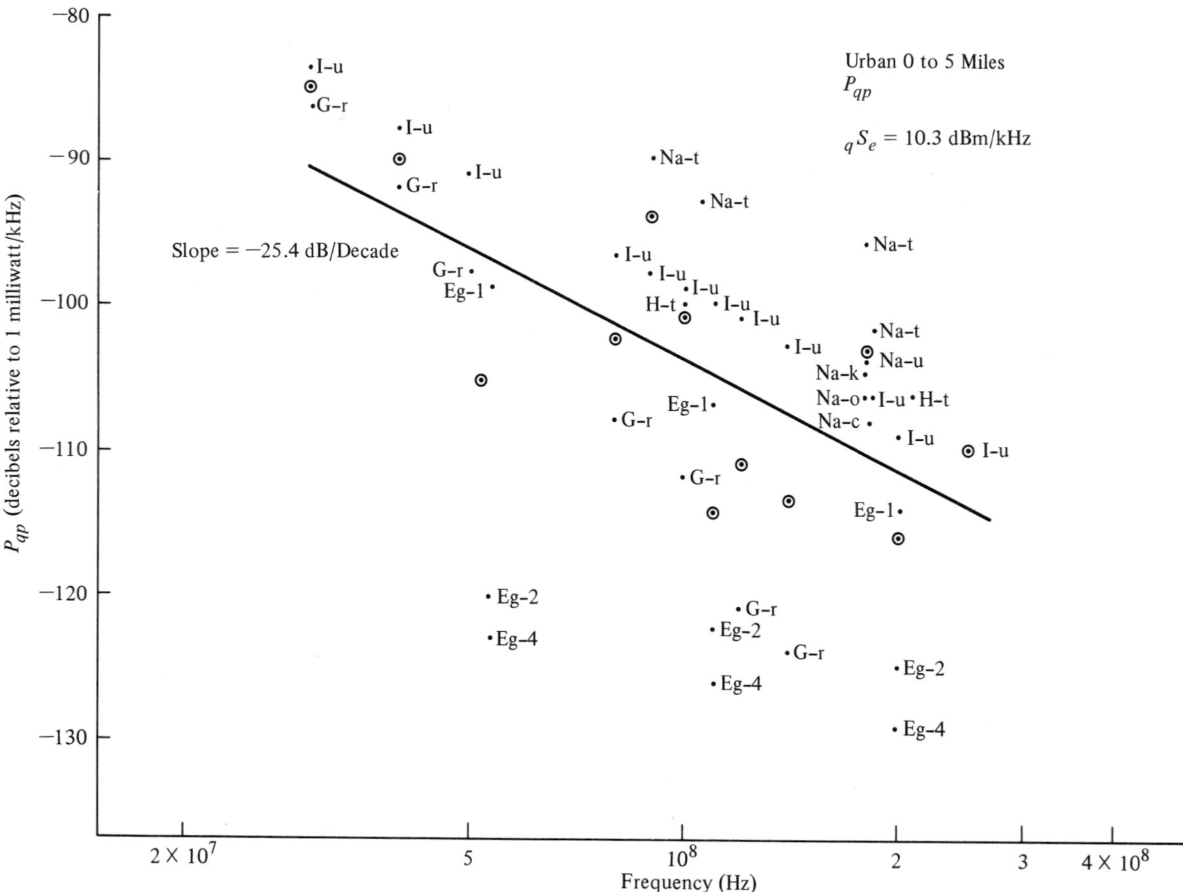

Fig. 6-25. P_{qp} for composite man-made noise within the urban zone, 0 to 5 miles.

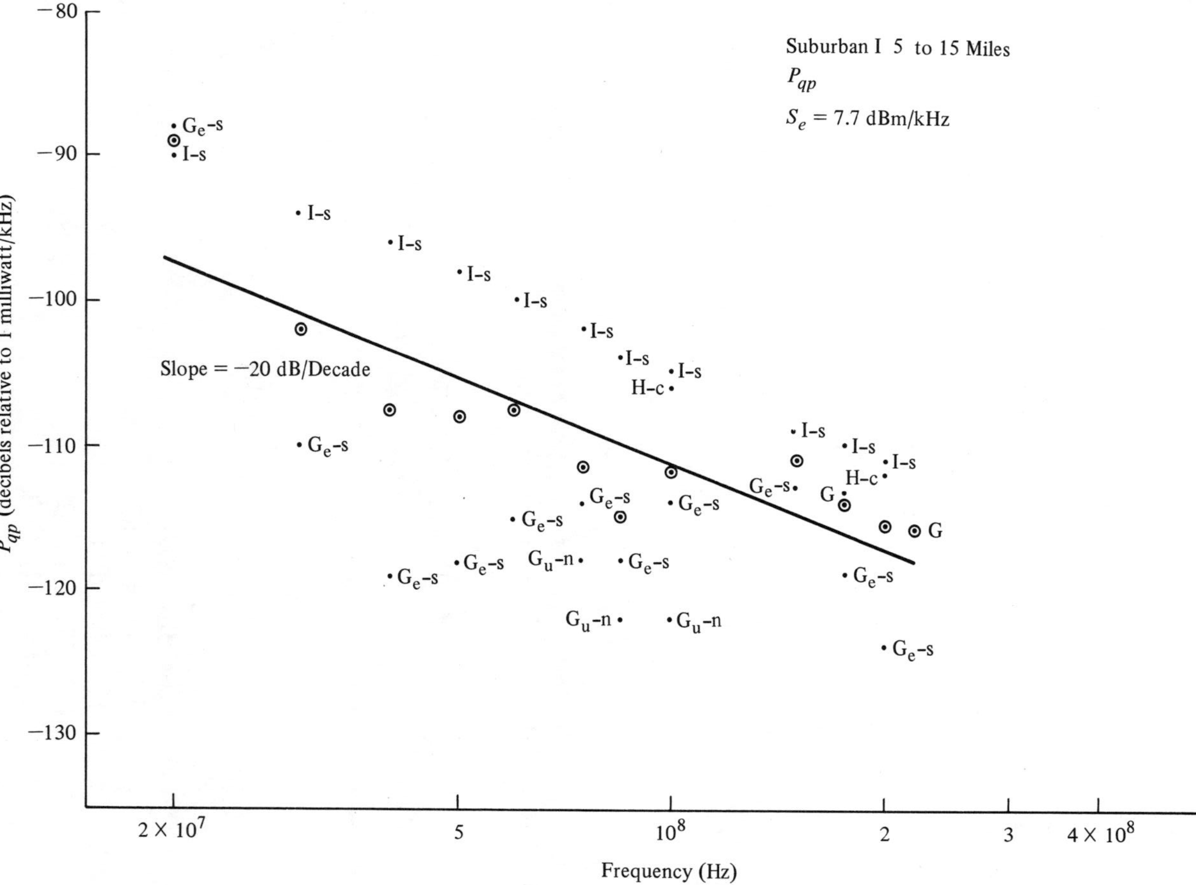

Fig. 6-26. P_{qp} for composite man-made noise within the suburban I zone, 5 to 15 miles.

6-22 through 6-26, the range variation may be represented by a polynomial of the form:[1]

$$P = \zeta_1 + \zeta_2 (d - 2.5) + \zeta_3 (d - 2.5)^2, \text{ in dBm/kHz} \qquad (6\text{-}5)$$

where $\zeta_i = a_i + b_i f$ and $i = 1, 2, 3$. In Equation 6-5, d is the distance in miles measured from the urban center, and f is the observation frequency. Each coefficient ζ_i is a linear function of frequency. The coefficients ζ_i are determined from three constaints placed upon Equation 6-5 at a specified frequency f:

1. at the point $d = 0$ the derivative of P with respect to d, $\left.\frac{\partial P}{\partial d}\right|_{d=0} = 0$, a condition that has been observed to occur for composite presentations of urban incidental-noise data.[1]
2. at $d = 2.5$ miles, Equation 6-5 at frequency f_m in MHz equals the value of the least-squares-regression line derived for the urban zone, i.e.,

$$P_a = -89.9 - 12.3 \log_{10} f_m \qquad (6\text{-}6)$$

$$P_{qp} = -53.2 - 25.4 \log_{10} f_m \qquad (6\text{-}7)$$

3. at $d = 10$ miles, Equation 6-5 at frequency f_m in MHz equals the value of the least-squares-regression line derived for the suburban I zone, i.e.,

$$P_a = -98.5 - 12.7 \log_{10} f_m \qquad (6\text{-}8)$$

$$P_{qp} = -71.3 - 20 \log_{10} f_m . \qquad (6\text{-}9)$$

Application of constraints 2 and 3 has the effect of requiring Equation 6-5 assume the least-squares-regression line values for the urban and suburban I zones at the respective zone centers. Typical values of ζ_i for P_a and P_{qp} at a frequency of 100 MHz are

P_a: $\zeta_1 = -114.5$ dBm/kHz
$\zeta_2 = -0.5$ dBm/kHz/mile
$\zeta_3 = -0.1$ dBm/kHz/mile^2
P_{qp}: $\zeta_1 = -104$ dBm/dHz
$\zeta_2 = -0.5$ dBm/kHz/mile
$\zeta_3 = -0.1$ dBm/kHz/mile^2.

A rendering of the range polynomial of Equation 6-5 as a volume of revolution centered over a metropolitan area is shown in Fig. 6-27.

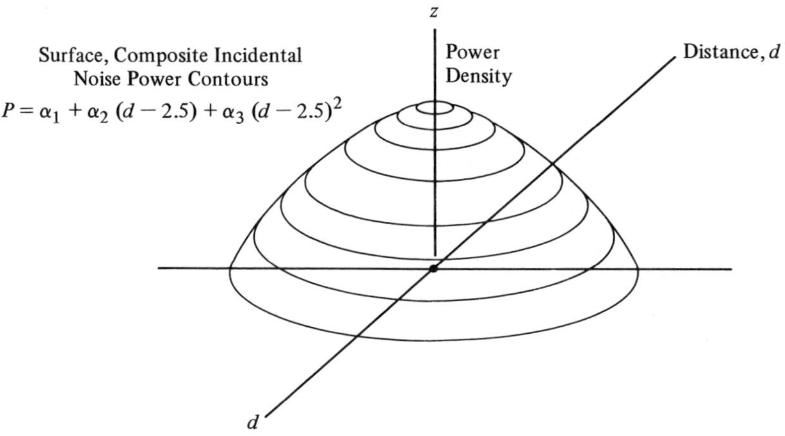

Fig. 6-27. Surface composite incidental-noise-power contours.

The power spectral density P is represented as increasing along the upward or positive z-axis. However, it must be recalled that Equation 6-5 is descriptive of measurements performed on or very near the surface and therefore, the positive z-axis of Fig. 6-27 represents the power variable and not elevation of the observation point.

Translation of the Circuit Reference Point for Mean Incidental-Noise Power. When employing mean values of composite incidental-noise power in studies of wireless systems maximum utility is achieved by referencing the measurements to the input terminals of the receiver, the same point in the receiving system at which receiver noise tem-peature and noise figure are conventionally referenced. In so doing, the contribution to receiving-system average noise power attributed to man-made external sources may be easily superimposed upon the receiver noise contribution. If the antenna-circuit losses, L_l, and mean line temperature, T_l, are known, the system-noise reference point may be translated to the output terminals of the observing antenna using:

$$T_a = T_l (L_l - 1) + T_x L_l \tag{6-10}$$

where

L_l = power input/power output
T_a = receiver-noise temperature referred to the antenna terminals
T_x = receiver-noise temperature referred to the input of the receiver.

T_a and T_x become identical when the antenna-line losses are zero, i.e., when $L_l = 1$. Use of Equation 6-10 provides an optional reference point in the receiving system—namely, at the antenna terminals, where the computed or measured noise temperature or power arising from external man-made noise sources may be stipulated.

INTERPRETATION OF THE FREQUENCY DEPENDENCE OF MEAN COMPOSITE INCIDENTAL RADIO-NOISE POWER

Theoretical Developments of the Noise-Power Function

An electrical circuit analog of the man-made noise-generation and radio-propagation process is shown diagramatically in Fig. 6-28. The model depicted consists of the following elements: a time-varying source, or sources, of incidental noise, $N(t)$, which are distributed within a metropolitan area and emit a spectral power density, $S(\omega)$. The radiated power density, $S(\omega)$, is coupled into the propagation medium from the sources by means of conducting attachments, such as automobile ignition leads, power-line conductors, and resistance welder electrodes, all of which act as antennas possessing a gain, G_T. In the propagation medium lying between the sources and observer, linear signal-processing occurs, which alters both the phase and amplitude of the noise-source signal in a manner that may be represented by a linear filter with a voltage-transfer function, $H(j\omega)$, or power-transfer function, $|H(j\omega)|^2$. At the observation point, the available power p_n appearing at the circuit terminals of the receiving antenna is coupled into the associated radio receiver. The observed noise power p_n referred to the output terminals of the receiving antenna is obtained from the circuit, analog representation of the pro-

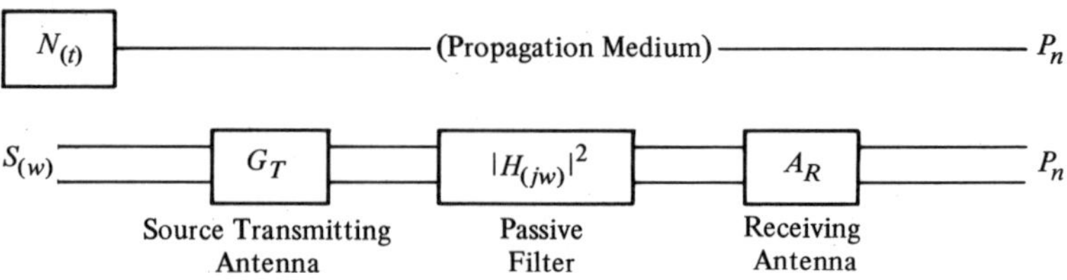

Fig. 6-28. Circuit analog of the man-made noise generation, propagation, and reception process.

cess by forming the product of the input spectral density of the noise sources $S(\omega)$ and the circuit quantities shown in Fig. 6-28:

$$p_n = S(\omega)\, G_T |H(j\omega)|^2\, A_R\,. \tag{6-11}$$

The product $G_T A_R$ appearing in Equation 6-11 may be expressed as:

$$G_T A_R = \frac{4\pi A_T}{\lambda^2} \cdot \frac{G_R \lambda^2}{4\pi} = A_T G_R = \text{constant}, \tag{6-12}$$

which is seen to be independent of frequency when the noise-source area, A_T, is fixed in size and the receiving-antenna gain, G_R, is constant. The latter condition is met for tunable monopoles, dipoles, and some broadband antennas. Therefore, Equation 6-11 for p_n may be rewritten as

$$p_n = S(\omega)|H(j\omega)|^2\, C, \tag{6-13}$$

where C designates a constant. Because the noise fields of many types of incidental sources are commonly superimposed at any metropolitan observation point and because great randomness in distribution of both the number and types of each category of sources exists within any metropolitan complex, it is both known as well as expected that strict sense-stationarity of the incidental-noise-field intensity does not occur. Nonetheless, in both limited locales and time periods, wide sense-stationarity of composite noise-source emissions is approached, if not exactly achieved. By definition, wide sense-stationarity of a random process is said to exist if the mean and variance of the random time-dependent functions are constant during a sampling period. To be of any general use in representing the noise process, wide sense-stationarity must occur over a time period of many minutes or hours and must predictably reoccur. For example, time-invariant mean noise power occurring for composite incidental-noise emissions during business-day rush hours in urban locations provides a practical encounter with the assumption of wide sense stationarity. In various measurements of composite metropolitan-area radio-noise emissions reported and examined for the purpose of preparing consolidated representations, time independence of mean noise power has been noted.[1, 6, 8, 10] Thus, the assumption of wide sense stationarity for composite noise emissions is consistent with existing evidence.

***Noise-Source Spectral-Power Density,* $S(\omega)$.** We begin with the assumption that at an observation point lying within a metropolitan area, the noise patterns received by a constant-gain antenna consist of a wavetrain of multiple independent pulses whose amplitudes are identically distributed and stationary in the wide sense. Therefore, the autocovariance function, $R[x_1(t), x_2(t)]$, of the noise may be written as

$$R(x_1, x_2) = \int_{-\infty}^{\infty} \int_{-\infty}^{\infty} x_1 \, x_2 \, p(x_1, t_1; x_2, t_2) \, dx_1 \, dx_2 \tag{6-14}$$

where x_1 and x_2 are the pulse amplitudes occurring at times t_1 and t_2, and $p(x_1, t_1; x_2, t_2)$ is the joint-probability-density function of pulse amplitude x_1 occurring at time t_1, followed by pulse amplitude x_2 occurring at time t_2. This wave pattern appears at the terminals of the observing antenna as a bipolar, amplitude-modulated radio-frequency signal, with $-x_0 \leqslant x(t) \leqslant x_0$, where limits $\pm x_0$ are finite. At any selected values of reference amplitude, $x_L \leqq x_0$, the joint probability, $p[x_1, t_1; x_2, t_2]$ can be written as the product of three terms,

$$p(x_1, t_1; x_2, t_2) = p(x_1) \cdot p(mt) \cdot p(x_2), \tag{6-15}$$

which exploits the assumed independence of the pulses comprising the observed noise waveform and the observation that amplitude $x_2 > x_L$ will follow the occurrence of $x_1 > x_L$ only if an even number of level x_L crossings of the pulse pattern, $x(t)$, have transpired in time $t_1 \leqq t < t_2$. The probability $p(m, t)$ that exactly m crossings of level x_L by signal $x(t)$ occurs in time t may be written as follows, if the signal level transitions are assumed to be Poisson-distributed with an average event or level crossing rate ν:

$$p(m, t) = \frac{(\nu|t|)^m}{m!} \, e^{-\nu|t|}. \tag{6-16}$$

The limiting value of the event count $p(m, t)_\infty$ is achieved by summing all possible even-numbered values of Equation 6-16:

$$p(m, t)_\infty = \sum_{m=0}^{M} p(m, t) \qquad \text{as } M \to \infty \quad \text{for } m \text{ even.} \tag{6-17}$$

Thus, by introducing Equation 6-16 into Equation 6-17 and using the series expansion for the exponential function, a convenient representation for $p(m, t)_\infty$ may be obtained as follows:

$$\begin{aligned} p(m, t)_\infty &= \sum_{m=0}^{M} \frac{(\nu|t|)^m}{m!} e^{-\nu|t|} \\ &= e^{-\nu|t|} \sum_{m=0}^{M} \frac{1}{2}\left\{\frac{(\nu|t|)^m}{m!} + \frac{(-\nu|t|)^m}{m!}\right\} \\ &= \frac{1}{2} e^{-\nu|t|} \{e^{+\nu|t|} + e^{-\nu|t|}\} \\ &= \frac{1}{2}\{1 + e^{-2\nu|t|}\} \quad \text{for } M \to \infty. \end{aligned} \tag{6-18}$$

Introducing Equation 6-18 into Equation 6-15 and the results into Equation 6-14 yields for the autocovariance function of the noise process:

$$R(x_1, x_2) = \frac{1}{2}(1 + e^{-2\nu|t|}) \int_{-\infty}^{\infty} \int_{-\infty}^{\infty} x_1 x_2 \, p(x_1) \, p(x_2) \, dx_1 \, dx_2. \tag{6-19}$$

By invoking the assumption of wide sense stationarity, the integrals appearing in the equation for $R(x_1, x_2)$ become independent of time during the epoch of observation. Therein, the signal amplitudes, x_1 and x_2, may be interpreted as threshold exceedence values $x > x_L$. Consequently, Equation 6-19 may be written in an abbreviated form with K representing a constant:

$$R(x_1, x_2) = \tfrac{1}{2} K(1 + e^{-2\nu|t|}). \tag{6-20}$$

It is the property of a wide-sense-stationary random process, established by the Wiener-Khintchein theorem, that the Fourier transform of the autocovariance function is equal to the spectral power density, $S(\omega)$; thus,

$$\begin{aligned} S(\omega) &= \mathfrak{F}R(x_1, x_2) = K \int_{-\infty}^{\infty} (1 + e^{-2\nu|t|}) \, e^{-j\omega t} \, dt \\ &= K\left[\delta(\omega) + \frac{4\nu}{4\nu^2 + \omega^2}\right]. \end{aligned} \tag{6-21}$$

The delta function appearing in the above expression for $S(\omega)$ represents a zero-frequency or direct-current contribution to the power observed at the antenna terminals.

The second term of Equation 6-21 contains two variables, the average radio-frequency waveform level crossing rate, ν, and the angular observation frequency, ω. The relative magnitudes of ν and ω will determine the frequency variation of $S(\omega)$ in any selected portion of the radio spectrum. Three ranges of values may exist for the ratio ν/ω, namely:

1. $\nu << \omega$. When this condition occurs, the limiting value of the second term in $S(\omega)$ becomes proportional to ω^{-2}, and the rate of decline of $S(\omega)$ with increasing observation frequency is quite rapid. The physical interpretation of this lower limit upon the range of ν may be stated as follows: in the span of time during which an observation of the mean noise power is performed, the average number of signal-amplitude crossings of level x_L per second is much smaller than the angular observation frequency.
2. $\nu \simeq \omega$. For the condition of approximate equality between the average level crossing rate and the angular frequency, the spectral power density displays an inverse dependence upon frequency, i.e., $S(\omega) \simeq \omega^{-1}$.
3. $\nu >> \omega$. At the upper limit of the ratio, ν/ω, the average level crossing rate appreciably exceeds ω, resulting in $S(\omega)$ = constant, independent of the observation frequency and a function only of the average number of level crossings per second. Clearly, for very high radio frequencies, this condition can exist only in man-made noise environments comprised of very large numbers of sources collectively radiating very short duration pulses at high emission rates.

Transfer Function of the Propagation Medium, $H(j\omega)$. The transfer function, $H(j\omega)$, of the noise-propagation medium can be computed for the simplest case, free space, and then by example extended to irregular surface-terrain propagation. In free-space propagation, the transmitted power per unit area in a plane wave emitted by an omnidirectional source is $p_T/(4\pi R^2)$. For a receiving antenna with aperture A_R and gain $G_R = 4\pi A_R/(\lambda^2)$, the received power, p_R, is given by

$$p_R = \frac{p_T A_R}{4\pi R^2} = \frac{p_T G_R \lambda^2}{(4\pi R)^2}$$

where R is the source-to-receiver distance. The transfer function, $|H(j\omega)|^2$, for free space is the ratio of p_R/p_T, i.e.,

$$|H(j\omega)|^2 = p_R/p_T = \frac{G_R \lambda^2}{(4\pi R)^2} = \frac{G_R c^2}{(2R\omega)^2}. \tag{6-22}$$

Therefore, for free-space transmission, one would expect the transfer function to vary inversely with the product of the squares of the frequency and the distance from the source.

A free-space dependence of $H(j\omega)$ will apply to observation points well elevated above a surface-noise-source distribution, but not in the general metropolitan-area situation. In determining the modification to Equation 6-22 needed to adequately represent the radio-propagation environment encountered during metropolitan-area noise investigations, it is essential to take note of three facts:

1. radiated noise arising from incidental sources is coupled to the surrounding space by means of short conducting members acting as antennas. In all cases, these effective antennas are short and located near the surface.
2. the propagation path intervening between the noise sources and any observation point on the surface of a metropolitan area is interrupted by the presence of surface irregularities whose origin may be either natural processes, as in the case of hills, or human efforts, as, for example, high-rise structures and towers. Either type of obstacle will scatter and may, as well, partially absorb the radiated energy:
3. most noise-measuring instrumentation and many types of radio receiving equipment employed in metropolitan areas use short, tuned antennas whose dimensions are comparable or smaller than the free-space wavelength of the radio observation frequency. As a consequence, it becomes admissible to exploit the results of propagation studies performed over irregular terrain using short, low-height antennas.[11-13] These studies, initially conducted in the HF and VHF bands and later extended to the UHF and SHF bands,[14] were used to determine the total path loss as a function of antenna height, an-

tenna polarization, path length, terrain roughness, and frequency. A large amount of data has become available, which permits determination of the basic transmission loss, l_b, under conditions representative of a radio-propagation environment similar to that encountered in metropolitan-area radio-noise investigations.

The basic transmission loss, l_b, is defined as the radio path loss occurring between two perfect omnidirectional antennas. In terms of the input power, p_T, and the received power, p_R, $l_b = p_T/p_R$. From Equation 6-13, the propagation space transfer function may be written as

$$|H(j\omega)|^{-2} = C\,S(\omega)/p_n.$$

The noise-power density transmitted from an area distributed source, $S(\omega)$, is identically equal to p_T, per unit bandwidth, appearing in the defining equation for l_b, and thus the power transfer function is observed to be inversely proportional to the basic transmission loss for the general case of radio propagation over irregular terrain.

The initial investigation of path-propagation loss for low-height antennas in irregular terrain were conducted at frequencies of 20, 50, and 100 MHz within two widely separated locales, one in Colorado and the other in Ohio.[12,13] The measurements have permitted the extraction of l_b as a function of various experimental and radio path conditions. Table 6-3[13] summarizes median values of L_b in decibels at the three observation frequencies. Examination of the table reveals

Table 6-3. Measured, median basic transmission loss, L_b, in decibels for two irregular terrain conditions. Antenna heights in meters and polarization used are indicated by 3V, 3H, 6H, etc. (After Barsis et al., 1964)

Frequency (MHz)	Location	Range (km)	Attenuation L_b for Various Antenna Heights (m) and Polarizations					
			3V	3H	6V	6H	9V	9H
20	Ohio	10	135.4	-	-	-	-	-
		20	137.2	-	-	-	-	-
50	Ohio	10	137.8	-	-	-	-	-
		20	143.4	-	-	-	-	-
100	Ohio	10	143.6	143.8	138.1	138.1	137.2	139.4
		20	144.5	144.9	143.1	142.5	141.2	141.5
100	Colorado	10	127.5	136.0	-	131.0	-	128.0
		20	140.4	145.4	-	138.4	-	137.5

that L_b increases with increasing frequency. In terms of $|H(j\omega)|^2$, increasing frequency causes the transfer function to decrease. Also evident from Table 6-3 is a dependence of L_b upon antenna height; however, notice that the frequency variation of L_b is essentially independent of antenna height for the height range of 3 to 9 m.

In the test areas of Ohio and Colorado from which the median basic transmission-loss data of Table 6-3 were obtained, the propagation paths traversed terrain whose interdecile elevation variations, Δh, lay within the range of 50 to 100 m. The interdecile terrain elevation variation, Δh, is defined as the difference in meters between the 90 and 10 percentile values of the terrain profile cumulative distribution, which is formed from sampling the elevation profile along the interterminal great-circle path at equally spaced intervals. The range of variation in Δh, which may be encountered in natural terrain, is 5 to 700 m, corresponding, respectively, to flat marshlands and the Rocky Mountains. Within a typical urban area composed of both natural and man-made obstacles, a representative interval for Δh is 25 to 100 m.

The median values of L_b from Table 6-3 may be used to compute the frequency decrement δ_F, which is defined as

$$\delta_F = 10 \log_{10} \frac{|H(j\omega_2)|^2}{|H(j\omega_1)|^2} = 10 \log_{10} \left(\frac{l_{b\,1}}{l_{b\,2}} \right). \qquad (6\text{-}23)$$

The notations $l_{b\,1}$ and l_{b2} designate the basic transmission loss observed at frequencies f_1 and f_2, respectively. In computing δ_F, values of L_b for each antenna-height group (3, 6, and 9 m) and for each polarization, vertical and horizontal, were processed separately, and the resulting values of δ_F were averaged for the ranges 10 and 20 km. Table 6-4 provides the mean values of the frequency decrement that arises from these reductions. Variation of the frequency decrement with frequency interval is observed to occur, possibly as a consequence of the location variability existing in the median values of L_b. The grand average of δ_F for all frequency and range intervals is -11.1 dB per decade.

Taking the antilog of Equation 6-23 and solving for the transfer function rato using the set mean value of δ_F from Table 6-3, one obtains:

$$\frac{|H(j\omega_2)|^2}{|H(j\omega_1)|^2} = \text{antilog}_{10}\ (-11.1/10). \qquad (6\text{-}24)$$

Table 6-4. Frequency decrement δ_f, in decibels per decade change in frequency, for irregular terrain at two values of radio-propagation range, 10 km and 20 km.

Frequency Interval (MHz)	Frequency Decrement δ_F dB/decade for Ranges:	
	10 km	20 km
20–50	-6	-15.5
50–100	-19.3	-3.7
20–100	-11.7	-10.4
Average	-12.3	-9.9

Average for all entries: 11.1 dB/decade
Average for 50-to-100-MHz interval: 11.3 dB/decade

After setting $|H(j\omega_2)|^2 = K\omega^\alpha$ in Equation 6-24 with $\omega_2 = 10\,\omega_1$, the exponent α is resolvable from the relationship:

$$10^\alpha = \text{antilog}_{10}\,(-1.11) = 10^{-1.11}$$

and results in $\alpha = -1.11$. Thus, the frequency dependence of the power-transfer function has been found to be representable by the function:

$$|H(j\omega)|^2 = K\omega^{-1.11}. \tag{6-25}$$

Comparison of the Computed and Observed Frequency Variation of Composite Incidental Noise

The frequency-dependent factors contained within the expression for composite metropolitan-area surface-noise power, p_n, of Equation 6-13 may now be replaced by their parametric representations, Equations 6-21 and 6-25, to yield

$$p_n = \overline{C}\left[\delta(\omega) + \frac{4\nu}{4\nu^2 + \omega^2}\right]\omega^{-1.11}. \tag{6-26}$$

Within each partitioning interval established by the limiting values of the ratio ν/ω, p_n assumes a distinctive dependence upon the observation angular frequency as follows:

A. $\nu << \omega \qquad p_n = K_1\,\omega^{-3.11}$ (6-27)
B. $\nu \simeq \omega \qquad p_n = K_2\,\omega^{-2.11}$ (6-28)
C. $\nu >> \omega \qquad p_n = K_3\,\omega^{-1.11}$. (6-29)

In these equations, $\overline{C}$, K_1, K_2, and K_3 represent frequency-independent factors. If the three partitioning intervals are respectively assigned to the radio-frequency spectrum as follows:

A. less than 20 MHz
B. 20 MHz to 200 MHz
C. 200 MHz to 1 GHz

then associations with composite noise power frequency groups I through IV from Figs. 6-12 through 6-18 may be performed in the following manner:

A. IA, IB, IIA, IIB
B. IIC, III
C. IVA.

The average frequency decrements for the experimental data associated with each partitioning interval may be computed from Figs. 6-12 through 6-18 and are found to be, in dB/decade:

A. $\delta_F = -27.6$
B. $\delta_F = -18.6$
C. $\delta_F = -13.2.$ (6-30)

The theoretically derived frequency variations of composite, surface metropolitan noise power, p_n, obtained from Equations 6-27 to 6-29 are shown by dashed lines in Fig. 6-29. In the same figure, experimentally determined frequency dependences for the three partitioning intervals given by Equation 6-30 are drawn as solid lines. In constructing Fig. 6-29, the p_n functions, Equations 6-27 to 6-29 have been normalized to unity within each interval at frequencies equal to 20 MHz, 200 MHz, and 1 GHz, respectively, for the intervals A, B, and C. The experimentally derived, composite metropolitan-area incidental-noise power from Figs. 6-12 through 6-18, grouped into partitioning intervals as noted previously, was constrained to pass through the three normalization frequencies with average slopes given by Equation 6-30.

Agreement between the average frequency decrements observed for composite metropolitan-noise data partitioned by values of the ratio ν/ω and the theoretically predicted frequency variation is seen to be quite close, up to the limit of adequate experimental data, pres-

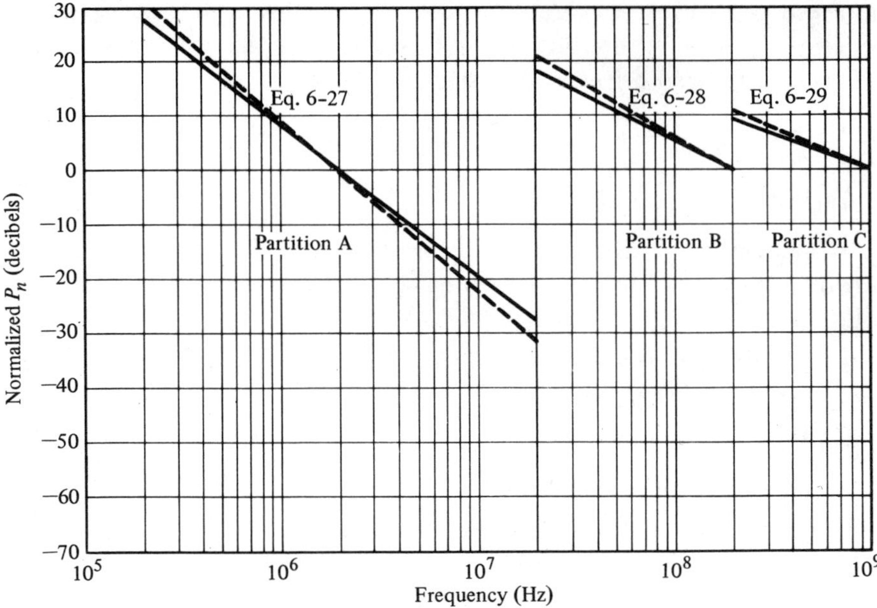

Fig. 6-29. Normalized values of P_n in decibels versus frequency for three partitioning intervals. Theory (- - -), experimental data (—).

ently approximately 1 GHz. There exists one set of composite noise data, Fig. 6-19, for the frequency interval 450 to 950 MHz, which displays a very large negative decrement. The cause of the large negative slope is attributed to the existence of a UHF-band resonance in the radiation spectrum of automotive ignition noise occurring in the interval 300 to 600 MHz. Noise levels observed at frequencies above this resonance will manifest larger negative slopes than data obtained at or below the median resonance frequency.

Above 1 GHz, composite urban noise levels diminish rapidly with increasing frequency as a direct consequence of a decreasing population of incidental and restricted radiation sources possessing appreciable spectral energy in the upper UHF and SHF bands.

TEMPORAL VARIATIONS IN COMPOSITE INCIDENTAL-NOISE POWER

The composite radio emission produced in a metropolitan area by the multiple types and numbers of incidental-noise sources varies diurnally with changes in business and industrial activities. Automotive traffic-noise-level variations correlate with alterations in traffic volume. The totality of radiated power from industrially and commercially employed incidental and restricted radiation devices is cyclically dependent upon the business day, while with electric power generation and transmission facilities, noise emissions are dependent upon the temporal changes in power demand. Two characteristics are evident in the weekly variations of composite metropolitan-area noise as a consequence of the three aforementioned time-dependent processes:

1. The onset of rush-hour activity at the beginning and end of the business day raises the business-day composite levels above the mean daytime average, primarily, but not necessarily exclusively, as a consequence of increases in automotive ignition noise radiation. Some portion of the beginning and end of business-day noise peaks arise from equipment startup and shutdown transients.

2. Nonworkday, daytime, mean composite noise-power levels do not achieve the business-day mean value because of reduced usage occurring for electrically operated industrial and commercial equipment and the attendant drop in power generation and transmission facilities load.

The interaction of these noise-contributing factors is displayed in the temporal variations of F_a presented in Fig. 6-30.[15] In performing

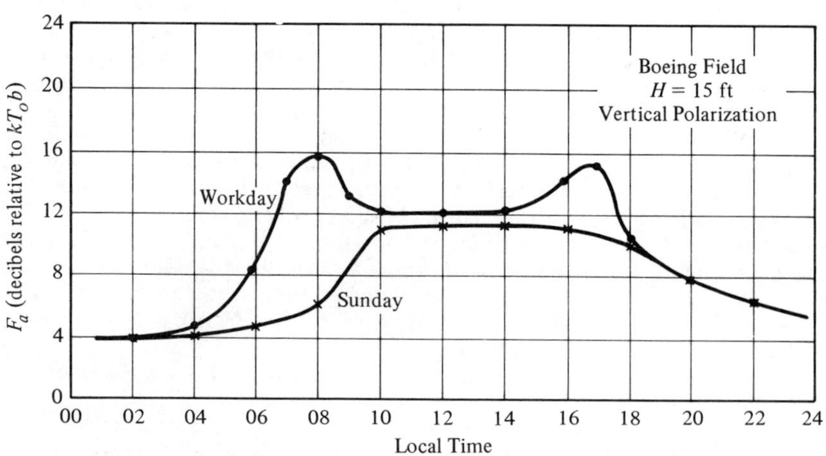

Fig. 6-30. Temporal variations of composite incidental noise power, Seattle area. (After Buehler et al., 1968)

the measurements presented in Fig. 6-30, the surface observation point was positioned several hundred meters from the nearest man-made incidental-noise source in an effort to record temporal changes in only the composite man-made incidental ambient noise level. The business-day maxima occurring in Fig. 6-30 at 8:00 A.M. and 4:30 P.M. contain some contribution from traffic ignition radiation. The observed excursions in the diurnal noise power for the urban Seattle location, Fig. 6-30, are 12 dB for a business day and 7 dB for Sunday. Two characteristics of a Sunday or holiday diurnal-noise-power profile are the absence of rush-hour peaks and a delayed rise to the midday mean value.

References

1. Skomal, E. N. Distribution and frequency dependence of unintentionally generated man-made VHF/UHF noise in metropolitan areas. *IEEE Trans. Electromagnetic Compatibility* **EMC-7** (3): 263–278 (1965).
2. Skomal, E. N. Analysis of airborne VHF/UHF incidental noise over metropolitan areas. *IEEE Trans. Electromagnetic Compatibility* **EMC-11** (2): 76–83 (1969).
3. Skomal, E. N., and Reed, K. C. Computer study of a diversity augmented, vehicular, radio communication system in a man-made noise environment. *IEEE Trans. Electromagnetic Compatibility* **EMC-8** (2): 67–73 (1966).
4. Spaulding, A. D., and Disney, R. T. Man-Made Radio Noise, Part I, Estimates for Business, Residential, and Rural Areas. U.S. Dept of Commerce, O.T. Report 74–38, June 1974.
5. CCIR. World Distribution and Characteristics of Atmospheric Radio Noise. Report 322, International Telecommunication Union, Geneva, 1964.
6. Skomal, E. N. Distribution and frequency dependence of incidental man-made HF/VHF noise in metropolitan areas. *IEEE Trans. Electromagnetic Compatibility* **EMC-11** (2): 66–75 (1969).
7. Skomal, E. N. Recent extensions of composite, incidental man-made radio noise data and their relevance to the hypothesis of the noise envelope statistics transformation. *1971 IEEE International Convention Electromagnetic Compatibility Symposium Record* 222–234 (July 13–15, 1971).
8. Skomal, E. N. An analysis of metropolitan incidental radio noise data. *IEEE Trans. Electromagnetic Compatibility*, **EMC-15** (2): 45–57 (1973).
9. Geselowitz, D. B. Response of ideal radio noise meter to continuous sine wave, recurrent impulses, and random noise. *IRE Trans. Radio Frequency Interference* **RFI-3**: 2–10 (1961).
10. Skomal, E. N. Man-made noise, Part II, experimental data. *Frequency* 12–17 (March 1967).
11. Barsis, A. P., and Rice, P. L. Predicition and Measurement of VHF Field Strength Over Irregular Terrain Using Low Antenna Heights. National Bureau of Standards Report 7962, October 14, 1963.

12. Barsis, A. P., McQuate, P. L., and Miles, M. J. Initial Results of VHF Field Strength Measurements Over Irregular Terrain Using Low Antenna Heights. National Bureau of Standards Report 8269, July 7, 1964.
13. Barsis, A. P., and Miles, M. J. Cumulative Distributions of VHF Field Strength Over Irregular Terrain Using Low Antenna Heights. National Bureau of Standards Report 8891, October 1, 1965.
14. McQuate, P. L., Harman, J. M., Johnson, M. E., and Barsis, A. P. Tabulations of Propagation Data Over Irregular Terrain in the 230- to 9200-MHz Frequency Range. ESSA Technical Report ERL 65-ITS 58-2, December 1968.
15. Buehler, W. E., King. C. H., and Lunden, C. D. VHF city noise. *1968 IEEE Electromagnetic Compatibility Symposium Record* 113–118 (1968).

7
Composite, Metropolitan-Area, Elevated and Airborne Incidental Noise

The surface incidental-radio-noise sources, which in a metropolitan area are both numerous and pervasive, generate within a large portion of the radio spectrum an ambient interference environment that envelops a populated region and extends from the urban core far into the rural surroundings. No surface enclave in an urban region is known to be free of incidental noise, even though pronounced minima occur and large temporal variations exist. Collective contributions from the many independently radiating sources create a composite noise-field intensity, which achieves a maximum value at or near the center of an urban area decreasing as the separation from the urban core enlarges.

Although the various surface and near-surface noise sources that contribute to the composite metropolitan-area noise field radiate interference in uniquely directional patterns, very few selectively favor the horizontal plane. Thus, much of the radiated energy is propagated above the surface either on line-of-sight paths or via surface and near-surface reflections. In this manner, an ambient noise envelope is created about each metropolitan complex. The presence of an incidental noise envelope enclosing an urbanized geographic area becomes a consideration in the design, use, and analysis of data derived from wireless receivers, which are either airborne or spaceborne.

It is possible to describe by means of a quadratic function the manner in which composite metropolitan-area surface noise power decreases with frequency and separation distance from the urban center (Chapter 6). With this formulation, certain general properties of metropolitan-area incidental noise can be identified and analytically represented. However, idiosyncratic variations of man-made noise intensity remain unrepresented by the general formulation. The latter are manifest by unique urban-noise signatures observable from both surface and airborne observation points. Noise-signature distinctions are evident as emission patterns associated with major trafficways, industrial complexes, and business centers. The noise-signature features, which can distinguish one metropolitan complex from another, are observable at various points along an elevated flight path intersecting an area. Typically, the smaller the illuminated surface area subtended by the observing antenna beam, the greater is the detectable noise signature detail. The localized variations in incidental noise power unique to a metropolitan area are superimposed upon a general noise background that varies in a predictable manner with altitude and distance from the urban center. This general noise background may be predicted for a typical urban area by employing the relationships that were used to describe the surface, composite, incidental noise power as a function of frequency and surface distance in Chapter 6.

Descriptions of the various characteristics of elevated incidental noise and the procedures available to compute the general noise background observable from either space or airborne platforms form the content of the present chapter.

EXPERIMENTAL AIRBORNE INCIDENTAL-NOISE DATA

Systematic investigations of the radio-spectrum emissions existing above the surface, arising from the urban area beneath, have been pursued exclusively with the use of aircraft. Specially conditioned aircraft engines and electrical systems have been found to be necessary to eliminate ignition and auxiliary electrical equipment interference during continuous flight measurements of metropolitan incidental noise. Observation of the noise levels existing above an urban complex requires the use of an antenna that provides lower hemispherical coverage when traverses of high-intensity surface zones are performed. Lower hemispherical coverage patterns, either linearly or circularly polarized, have also been employed in noise studies using off-center urban traverses when they have replicated the beam directivities and orientations of aircraft antennas used in navigation, position fixing, or landing approach systems. However, more commonly forward pointing or occasionally laterally directed patterns are used in airborne off-center and urban-approach noise measurements.

No evidence has been recorded to indicate that airborne antenna noise temperature is dependent upon incidental noise polarization. Viewed from the air, the radiation field arising from an area distribution of independent noise sources should be unpolarized. Downward-directed, hemispherical coverage, circularly polarized antennas have not yielded significantly greater antenna-noise temperature than linearly polarized antennas. This may be because antennas admit a significant fraction of the detected surface noise radiations through their side lobes, which generally do not preserve the polarization selectivity of the main beam.

Measurement of antenna-noise temperature, T_a, may be transformed to F_a by the operation,

$$F_a = 10 \log (T_a/290), \tag{7-1}$$

which is convenient for comparing airborne data with surface-noise-power measurements. It would also be desirable to transform T_a into surface-noise-brightness temperature, T_b, to facilitate comparisons with radio-noise emission data from extraterrestrial and galactic sources. When the surface distribution of noise sources at a constant temperature T_b covers an area equal to or greater than the beam intercept area, the surface-brightness temperature is equal to the antenna temperature. However, if the surface distribution of noise sources is nonuniform or if it is smaller than the beam-intercept area, conversion of T_a or F_a to T_b requires a knowledge of both the surface area subtended by the antenna beam, Ω, and the noise-emitting area, Ω_s, at temperature T_b, i.e.,

$$T_b = T_a(\Omega_s/\Omega), \tag{7-2}$$

neither of which may be available, but for different reasons. Consequently, surface-brightness temperature is rarely obtainable. The typically encountered difficulties preventing transformation of T_a into T_b arise from

1. distortions in an airborne antenna pattern created by the presence of aircraft structures
2. the attendant complexities of executing gain and pattern measurements on aircraft antennas providing horizontal plane and lower hemispherical coverage
3. uncertainties in describing the geographical perimeter and thus the radiating area of a metropolitan distribution of surface noise sources.

Airborne Incidental-Noise Profiles

Measurements of T_a or F_a for incidental-radio-noise power as a function of aircraft location have been performed at altitudes between 2000[1,2] and 40,000[3] ft over several major North American cities and at intermediate altitudes above several English cities.[4] Incorporated in these and other airborne-noise studies has been the frequency interval of 49 MHz to 1 GHz[5,6] and attempted extensions to 3 GHz[6], which failed to yield incidental-noise levels exceeding the internal noise of the instrumentation for most of the flight profiles.

Forward-directed antennas such as those employed in extended-range airborne communication systems are able to detect surface-noise emissions from urban complexes at a minimum range D_m whenever an aircraft rises above the local optical horizon of a metropolitan area. This event occurs at a range of $D_m = 1.17\, h^{1/2}$, in which h is the aircraft altitude in feet and D_m is in nautical miles. When flying at an altitude of 40,000 ft, a plane equipped with a forward-directed, high-gain antenna and low-noise receiver tuned to a frequency below 1 GHz will begin to detect incidental noise emissions from a large metropolitan complex at a range of approximately 214 nautical miles. This effect is evident in Fig. 7-1, which plots for 130 MHz the observed change in antenna temperature, expressed as F_a using Equation 7-1, for an aircraft approaching the Seattle metro-

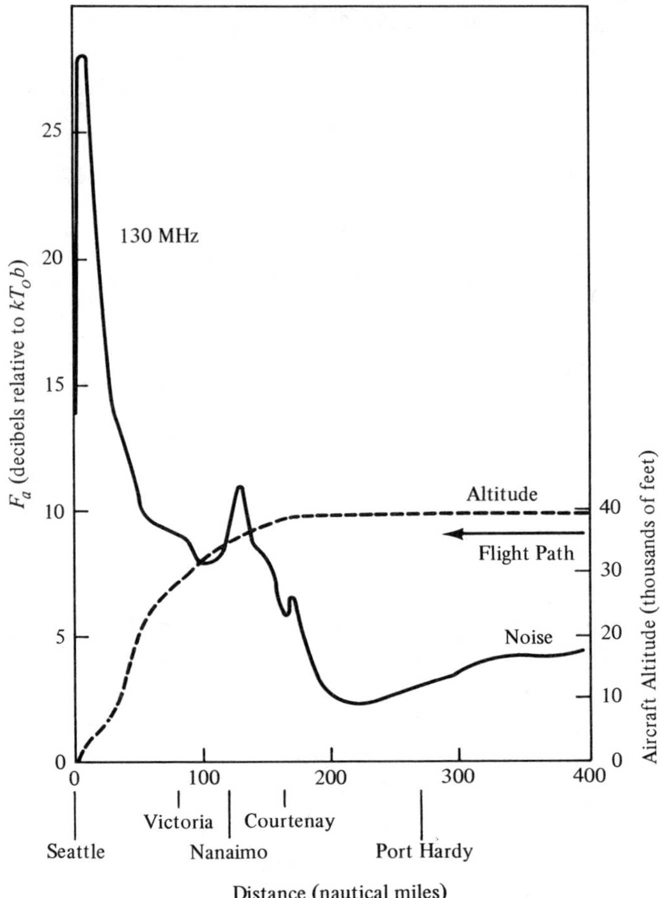

Fig. 7-1. Metropolitan-area incidental-noise profile observed by a forward-directed antenna mounted on an aircraft approaching Seattle, Washington. Frequency: 130 MHz. (After King and Lunden, 1964)

politan area from the northwest.[3] The measurements of Fig. 7-1 were obtained using a forward-directed, linearly polarized Yagi antenna possessing a gain of approximately 9 dB relative to an isotropic radiator. The first indication that the Seattle metropolitan complex is being approached appears as a decrease in F_a at a range of from 220 to 280 nautical miles. The decrease in noise power is caused by destructive interference (multipath fading) between the noise signals reaching the antenna along two paths:

1. through the atmosphere but very close to the surface
2. via grazing incidence reflection at the surface where a 180° phase reversal occurs at the reflection point in the vertically polarized component of the noise signal.

Notice the observed noise power during approach begins to increase in Fig. 7-1 at a range of about 210 nautical miles as the surface-reflected, vertically polarized signal decreases and its phase relative to the direct-path component alters from 180°. Thereafter, the observed signal power rises rapidly as the elevation angle of the plane above the urban horizon increases, reaching a maximum at a distance determined by (1) the aircraft antenna pattern, (2) the dimensions of the urban area, (3) the descent flight profile, and (4) the location of the airfield. The precipitous reduction in F_a commencing at a distance of less than 5 nautical miles from touchdown results, primarily, from the diminished horizon of the aircraft antenna. Superimposed upon the ascending portion of the noise-temperature profile occurring during the urban center approach are subsidiary maxima; contributed by intense, localized surface-noise sources possibly generated in several smaller cities of the Seattle metropolitan area lying on or near the flight path.

The predominant interest motivating airborne-noise studies over and in the vicinity of urban areas has been the expectation that they represent a more rapid and less costly method of determining surface incidental-noise-power levels than do surface studies. To perform aerial mapping of surface, composite, incidental noise power, downward-directed antennas are required. When utilizing antennas whose main beam is pointed vertically downward, the mapping aircraft must follow a flight profile, which passes above the surface-noise source distribution of interest. If a low-directivity antenna is used—that is, one with a large solid angle coverage pattern—and if the flight altitude is large, the incidental-noise-power profile will reveal only slight changes in magnitude with aircraft position along the flight path. Few, if any, of the existing localized surface noise sources will be individually resolved. A noise profile plotted as F_a for Seattle and obtained with a low-directivity antenna is shown in Fig. 7-2(a).[5] The flight path intersected the center of Seattle at an altitude of 5000 ft, on an east-west traverse. A trailing-wire vertically polarized dipole antenna was used for these measurements. The combination

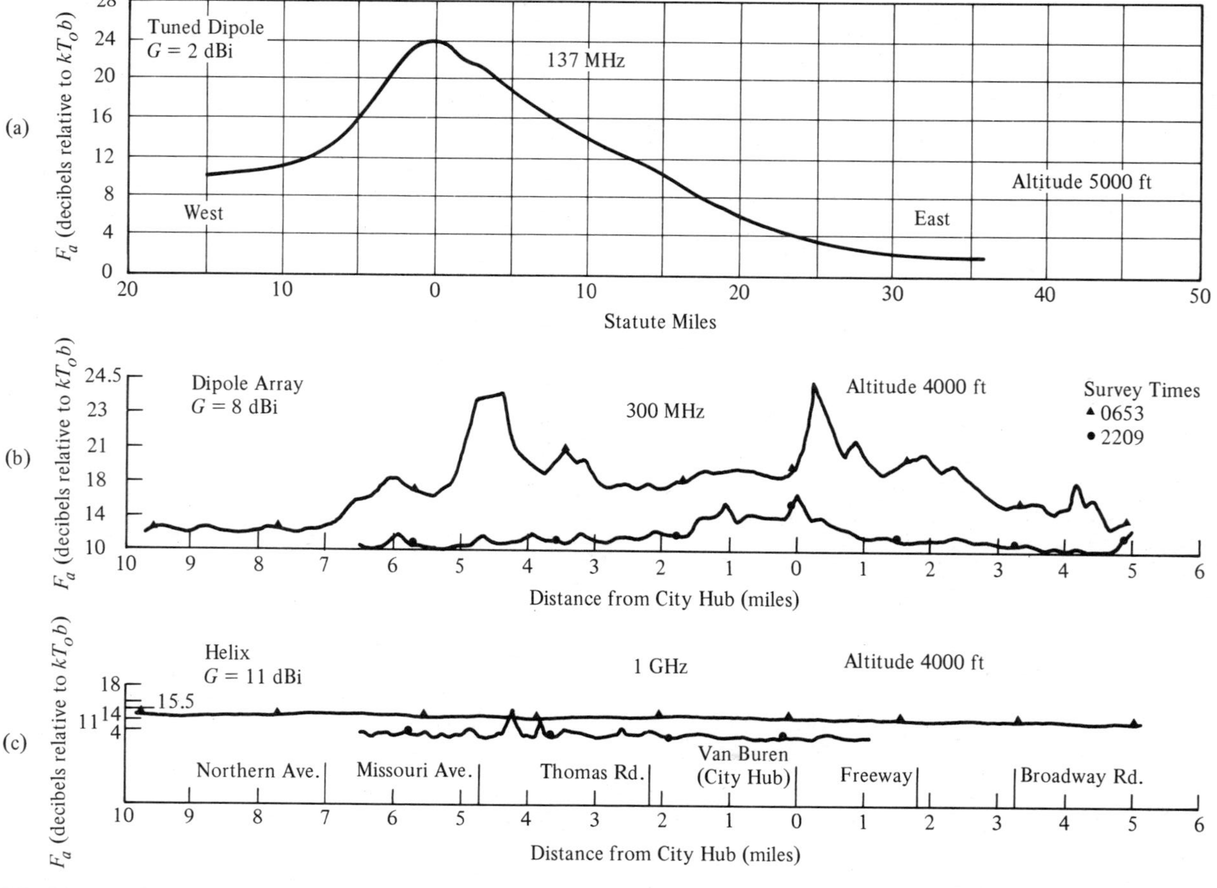

Fig. 7-2. Metropolitan-area airborne-noise profiles observed by downward-directed antennas. **a**, Seattle center traverse at 5000 ft for 137 MHz during a business day. Tuned dipole antenna. (After Buehler et al., 1968.) **b** and **c**, Phoenix center traverse at 4000 ft for 300 MHz and 1 GHz. Medium-directivity antennas used for business-day noise profiles at two periods 6:53 A.M. and 10:09 P.M. (After Anzic and May, 1971.)

of low-antenna directivity and moderate altitude prevented resolution of most detailed surface-noise variations at 137 MHz. Deviation of the noise profile from east-west symmetry in the western portion is attributable to noise contributions arising from the populated areas on the western shore of Puget Sound.

When antenna resolution is increased without substantially altering the flight altitude, a fine structure existing in the surface-noise-power profile becomes apparent. Individual maxima and mimima in surface emission level per unit area are then observed. The profile of Fig. 7-2(b) exemplifies this point.[6] Some of the evident increase in fine structure arises from the higher observation frequency, 300 MHz; however, the principal portion derives from the narrower beam. For these measurements, a circularly polarized dipole array antenna with a gain of 8 dB above isotropic was used.

Two periods of a business day are represented by the pair of incidental-noise profiles plotted with ● and ▲ in Figs. 7-2(b) and (c). Advent of business activities at 6:53 A.M. stimulated an increase in the airborne noise levels at 300 MHz, which over the central Phoenix business district exceeded the late evening intensity at 10:00 P.M. by 10 dB. Lesser diurnal changes in magnitude appeared above those surface locations possessing a reduced source density. All noise powers observed at 300 MHz were appreciably larger than the effective receiving system noise figure of 3.5 dB. The nighttime measurements of antenna-noise temperature at 1 GHz, presented as F_a in the lower trace of Fig. 7-2(c), only barely exceed the internal noise of the 2.5-dB noise figure receiver for most of the flight path. The vertical scale compression in Fig. 7-2(c) tends to conceal the diurnal variation in noise level, the maxima of which are comparable to those observed at the lower frequency in Fig. 7-2(b).

Two additional profiles of urban-area airborne noise are shown in Fig. 7-3 for altitudes of 2000 and 18,000 ft.[1] Observed antenna noise temperatures are presented in °K, which may be converted to F_a by means of Equation 7-1, and yields 19.5 and 22.5 dB, respectively, for Figs. 7-3a and b. The applicable surface-distance scales are obtainable from the flighttime-versus-distance inset on each figure. Some smoothing was affected in processing the Philadelphia data suppressing a portion of the detail; even so, the decrease in surface-noise-source resolution resulting from a ninefold increase in altitude

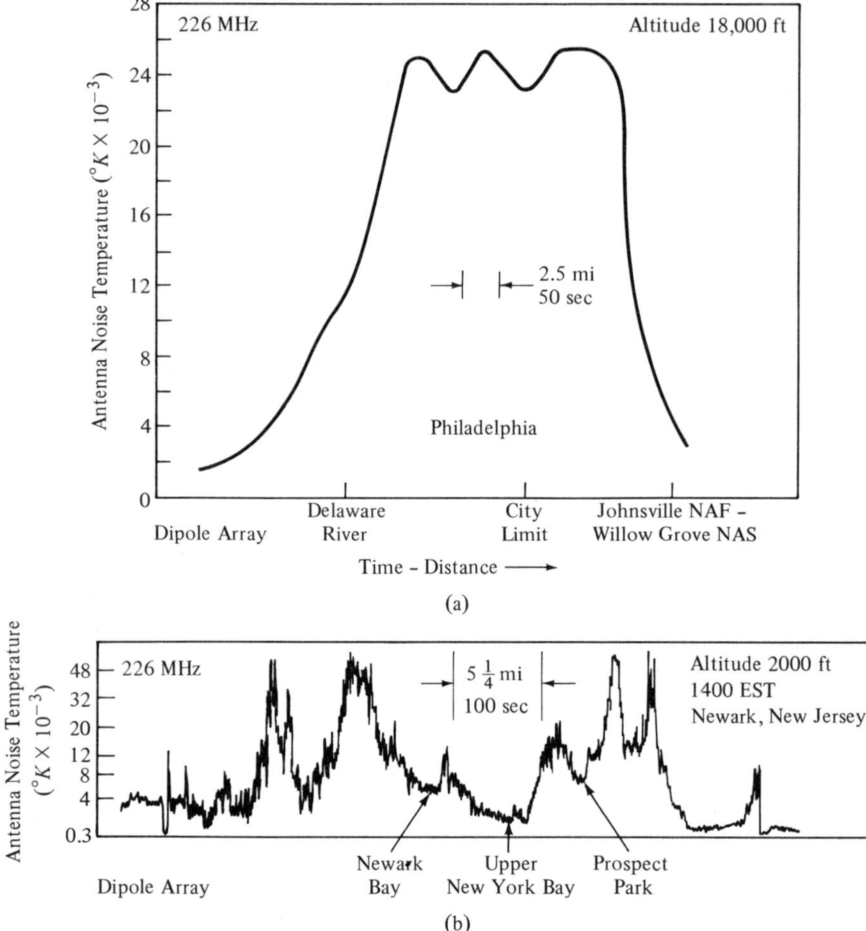

Fig. 7-3. Metropolitan-area airborne noise profiles for Philadelphia and Newark observed at a frequency of 226 MHz and altitudes of 18,000 and 2,000 ft. (After Ploussios, 1968)

is striking. Identical antennas and receiving equipment were used for both series of measurements, which consisted of a downward-directed, linearly polarized dipole and a three-channel receiver with a 300°K excess noise temperature (noise figure of 3.1 dB). The maximum noise factors F_a of Figs. 7-2(b) and 7-3(b), which were recorded with comparable antennas and receivers, are 24 and 22.5

dB, respectively, differing only in signature detail, which is attributable primarily to differences in flight altitude.

The geographical areas of major cities in North America and Europe are sufficiently great that even for dipole antennas with low pattern directivity, the beam-intercept area, Ω_s, will not exceed the metropolitan surface boundaries while the aircraft is near the regional center until altitudes of 10 miles or more are attained. Because of unequal surface-noise-source distributions and dissimilar emission levels arising from the multiple types of incidental sources that occur, surface-brightness temperature may be uniform only over subportions of a metropolitan complex. Thus, only when narrow-beam antennas and low flight altitudes are used will the observed antenna-noise temperature approximate surface-brightness temperature and only in localized sectors of an urban region, such as the core, or in perimeter zones free of traffic corridors. By selecting the maximum values of antenna noise temperature observed with limited coverage antennas during aircraft flights across the urban centers of various metropolitan areas, a common bases for comparison is attained. Figure 7-4 provides a comparison of maximum observed antenna-noise temperatures expressed as F_a for urban-center traverses recorded for a range of frequencies, altitudes, and metropolitan areas. Only surface incidental and restricted noise-source emissions are represented; coherent signals have been eliminated. In the notation used, the altitude of the measurement in thousands of feet and a descriptor indicating the size of the area studied are shown as a symbol pair, e.g., point 5L, 5000 ft above a large city. Single values of maximum F_a are plotted as one point. A range of maximum daily values of F_a is graphed as a vertical line. Broken vertical lines mark late evening maxima, as does the letter *N* placed after the urban-size designator. On single point entries, business-day maxima are shown without *N*. All data were obtained using linearly polarized antennas with vertical, downward-pointed beams, unless *O*, indicating circular polarization, follows the designation symbol. Multiple sources have been drawn upon to construct Fig. 7-4.[1,2,4-8] Notice the business-day incidental-noise maxima for comparable-size metropolitan areas and aircraft altitudes decrease with frequency, reflecting the trend observed in Chapter 6 for composite surface-noise data. Further, the general trend observed for incidental-noise radiations is a decrease in

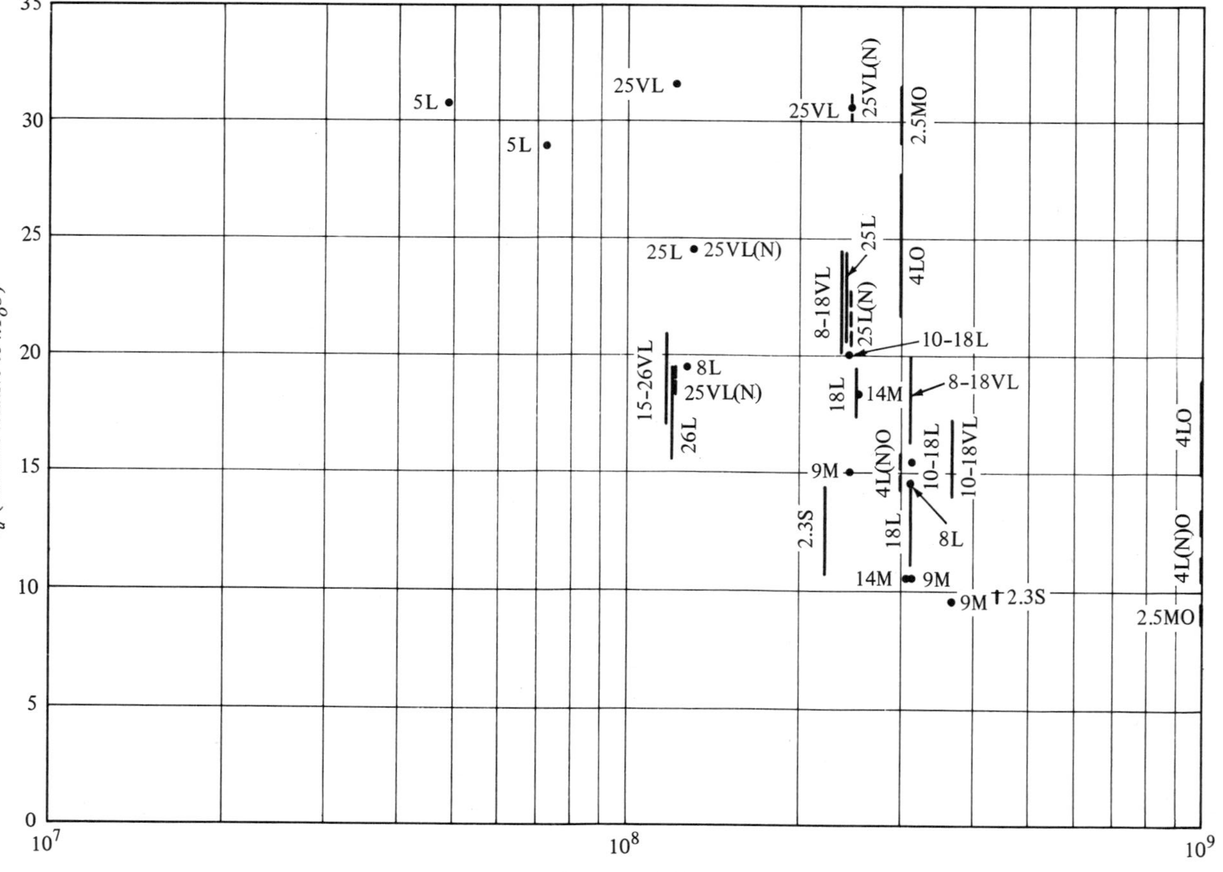

Fig. 7-4. Maximum values of airborne antenna-noise temperature, F_a, in decibels as a function of frequency. Symbol key: aircraft altitude in kilofeet; city size indicators (*S*, *M*, *L*, *VL*) are initials for small, medium, etc.; (*N*) is the nighttime maxima; *O*-circularly polarized antenna. Solid lines represent the range of business-day maxima; dashed lines represent the range of the night-time maxima. In the absense of a suffix *O*, the antenna used was linearly polarized.

power with increasing altitude above cities of comparable size. Daytime maxima are rarely seen to exceed the nighttime maxima by more than 5 dB, irrespective of the urban size. In some recorded cases, no detectable difference exists—e.g., the 243-MHz entry at 25,000 ft for a very large city (New York).[8] No evidence appears in these data that a greater noise temperature occurs for circularly polarized antennas than for linear antennas. In this respect, the airborne results are comparable to surface observations of composite incidental noise, which have consistently yielded no evidence of a polarization differential.

THEORETICAL ANALYSES OF AIRBORNE INCIDENTAL RADIO NOISE

Representations of Composite Surface-Noise Power

The analytical foundation for a calculation of the available power at the output terminals of an airborne antenna, p_h, rests upon a knowledge of the surface composite noise-power function, $p_s(f, d)$, which is dependent upon both frequency (f) and distance (d) measured along the surface from the center of an urban area. The surface-noise-power function, $p_s(f, d)$, developed in Chapter 6, was shown to display an inverse dependence upon frequency and distance, which, when expressed in decibels relative to 1 mW per detection bandwidth, B, may be written as

$$P_s(f, d) = \xi_1 + \xi_2(d - k) + \xi_3(d - k)^2, \text{ in } dBm/B, \qquad (7\text{-}3)$$

where k is a constant, which when set equal to 2.5 miles provides a good representation of VHF- and UHF-band composite, surface, incidental-radio-noise-power data. The coefficients ξ_i of Equation 7-3 contain the frequency dependence of the surface noise power function.

Representation of the surface, composite incidental-noise power in the units of power, $p_s(f, d)$, rather than in decibels with respect to a reference, may be obtained from Equation 7-3 by noting that the ξ_i's appearing therein are always negative. Transformation of Equation 7-3 to the exponential form with the reference power and conversion constant accumulated in the constant coefficient K_1 yields for $p_s(f, d)$:

$$p_s(f, d) = K_1 \exp(\alpha_1 - \alpha_2 k + \alpha_3 k^2) \exp[d(\alpha_2 - 2k\alpha_3) + d^2\alpha_3]$$
$$\cong K_1 \exp(\alpha_1) \exp(dC_1 + d^2\alpha_3) \quad (7\text{-}4)$$

since

$$|\alpha_1| \gg |\alpha_2 k|; \quad |\alpha_1| \gg |\alpha_3 k^2|$$

and

$$C_1 = (\alpha_2 - 2k\alpha_3) \cong 0.$$

Examination of the coefficient α_1 reveals that it is approximately equal to $P_s(f, 0)$—i.e., the surface noise power at the urban center, expressed in decibels. Furthermore, $|\alpha_1|$ increases with increasing frequency and may be expressed as $|\alpha_1| \propto f^\gamma$ for $\gamma > 0$. Inserting this approximation for $|\alpha_1|$ into Equation 7-4, recalling ξ_i's are negative, and using a power-series representation for the exponentials allows $p_s(f, d)$ to be approximated for $d > 0$, $\omega > 0$ by:

$$p_s(f, d) = K_1(1 + \eta_1 f + \eta_2 f^2 \ldots)^{-1}(1 + \mu_1 d + \mu_2 d^2 \ldots)^{-1} \quad (7\text{-}5)$$

in which η_j and μ_j, $j = 1, 2, \ldots n$ are positive, series expansion constants. The general functional form of $p_s(f, d)$ indicated by the expansion, Equation 7-5, may be written:

$$p_s(f, d) = K\omega^{-\beta} d^{-\alpha} \quad (7\text{-}6)$$

in which K is a constant, $\omega = 2\pi f$, and the exponents α and β are positive and constant over subportions of the intervals $0 \leqq \omega < \infty$, $0 \leqq d < \infty$. Equations 7-3 and 7-6 are alternate forms available for representing the surface, composite, incidental noise power and will be incorporated into later analyses as convenience dictates.

Integral Representation of Airborne Noise Power

Calculation of P_h, the incidental noise power at altitude h, is dependent upon a knowledge of:

1. the pattern function and losses of the measuring antenna
2. the distribution of the surface incidental-noise sources as a function of position and frequency
3. the transmission path losses

4. the degree of correlation existing between the noise emissions of neighboring surface areas or, equivalently, the correlation length of the surface noise distribution.

Given this information, p_h may be represented as an integral, which in the general case, requires numerical evaluation. The degree of difficulty encountered in performing the integration varies with the power pattern of the antenna, the form of the function representing the surface-noise distribution and the ground projection of the observation point. For a few select antennas, the power patterns are reasonably simple functions of the polar angles. Such is the case for vertically and horizontally polarized electric and magnetic dipoles and monopoles. Since the integrand of the function representing p_h contains the product of the power pattern of the receiving antenna and the distribution function of the surface-noise sources, the presence of angular symmetry in the range of integration markedly simplifies evaluation. The simplification becomes dramatic when the observation point is positioned above the center of a symmetrical surface-noise distribution. Commonly, central positioning of an airborne observer above a metropolitan noise distribution is of primary importance in aerial noise surveys, representing, as it does, the locus of the maximum observable incidental-noise level. Consequently, several sets of elevated incidental-noise measurements exist, which include urban-centered observer locations. In the VHF- and UHF-band studies discussed earlier, dipole and monopole antennas were often used; in addition, the loci of measurement permit exploiting several analytical symmetries.

It is possible, without loss of accuracy, to assume in the following work that the atmospheric radio-path losses are zero, as the frequencies for which airborne data are available are less than 3 GHz and the observation altitudes less than 10 miles.

The issue of possible correlation occurring between surface incidental-noise emissions is presently open. Insufficient temporal variation data are available to resolve the question. For most point-noise sources, such as vehicles and manufacturing equipment, the radiated fields are uncorrelated. Power-transmission lines may yield correlated noise patterns along some portion of their runs if the incidental noise is propagated with low attenuation from single-point sources. However, the majority of incidental noise sources found in principal portions of a metropolitan area will radiate independently, and thus the simplest assumption is the most probable, i.e., the surface noise source correlation length is zero. This assumption permits the treatment of incidental noise power from single emitters as additive quantities.

The dominant incidental radio-noise sources in metropolitan areas are known to be located on or comparatively near the surface. Of the dominant sources of incidental noise in the VHF band, power-transmission and power-distribution lines possess the greatest average surface separation, although electrical equipment such as elevator motors located on upper building levels are occasionally found at higher altitudes. Most heavy electrical machinery and automobiles are surface-located noise emitters. Collectively, all incidental-noise sources can be considered to be distributed two-dimensionally on a plane that is parallel to and near the surface. Because all incidental noise sources may be present in any proportion within an urbanized area and because noise from these sources radiates without pronounced directivity, the noise power emission from any unit area of a two-dimensional distribution may be assumed to be uniform. Implicit in this assumption is the point that the earth and the physical surroundings of the sources are either good noise-absorbing or noise-scattering media or both.

At altitude h, the observed noise power density, p_h, per unit detection bandwidth, B, received by an antenna of aperture area, A_r, is given by:

$$p_h = A_r \int \frac{p_s(f, d)\, F(\gamma_1, \gamma_2)}{4\pi R^2}\, dA, \text{ in watts}/B, \tag{7-7}$$

where $F(\gamma_1, \gamma_2)$ is the normalized (to unity) power pattern of the receiving antenna in terms of the generalized angle variables (γ_1, γ_2). The angle variable pair (γ_1, γ_2) must be chosen to exploit the existing symmetries of the antenna pattern and the surface distribution of noise sources. The domain of integration expressed by Equation 7-7 is understood to extend over the surface area containing the incidental noise sources.

When the observing antenna is positioned above an urban area at height, h, from the surface, a separation distance, R, exists between the antenna and any differential element, dA, of the surface contain-

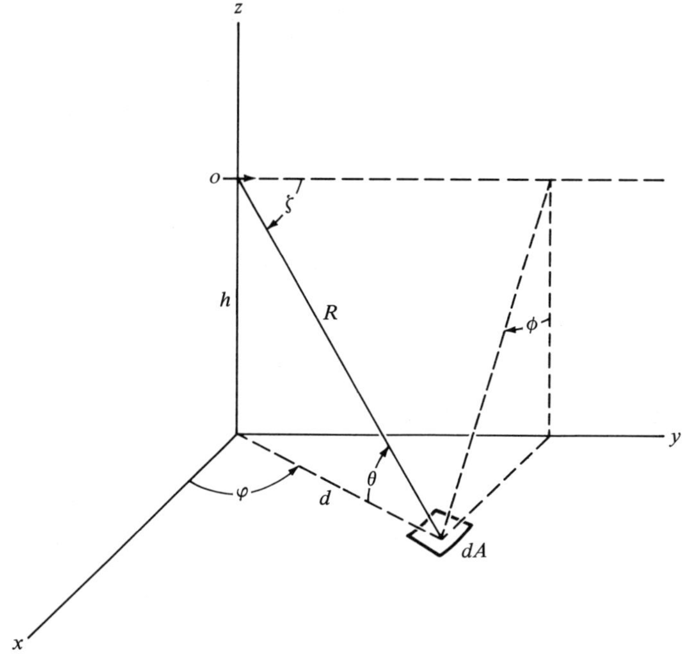

Fig. 7-5. Coordinate system for an observing antenna located at an altitude, h, above a surface distribution of incidental noise sources.

ing the incidental noise sources. The principal coordinate angles are shown in Fig. 7-5. The generalized coordinate pair (γ_1, γ_2) is selected from the four angles in Fig. 7-5 when the antenna pattern is identified, whereupon one or more of the pair will also appear in the representation for the differential surface area, dA.

The surface-noise-power density, $p_s(f, d)$, appearing in Equation 7-7, is related by the effective aperture area of the surface test antenna, A_m, to the surface noise power per unit bandwidth, $p_n(f, d)$ as:

$$p_s(f, d) = p_n(f, d)/A_m, \text{ in watts}/B. \tag{7-8}$$

Both $p_s(f, d)$ and $p_n(f, d)$ are functions of frequency and distance, d, measured radially from the center of an urban area. $P_n(f, d)$ has been compiled in Chapter 6 from data acquired by two types of measuring instruments: average-power and quasi-peak detectors. When average-power detectors are used to execute the measurement of p_h

and when sufficient surface-power measurements are available to formulate a representation of $P_n(f, d)$ of the form of Equation 7-3 for the general metropolitan complex, substitution of Equation 7-8 into Equation 7-7 and inclusion of the correct antenna-power pattern function, $F(\gamma_1, \gamma_2)$, completes the formal representation of $p_n(f, d)$:

$$p_n(f, d) = \frac{A_r}{4\pi A_m} \int \frac{p_n(f, d)\, F(\gamma_1, \gamma_2)\, dA}{R^2}. \tag{7-9}$$

In affecting an evaluation of the integral, Equation 7-9, a closed-form representation is possible if sufficient symmetries exist; otherwise, numerical quadratures must be employed.

Situations will occur when either the noise field intensity at altitude h has been measured with quasi-peak detectors or when $p_n(f, d)$ is expressible for a general metropolitan area only in terms of ${}_{qp}p_n(f, d)$. If either condition applies, then the quantity

$$({}_{qp}E^2 A_m)\, Z_0^{-1},$$

plotted in Figs. 6-25 and 6-26, is of importance and is related to the available noise power, $p_n(f, d)$, through a multiplying factor, $\mathfrak{F}$, introduced in Chapter 6 as:

$$p_n(f, d) = {}_{qp}E^2 A_m Z_0^{-1} \mathfrak{F}^{-1}, \text{ in watts}/B. \tag{7-10}$$

As previously noted, $\mathfrak{F}$ is a function of Δf, τ, and T_c/T_d, which are defined as follows:

Δf = noise bandwidth of a quasi-peak detector
τ = average period of the noise impulses
T_c and T_d = charge and discharge time constants, respectively, of a quasi-peak detector.

Since $\mathfrak{F}$ is unitless, it is dimensionally correct to label the righthand side of Equation 7-10 a power, exclusive of $\mathfrak{F}$, and to denote it by the subscript qp to indicate the type of detector used in the measurement of the field quantity E_{qp}, which appears in the numerator, i.e.,

$${}_{qp}p_n(f, d) = {}_{qp}E^2 A_m Z_0^{-1}, \text{ in watts}/B. \tag{7-11}$$

Thus, the ratio of quasi-peak to available noise power is equal to

$$\mathfrak{F} = {}_{qp}p_n(f, d)/p_n(f, d) \tag{7-12}$$

and depends upon the quasi-peak detector characteristics as well as the average-noise-pulse period, τ. Large variations in the value of $\mathcal{F}$ may occur due to changes in τ or the average individual source radiated power when either the mixture of incidential-noise-source types or the emission characteristics of the noise radiators within an urban area are altered. As an example of the effect upon $\mathcal{F}$ produced by a change of the observed emission characteristics of composite, urban-area incidental-noise sources, one sees in Figs. 6-23 to 6-26 that the ratio of powers, Equation 7-11, varies approximately 10 dB as the observation frequency is changed from 20 to 200 MHz. The cause of this variation is ascribed to changes in the individual noise signals and not to propagation effects, which should identically affect both factors on the right of Equation 7-12. When both the types and the relative mixture of the incidental-noise sources remain fixed within an urban area, the ratio $\mathcal{F}$ at each observation frequency is a constant, independent of the observer location. The invariability of $\mathcal{F}$ may be assumed in the analyses of airborne composite metropolitan-area incidental noise during epoch durations of a few hours and for periods of months, provided results from similar portions of diurnal and weekly cycles are compared.

When an observation of airborne incidental radio-noise level is performed using a quasi-peak instrument, then, subject to the preceding constraints, the left-hand side of Equation 7-7 may be written as:

$$p_h = \mathcal{F}^{-1}\,{}_{qp}p_h, \text{ in watts}/B \tag{7-13}$$

in which

$${}_{qp}p_h = {}_{qp}E_h^2 A_m Z_0^{-1}, \text{ in watts}/B \tag{7-14}$$

and ${}_{qp}E_h$ is the observed quasi-peak electric field strength at altitude h in detection bandwidth B. Substituting Equation 7-12 into Equation 7-8 yields $p_s(f, d)$ in terms of the quasi-peak noise power, which with Equation 7-13 is substituted into Equation 7-7 to obtain the altitude dependence of ${}_{qp}p_h$ in integral form as:

$${}_{qp}p_h = \frac{A_r}{4\pi A_m} \int \frac{{}_{qp}p_n(f, d)\, F(\gamma_1, \gamma_2)\, dA}{R^2}, \text{ in watts}/B. \tag{7-15}$$

The ${}_{qp}p_n(f, d)$ appearing in the above integrand is obtained, by conversion from Equation 7-3, which expresses the surface distribution of composite incidental noise measured with quasi-peak field-intensity instrumentation. The concluding step requires the representation of the normalized airborne-antenna-power pattern, $F(\gamma_1, \gamma_2)$, in terms of the appropriate coordinate angles chosen to exploit any existing pattern symmetries.

Approximation to the Surface-Noise-Power Function. The representations for surface-noise power, either $p_n(f, d)$ in Equation 7-9 or ${}_{qp}p_n(f, d)$ in Equation 7-15, have units of power in watts, while the formulating function, Equation 7-3, is established by representing the observed composite incidental-surface noise as a ratio converted to decibels. It is possible, although not convenient, in affecting the evaluation of the integrals, Equations 7-9 and 7-15, to convert Equation 7-3 obtained by either average power or quasi-peak field-intensity instrumentation into watts using an exponential expression for either $p_n(f, d)$ or ${}_{qp}p_n(f, d)$ that is parametric in variables f and d. A much simpler procedure for performing the integration of Equations 7-9 and 7-15 employs a stepwise linearization of the integrand. A series of line segments closely matching the quadratic series of Equation 7-3 is chosen:

$$\begin{aligned}\tilde{p}_n(f, d) &= a_i - B_i(d - d_i)\\ &= A_i - B_i(R^2 - h^2)^{1/2}, \text{ in watts}/B \qquad (7\text{-}16)\end{aligned}$$

where $A_i = a_i + B_i d_i$, for $d_i \leqslant d \leqslant d_{i+1}$, $d_1 = 0$, $i = 1, 2, \ldots, N$. The expression $\tilde{p}_n(f, d)$ represents either $p_n(f, d)$ or ${}_{qp}p_n(f, d)$. The closeness of fit between Equations 7-16 and 7-3 is controlled by the selection of the coefficients, a_i and B_i, and N, the number of linear segments, or equivalently, the number of annular regions used to subdivide the metropolitan surface area containing the noise sources.

Determination of a_i and B_i appearing in Equation 7-16 is performed by converting Equation 7-3 into power per bandwidth B as a function of d for a fixed frequency f. The series of segments $i = 1, 2, \ldots, N$ of $\tilde{p}_n(f, d)$ are successively fitted to either the replot or the polynomial representation of Equation 7-3 with the initial segment, $i = 1$, assigned to the interval $0 \leqslant d \leqslant d_1$. Segment endpoints d_i are adjusted until the maximum value of

$$|p_n(f, d)/A_m - \text{antilog } P_s(f, d)/10| \leqslant \delta p_n, \qquad i = 1, 2, \ldots, N, \tag{7-17}$$

where δp_n is the maximum error of approximation in watts/B. Series coefficients, a_i and B_i, are then determined by setting $p_n(f, d) = \tilde{p}_n(f, d)$ and $d = d_i$ for $i = 1, 2, \ldots, N$, pairing and solving the resulting set of equations.

The use of a linearized representation of $p_n(f, d)$ in the integrals for airborne noise power (Equations 7-9 or 7-15) sectorizes the integration range of the angular variables (γ_1, γ_2) appearing in both the antenna pattern function $F(\gamma_1, \gamma_2)$ and in the incremental surface area, dA. The integration is then performed over circular annular regions centered in the metropolitan area, with the formal representation of Equations 7-9 and 7-15 becoming:

$$p_h = \frac{A_r}{4\pi A_m} \sum_{i=1}^{N} \int \frac{[A_i + B_i(R^2 - h^2)^{1/2}] F(\gamma_1, \gamma_2)\, dA}{R^2}, \text{ in watts}/B, \tag{7-18a}$$

or using the alternate form in Equation 7-16:

$$p_h = \frac{A_r}{4\pi A_m} \sum_{i=1}^{N} \int \frac{[a_i - B_i(d - d_i)] F(\gamma_1, \gamma_2)\, dA}{R^2}, \text{ in watts}/B. \tag{7-18b}$$

Similar results are obtained for the quasi-peak power expressions:

$${}_{qp}p_h = \frac{A_r}{4\pi A_m} \sum_{i=1}^{N} \int \frac{{}_{qp}[A_i + B_i(R^2 - h^2)^{1/2}] F(\gamma_1, \gamma_2)\, dA}{R^2}, \text{ in watts}/B \tag{7-19a}$$

$${}_{qp}p_h = \frac{A_r}{4\pi A_m} \sum_{i=1}^{N} \int \frac{{}_{qp}[a_i + B_i(d - d_i)] F(\gamma_1, \gamma_2)\, dA}{R^2}, \text{ in watts}/B. \tag{7-19b}$$

Conversion of the Area Distributed Incidental Noise Envelope Distribution Function By Radio-Propagation Processes

Before completing evaluation of the integral representations of airborne noise power, it is necessary to introduce and develop a corollary concept pertaining to the envelope statistics of composite incidental noise. The noise process to be addressed is a manifest transformation in the composite noise-envelope distribution, which becomes evident with changes in either the observation frequency or source-separation distance. This effect is not uniquely associated with airborne composite incidental noise, but may be manifest in surface-noise data as well.

Representation of the Observed Incidental-Noise Voltage

The observed incidental noise voltage, V_n, at the output terminals of a matched receiving antenna, is representable as a series of terms, each resulting from emissions, $\hat{V}_n$, of many independent random sources, N:

$$V_n = \sum_{n=1}^{N} \hat{V}_n .$$

Each term $\hat{V}_n$ arising from the nth random source, may be described by a series expansion:

$$\begin{aligned}
\hat{V}_n &= \sum_{m=0}^{M} a_m \cos(\omega_m t + \phi_m) \\
&= \cos\omega_0 t \sum_{m=0}^{M} a_m \cos(\omega_m t + \omega_0 t + \phi_m) \\
&\quad + \sin\omega_0 t \sum_{m=0}^{M} a_m \sin(\omega_m t + \omega_0 t + \phi_m) \\
&= \cos\omega_0 t \sum_{m=0}^{M} C_m + \sin\omega_0 t \sum_{m=0}^{M} S_m \qquad (7\text{-}20)
\end{aligned}$$

where ω_0 represents the center angular frequency of the observation band. Equation 7-20 describes a random process in which the stochastic independent variable is the random signal phase, ϕ_m, which may be rewritten and transformed as shown, using cosine and sine relationships. The resulting transformed random variables become sine and cosine functions, abbreviated by the symbols S_m and C_m with means $\overline{S}_m$ and $\overline{C}_m$.

When the representation for $\hat{V}_n$, Equation 7-20 is used to describe each of M signals arising from N independent random sources, a double summation for the total observed noise voltage, V_n, results, i.e.,

$$V_n = \cos \omega_0 t \sum_{n=1}^{N} \sum_{m=0}^{M} C_{n,m} + \sin \omega_0 t \sum_{n=1}^{N} \sum_{m=0}^{M} S_{n,m}.$$

Furthermore, if the emissions from each of the N sources are anharmonic, the number of terms represented by the double summation $(N + M)$ increases linearly with N. And finally, the sufficient condition that the central limit theorem applies to the random variables, $C_{n,m}$ and $S_{n,m}$, namely,

$$\lim_{(N+M)\to\infty} \frac{\sum_{n=1}^{N} \sum_{m=0}^{M} (C_{n,m} - \overline{C}_{n,m})}{\left(\operatorname{var} \sum_{n=1}^{N} \sum_{m=0}^{M} C_{n,m}\right)^{1/2}} \to N(0, 1)$$

$$\lim_{(N+M)\to\infty} \frac{\sum_{n=1}^{N} \sum_{m=0}^{M} (S_{n,m} - \overline{S}_{n,m})}{\left(\operatorname{var} \sum_{n=1}^{N} \sum_{m=0}^{M} S_{n,m}\right)^{1/2}} \to N(0, 1) \qquad (7\text{-}21)$$

where $N(0, 1)$ is a normally distributed random variable of zero mean, and unity variance, is the convergence, in distribution, of the ratio:[9]

$$\lim_{(N+M)\to\infty} \frac{\mu^{(2+\epsilon)}}{(\sigma_{c,s}^2)^{[1+(1/2)\epsilon]}} \to 0 \qquad (7\text{-}22)$$

for any integer, $\epsilon > 0$. μ in equation 7-22 is the $(2 + \epsilon)$th root of the sum of the $(2 + \epsilon)$th central moments of either random variable $C_{n,m}$ or $S_{n,m}$, i.e.,

$$\mu^{(2+\epsilon)} = \sum_{n=1}^{N} \sum_{m=0}^{M} E[|C_{n,m} - \overline{C}_{n,m}|]^{(2+\epsilon)}$$

$$\sigma_c^2 = \sum_{n=1}^{N} \sum_{m=0}^{M} \text{var}\, C_{n,m}$$

and similarly for $S_{n,m}$. When the preceding conditions on the second and higher central moments of $C_{n,m}$ and $S_{n,m}$ are met, and as the number of anharmonic, independent random sources, N, becomes very large, the observed noise-voltage envelope distribution approaches what is recognized to be a Rayleigh distribution.[10]

Conditions Affecting an Increase in the Number of Noise Sources, N, Within the Observation Range. The central limit theorem has been shown to apply to composite noise emissions when the number, N, of anharmonic, independent random noise sources lying within the detection range of an observer increases without limit. An unbounded increase in N is dependent upon, and must accompany, a variation in a primary independent variable such as frequency or range. To prescribe a proper parametric dependence for N, it is necessary to define and examine the spatial interval within which the N random sources exist.

A primary measure of random incidental noise voltage is the probability density function of the noise envelope. Determination of the noise-envelope voltage-density function may be experimentally accomplished by using either a linear voltage quantizer with equal voltage increments or a logarithmic quantizer with constant voltage ratio increments. The levels are derivable by observing the average received signal as a function of range or frequency using a signal generator whose output produces at the observation point the parametric dependences represented by Equation 7-6. For the two quantizers, the following equations hold for two independent random noise sources whose observed powers, $\hat{P}_1(R_1, \omega_1)$ and $\hat{P}_2(R_2, \omega_2)$, are proportional to the respective squares of the two quantizer voltage levels:

$$[\hat{P}_1(R_1, \omega_1)]^{1/2} - [\hat{P}_2(R_2, \omega_2)]^{1/2} = K_1,$$

a constant for a linear quantizer, (7-23a)

or

$$\hat{P}_1(R_1, \omega_1)/\hat{P}_2(R_2, \omega_2) = K_2,$$

a constant for a logarithmic quantizer. (7-23b)

R_1 and R_2 are the distances between the noise source and the observer, and $\omega_1/2\pi$, $\omega_2/2\pi$ are the observation frequencies.

Using Equation 7-6 in Equation 7-23 results in the following pair of constraints on the respective quantizers, linear and logarithmic:

$$2K_1(K)^{-1/2} = \frac{\alpha \Delta R}{R_1^{1+(1/2)\alpha}\omega_1^{(1/2)\beta}} + \frac{\beta \Delta \omega}{R_1^{(1/2)\alpha}\omega_1^{1+(1/2)\beta}}$$

$$= \frac{1}{R_1^{(1/2)\alpha}\omega_1^{(1/2)\beta}}\left(\frac{\alpha \Delta R}{R_1} + \frac{\beta \Delta \omega}{\omega_1}\right) \quad \text{(7-24a)}$$

$$K_2 - 1 = \frac{\alpha \Delta R}{R_1} + \frac{\beta \Delta \omega}{\omega_1}. \quad \text{(7-24b)}$$

In obtaining Equation 7-24, $R_2 = R_1 + \Delta R$ and $\omega_2 = \omega_1 + \Delta\omega$ have been used, subject to the conditions that $\Delta R/R_1 << 1$ and $\Delta\omega/\omega_1 << 1$. The incrementals ΔR and $\Delta\omega$ may be interpreted as the range and frequency excursions, which, when separately varied, produce an average observed signal level change equal to the difference between a pair of quantization levels. Examination of Equation 7-24 reveals that, for a fixed $\Delta\omega$:

1. ΔR increases with increasing R_1 when the center observation frequency $\omega_0 = (2\omega_1 + \Delta\omega)/2$ is fixed
2. ΔR increases with ω_0 when the range R_1 to the nearest source is fixed.

When it may be assumed that the individual incidental-noise sources are distributed without large intervening voids in the geographical area contributing to the total detectable noise signal, then the number of independent random noise sources, N, contributing to the signal within a detector quantization level:

1. increases with R for a surface observer (or as $R \cdot \cos\theta$ for an observer at an elevation angle θ) if ω_0 is constant, or
2. increases with ω_0 for a constant R.

Situations (1) and (2) will occur provided that:

a. the frequency interval, $\Delta\omega$, remains constant
b. the receiver gain is linear
c. the internal noise of the receiver is negligible compared to the external incidental-noise levels.

The existence of large voids between surface incidental-noise sources will not permit the assumption that N increases in proportion to ΔR over a range of the variable R and consequently over a set of quantization intervals. An estimate of the maximum permissible void dimension that will allow the proportionality assumption between N and ΔR is obtained by the following computation.

Let L represent the average separation distance between surface incidental-radio-noise sources. An upper bound on the average area containing one source, L^2, is obtained by equating L^2 to the area of the annular region between two concentric circles of radii R_1 and R_2, which represent the separations between two signal sources of observed powers $\hat{P}_1$ and $\hat{P}_2$.

The resulting relationship is:

$$L^2 \leqslant \pi(R_2^2 - R_1^2).$$

By selecting small signal quantization intervals, the approximation, $R_2 = R_1 + \Delta R$, may be introduced when $\Delta R/R_1 << 1$. The expression for L^2 now becomes

$$L^2 \leqslant 2\pi R_1(\Delta R).$$

Using Equation 7-23 to obtain ΔR, the bounds on L^2 that apply to linear and logarithmic quantizers are

1. Linear quantizers:

$$L^2 \leqslant \frac{2\pi R_1^2}{\alpha}\left(\frac{2K_1}{K^{1/2}} R_1^{(1/2)\alpha} \cdot \omega_1^{(1/2)\beta} - \frac{\Delta\omega}{\omega_1}\right) \tag{7-25a}$$

2. Logarithmic quantizers:

$$L^2 \leqslant \frac{2\pi R_1^2}{\alpha}\left(K_2 - 1 - \beta\frac{\Delta\omega}{\omega_1}\right) \tag{7-25b}$$

where $\Delta\omega/\omega_1 << 1$. In Equation 7-25, R_1 is the smallest separation distance between the observer and a signal source used to establish the quantization levels. This separation should be interpreted as the smallest distance between an observer and a noise source when evaluating the noise-void dimensions.

Equations 7-25a and b apply when the frequency-dependent transformation of the noise envelope is evaluated. If the observation frequency is fixed and the range-dependent conversion of the noise-envelope distribution function is assessed, the expressions for L^2 become

1. Linear quantizer:

$$L_R^2 \leqslant \frac{4\pi K_1}{\alpha K^{1/2}} R_1^{2+(1/2)\alpha} \omega^{(1/2)\beta} \tag{7-26a}$$

2. Logarithmic quantizer:

$$L_R^2 \leqslant \frac{2\pi R_1^2}{\alpha}(K_2 - 1). \tag{7-26b}$$

It is evident by comparing Equation 7-25 and 7-26 that the maximum permissible value of L is greatest for detection of a range-dependent effect. Enlargement of the incremental annular surface region ΔR with either range or frequency, subject to the constraints on the dimensions of a surface-noise-void size (Equations 7-25 and 7-26), produces a monotonic increase in the number of independent noise sources contributing to the total incidental noise observed between specified pairs of levels. Thus, the sufficiency condition for invoking the central limit theorem is satisfied. Consequently, the envelope of incidental man-made noise approaches a Rayleigh distribution as either range, R, or observation frequency, $\omega/2\pi$, increases.

Implicit in this development are the assumptions that receiver gain is linear and that internal receiver noise is negligible. Notice that the conversion to Rayleigh of the incidental noise-envelope distribution is hypothesized, in limit, with either increasing range or observation frequency, but the rate of convergence is not predicted by the analysis.

Observed Dependence of Man-Made Noise-Envelope Statistics upon Range and Frequency

Range Dependence. Above a metropolitan area noticeable alterations in man-made incidental-radio-noise characteristics have been observed to accompany altitude changes. When UHF-band incidental noise was observed with a linearly polarized, vertically downward directed, twin-dipole antenna array of approximately 10 dB directivity,[1] it was noted that the ensemble of impulsive emissions originating from the surface sources underwent progressive modification as the observation altitude increased. Observed with a broadband visual display, the appearance of the ensemble was seen to approach a randomness in form similar to thermal noise. This change in characteristic became pronounced at flight altitudes greater than 5000 ft. It was concluded that an alteration of the surface noise density per unit solid angle had increased in proportion to the altitude increase.

Subsequently, similar observations were noted during in-flight studies of man-made incidental HF/VHF-band noise above Seattle.[5] The latter tests were performed with an airborne antenna having a directivity 8 dB lower than the initial study,[1] which was probably the major reason thermalization of the observed incidental noise appeared to occur at altitudes lower than 5000 ft.[5] The simultaneous availability of both quasi-peak and field-intensity detector measurements permitted acquiring in the second study a measure of airborne incidental-noise thermalization, namely, the ratio of quasi-peak to field intensity (or average voltage). This ratio was found to range from 8 to 10 dB, allowing the conclusion that composite, incidental noise had substantially altered from the impulsive distribution existing on the surface approaching thermalization with increasing flight altitude.

Frequency Dependence. Measurement of the quantity V_d representing the ratio, in decibels, of the rms to average noise-signal-envelope voltage as a function of frequency provides a method for determining the degree of progressive thermalization occurring in impulsively distributed noise. For impulsive composite incidental noise observed in business areas with receivers possessing either 4 or 10 kHz detection bandwidths, V_d has been found to range from 3 to 7 dB (Figs. 6-8 and 6-9). Notice the average observed values of V_d decrease as the observation point moves to quieter sites. The frequency range displayed in Figs. 6-8 and 6-9 is the lower middle portion of the radio spectrum. It is not evident from these data that a systematic variation of V_d with frequency is occurring.

Investigations extending the measurement of V_d for composite incidental surface noise in central urban areas beyond 250 MHz have been performed using linearly polarized antennas and a linear receiver with a 10-kHz detection bandwidth followed by signal-processing circuitry for obtaining the rms and average noise-envelope voltage.[11] Experimentally determined values of V_d for three urban surface sites in Ottawa, Canada, are presented in Fig. 7-6 for the frequency range of 160 to 500 MHz. At each of the three locations, 100 independent data samples were taken during a 10-day period. For each location, the mean value of V_d has been plotted and, in addition, the mean value for the combined data of all five sites. Range bars designating ±1 standard deviation about the means are also shown. Notice that, between 200 and 500 MHz, V_d for each site and for the combined data of all sites varies inversely with frequency. Within this interval, the noise figure of the receiving system rose by 1.8 dB, which accounts for some but not all of the measured decrease in noise-distribution impulsiveness. The data of Fig. 7-6 show a distinct decrease in noise impulsiveness, manifesting the

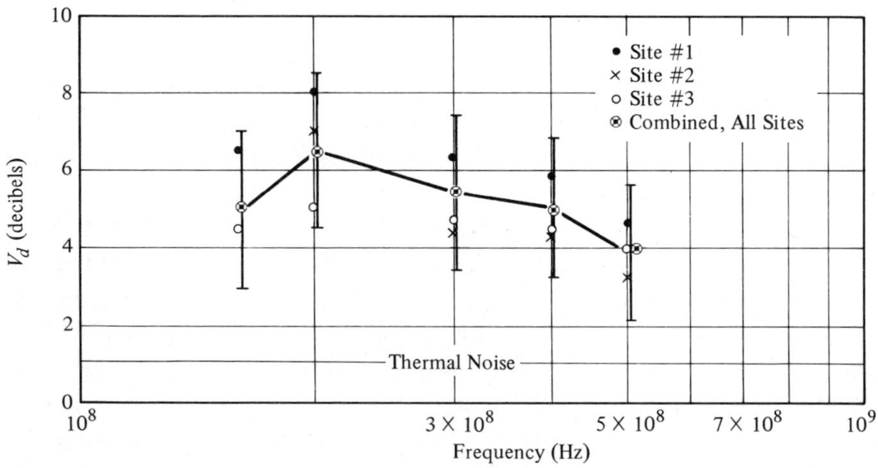

Fig. 7-6. V_d versus frequency for composite, incidental noise at three surface sites. (After Lauber and Bertrand, 1977)

conversion of the noise-envelope distribution to Rayleigh form with increasing frequency.

A second set of data revealing a transformation of composite incidental impulsive noise toward a Rayleigh envelope distribution with increasing frequency has been recorded.[12] At each of 19 surface observation points located between 5 and 35 miles from the centers of Laramie and Cheyenne, Wyoming, measurements were made of V_d at several frequencies lying between 250 kHz and 250 MHz.

Table 7-1 presents V_d averaged for frequencies of 250 kHz, 500 kHz and 102 MHz, and 250 MHz, plus the standard deviations of the averages. The four frequencies represent the upper and lower limits of measurement and the adjacent selections. In preparing Table 7-1, paired averages were constructed for each site to reduce the scatter occurring in the data. All averaging was performed in the units of decibels. The column headed "Note" contains a coded comparison of the results. The absence of an entry in this column opposite a site designation denotes that the averaged values of V_d decreased with increasing frequency by a statistically significant amount. An entry of *N* in the column indicates the averaged V_d value for the higher pair of frequencies was larger than the averaged V_d for the lower pair. An entry of *?* in the Note column indicates that although the averaged value of V_d for the higher-frequency pair was less than the averaged V_d for the lower-frequency pair, the difference did not exceed the largest standard deviation occurring for either pair. Of the 19 test locations, 12 display averaged V_d values, which decreased with increasing frequency by a statistically significant amount. This behavior is consistent with the hypothesis that the composite incidental-noise-distribution function approaches a Rayleigh envelope distribution with increasing frequency of observation. Another three locations display a decrease in averaged V_d with increasing frequency but these cases also display large data variances. Only 4 of the 19 locations show increases in averaged V_d with frequency, but in 2 of the 4 cases the differences in averaged V_d are less than the greatest standard deviation of either average. Also for three of the four sites—sites 7, 9, and 10—the primary data sets were incomplete, lacking information for the 102-MHz entry.

The preponderance of cases presented in Table 7-1 display a frequency variation of averaged V_d at a fixed noise-source separation

Table 7-1. Measured values of V_d for surface composite incidental man-made noise. (After Essa Laboratories, 1968)

Site	Average V_d for 250 and 500 kHz	$(\text{Variance})^{1/2}$ of Average V_d	Average V_d for 102 and 250 MHz	$(\text{Variance})^{1/2}$ of Average V_d	Note
Under high-voltage line	4.3	9.4	4.65	6.45	N
Along high-voltage line	8.65	8.2	4.65	5.0	?
0.25 Mile from high-voltage line	4.5	3.0	1.2	0.68	–
Q-5	6.3	1.2	1.1	0.18	–
No. 1	3.7	0.85	0.95	1.3	–
No. 2	3.05	0.33	2.05	0.35	–
No. 3	4.15	0.32	2.6[a]	0.2[a]	–
No. 4	5.3	1.2	0.6	0.18	–
No. 5	5.9	1.2	1.5	0.33	–
No. 6	1.95	0.16	0.75	0.18	–
No. 7	2.65	0.4	3.6[a]	2.0[a]	N
No. 8	2.85	0.3	2.45	1.45	?
No. 9	2.2	0.3	5.3[a]	1.5[a]	N
No. 10	2.95	0.36	5.4[a]	0.7[a]	N
No. 11	6.1	1.2	1.1	0.2	–
No. 12	12.8	42	11	61	?
No. 13	7.0	0.73	0.9	0.16	–
No. 14	5.5	0.6	0.9	0.16	–
No. 15	4.15	0.4	1.45	0.11	–

[a]Indicates the measurement set was incomplete.

distance, which is consistent with the concept of thermalization of composite impulsive incidental noise with increasing frequency.

COMPARISONS OF THEORETICAL COMPUTATIONS OF AIRBORNE INCIDENTAL NOISE WITH EXPERIMENTAL DATA

Insertion of the appropriately normalized antenna function $F(\gamma_1, \gamma_2)$ into Equations 7-18 and 7-19 followed by integration and summation yields the incidental noise power at altitude h. The observation point may be positioned above any surface point either centered or laterally displaced from a metropolitan area. The integrals, however, require numerical evaluation unless the observation point is located above the urban center. When centrally positioned, the symmetry existing in the representation of composite surface incidental-noise power (Equation 7-3), permits closed form evaluation of Equations 7-18 and 7-19 for vertical dipole or monopole antennas. Centering of the observation point does not assure that integration of Equations 7-18 and 7-19 can proceed without the use of numerical techniques even though the antenna employed is representable by a simple pattern function. A horizontally oriented dipole is an example of this situation; it will be shown later that numerical treatment is required even though airborne incidental-noise observations are peformed above an urban center.

Nonetheless, when the observation point is positioned above an urban center, some simplification is usually possible in the evaluation of expressions for p_h and, in addition, the analytical results obtained may be readily compared with an appreciable body of existing airborne noise data. Reported investigations of airborne incidental radio noise have, in virtually all instances, recorded the maximum airborne incidental-noise level, which is normally found above the center of the metropolitan complex.

In the subsequent portions of this section, both vertically and horizontally oriented dipoles and monopoles are examined as airborne observing antennas, each positioned to project a beam vertically downward.

Vertically Polarized Dipole or Monopole

Computation of Airborne Noise Power for Urban-Centered Locations. Referring to Fig. 7-5, the differential area dA is seen to be representable as

$$dA = R^2 \cot \theta \, d\theta \, d\varphi, \tag{7-27}$$

where θ is the elevation angle of the observation point. When the airborne antenna is a dipole, its effective area, A_r, becomes identically equal to the area of the antennas used for most surface noise measurements and to which the composite incidental surface noise plots have been referenced; thus, $A_r = A_m$ and

$$\frac{A_r}{4\pi A_m} = \frac{1}{4\pi}. \tag{7-28}$$

The normalized power pattern $F(\gamma_1, \gamma_2)$ may be written for a vertical dipole or monopole antenna in terms of the angle variables of Fig. 7-5 as:

$$F(\gamma_1, \gamma_2) = \frac{\cos^2(\frac{1}{2}\pi \sin \theta)}{\cos^2 \theta}. \tag{7-29}$$

Inserting Equations 7-27 through 7-29 into the representation for average power p_h at altitude h, and integrating with respect to φ over $0 \leqq \varphi \leqq 2\pi$ provides

$$_c p_h = \frac{1}{2} \sum_{i=1}^{N} \left[(a_i - B_i d_i) \int_{\theta_i}^{\theta_{i+1}} \frac{\cos^2(\frac{1}{2}\pi \sin \theta)}{\cos \theta \sin \theta} \, d\theta \right.$$

$$\left. + h \cdot B_i \int_{\theta_i}^{\theta_{i+1}} \frac{\cos^2(\frac{1}{2}\pi \sin \theta)}{\sin^2 \theta} \, d\theta \right],$$

$$\text{in watts}/B, \qquad \theta_i > \theta_{i+1}. \tag{7-30}$$

The second form of the integrand, Equation 7-18b, has been selected in writing Equation 7-30 to facilitate later reduction. The resulting expression represents the available incidental-noise power observable above the center of a metropolitan area. Exploitation of the symmetries existing in the antenna pattern function (Equation 7-29) and in the composite surface noise representation (Equation 7-3) becomes explicit in Equation 7-30. A comparable expression exists for ${}_{qp}p_h$. Letting ${}_{\alpha}I_i$ and ${}_{\beta}I_i$ denote the pair of integrals in Equation 7-29 permits a compact notation as follows:

$$p_h = \frac{1}{2} \sum_{i=1}^{N} [(a_i - B_i d_i)_{\alpha} I_i + hB_{i\beta} I_i], \text{ in watts}/B. \tag{7-31}$$

The integrals of Equation 7-31 can be evaluated by expanding the integrands in power series of sine and cosine θ and subsequently integrating term by term. The resulting series converges rapidly, requiring retention of at most five terms:

$$\begin{aligned}
{}_{\alpha}I_i &= \int_{\theta_i}^{\theta_{i+1}} \frac{\cos^2 (\frac{1}{2}\pi \sin \theta)}{\cos \theta \sin \theta} \, d\theta \\
&= \frac{1}{\pi^2} \ln \left| \frac{\sin^2 \theta_{i+1}}{\sin^2 \theta_i} \right| \\
&\quad - 0.149(\sin^2 \theta_{i+1} - \sin^2 \theta_i) \\
&\quad + 2.84 \times 10^{-2} (\sin^4 \theta_{i+1} - \sin^4 \theta_i) \\
&\quad - 3.45 \times 10^{-3} (\sin^6 \theta_{i+1} - \sin^6 \theta_i) \\
&\quad + 3.24 \times 10^{-4} (\sin^8 \theta_{i+1} - \sin^8 \theta_i) \ldots
\end{aligned} \tag{7-32a}$$

$$\begin{aligned}
{}_{\beta}I_i &= \int_{\theta_i}^{\theta_{i+1}} \frac{\cos^2 (\frac{1}{2}\pi \sin \theta)}{\sin^2 \theta} \, d\theta \\
&= \cot \theta_i - \cot \theta_{i+1} + 1.67(\theta_i - \theta_{i+1}) \\
&\quad + 0.405(\sin 2\theta_i - \sin 2\theta_{i+1}) \\
&\quad - 0.142(\sin^3 \theta_i \cos \theta_i - \sin^3 \theta_{i+1} \cos \theta_{i+1}) \\
&\quad + 1.96 \times 10^{-2} (\sin^5 \theta_i \cos \theta_i \\
&\quad - \sin^5 \theta_{i+1} \cos \theta_{i+1}) \ldots \\
&\qquad \theta_i > \theta_{i+1}
\end{aligned} \tag{7-32b}$$

The indexing order of the segmental distances d_i and associated elevation angles θ_i are shown in Fig. 7-7.

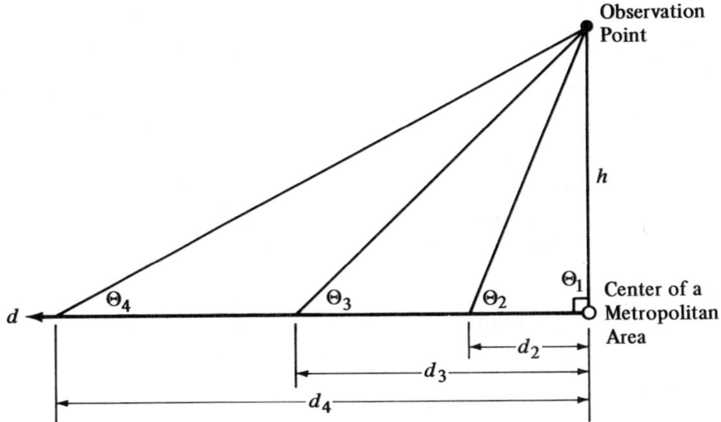

Fig. 7-7. Indexing order of surface segmental distances, d_i, and associated elevation angles, θ_i.

Evaluation of the integrals of Equation 7-32 follows from the selection of the frequency, which then permits determination of the fitted line-segment coefficients, a_i and B_i of Equation 7-16, using the polynomial form for $P_s(f, d)$ of Equation 7-3, converted into watts per unit of bandwidth. The number of segments and the segment endpoints, d_i, are established by the closeness of fit constraint represented by the inequality (7-17), which typically is set at 1 dB. Elevation angles θ_i, shown in Fig. 7-7, are then obtained from the values d_i and the observation altitude, h. Together these permit computation of ${}_{\alpha}I_i$ and ${}_{\beta}I_i$ and, finally, of p_h (Equation 7-31).

Comparisons of Measured and Computed Airborne Incidental-Noise Power for Urban-Centered Locations. Two experimental sources exist of available noise-power data observed as a function of altitude above the center of urban areas. The study encompassing the greatest diversity of measuring frequencies was performed by Buehler et al.[5] using a tuned, vertical-dipole antenna trailed from a light plane at altitudes from 1500 to 10,000 ft. The investigation was performed in the Seattle area and included flights at an altitude of 1 mile across the metropolitan center and on various laterals in the vicinity of the city. In addition, the height dependence of p_h was determined for the altitude interval of 1500 to 10,000 ft at an offset location 16 miles east of the city.

Traverses of the central Seattle area produced noise data at 49, 73, and 137 MHz. The available noise power was observed to attain its maximum value above the center of the city. In units of dBm/kHz, the observed maxima at an altitude of 1 mile, ${}_cP_h$, are listed in Table 7-2.

A second source of VHF airborne noise data measured with vertical-monopole antennas was obtained by Barnard.[4] This series of studies was performed in central England at 118 MHz. Flights were made over the centers of Birmingham, Manchester, Liverpool, and London at an altitude of 26,000 ft and over London at 15,000 ft. As in the preceding study, the receiver was calibrated to yield available power at the antenna terminals. Except for Liverpool, which is separated by less than 5 miles from the Irish Sea lying to the west, the English urban areas may be considered symmetrical surface distributions of incidental noise. The maximum values of ${}_cP_h$ obtained by Barnard over the centers of each metropolitan area are listed in Table 7-2 at a frequency of 118 MHz.

The fourth column of Table 7-2 contains the theoretically predicted values of ${}_cP_h$ using Equation 7-31. The coefficients a_i and B_i appearing in Equation 7-31 were obtained from the surface com-

Table 7-2. Comparison with theoretical predictions of observed incidental radio-noise power as a function of altitude above the center of several metropolitan areas.

		Incidental Noise Power, ${}_cP_h$ (dBm/kHz)			
Frequency (MHz)	Altitude	Observed	Calculated Equation 7-31	Difference (dB) (Observed–Calculated)	Measurement Location and Reference
49	0.95	-112.7	-114	1.3	Seattle[5]
73	0.95	-116.1	-115.4	-0.7	Seattle[5]
118	2.84	-123.8	-123.1	-0.7	London[4]
118	4.83	-122	-125.1	3.1	London[4]
118	4.83	-124	-125.1	1.1	Birmingham[4]
118	4.83	-126	-125.1	-1.3	Manchester[4]
118	4.83	-128	-125.1	-2.9	Liverpool[4]
137	0.95	-120	-119	-1.0	Seattle[5]
				-0.1 Average	

posite noise curves of Figs. 6-22 and 6-23 at the frequencies employed in the experimental investigations. The fifth column of Table 7-2 presents the differences in decibels between measurements and computations. The average error is approximately −0.1 dB.

Variation of Airborne Incidental-Noise Power with Altitude for Urban Offset Locations. The general representation of airborne noise power p_h for a vertical dipole antenna may be obtained from either Equation 7-7 or 7-15 with the aid of Equation 7-8 and Equation 7-29 and is:

$$p_h = \frac{A_r}{4\pi A_m} \int_{\varphi_1}^{\varphi_2} d\varphi \int_{\theta_1}^{\theta_2} \frac{p_n(f, d) \cos^2 (\frac{1}{2}\pi \sin \theta)}{\sin \theta \cos \theta} \, d\theta. \tag{7-33}$$

When the observation point is displaced to one side of an urban center (Fig. 7-8), the incidental noise emissions may be treated as arising from a centrally located core area and represented by an average level:

$$\bar{p}_n(f) = p_n(f, d), \tag{7-34}$$

which is a function of frequency only. Inserting Equation 7-34 into Equation 7-33 and representing the angle and distance independent factors by a constant, C, yields:

$$p_h = C \int_{\varphi_1}^{\varphi_2} d\varphi \int_{\theta_1}^{\theta_2} \frac{\cos^2 (\frac{1}{2}\pi \sin \theta)}{\sin \theta \cos \theta} \, d\theta.$$

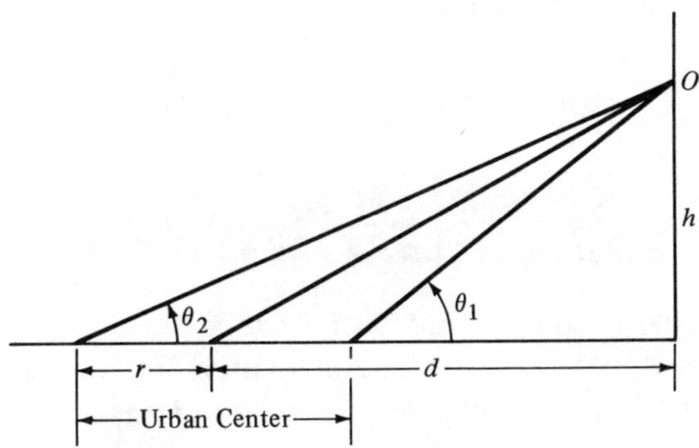

Fig. 7-8. Urban center offset geometry.

Taking the partial derivative of p_h in the above with respect to h, one obtains

$$\frac{\partial p_h}{\partial h} = C \int_{\varphi_1}^{\varphi_2} d\varphi \left[\frac{\cos^2 (\frac{1}{2}\pi \sin \theta_2)}{\sin \theta_2 \cos \theta_2} \cdot \frac{\partial \theta_2}{\partial h} - \frac{\cos^2 (\frac{1}{2}\pi \sin \theta_1)}{\sin \theta_1 \cos \theta_1} \cdot \frac{\partial \theta_1}{\partial h} \right].$$

Using Fig. 7-8, one obtains $h = (d - r) \tan \theta_1 = (d + r) \tan \theta_2$, and thus the partials of θ_1 and θ_2 with respect to h may be expressed as:

$$\frac{\partial \theta_2}{\partial h} = \frac{\cos^2 \theta_2}{d + r} \quad \text{and} \quad \frac{\partial \theta_1}{\partial h} = \frac{\cos^2 \theta_1}{d - r},$$

permitting the writing of $\partial p_h / \partial h$ as:

$$\frac{\partial p_h}{\partial h} = C \int_{\varphi_1}^{\varphi_2} d\varphi \left[\sin^2 \left(\frac{1}{2} \pi \sin \theta_1 \right) - \sin^2 \left(\frac{1}{2} \pi \sin \theta_2 \right) h^{-1} \right.$$

$$\cong C' \int_{\varphi_1}^{\varphi_2} h^{-1} \, d\varphi \left\{ \frac{1}{2} \pi (\sin \theta_1 - \sin \theta_2) + \frac{\pi^3}{48} (\sin^3 \theta_2 - \sin^3 \theta_1) \right.$$

$$\left. + \left(\frac{1}{2} \pi \right)^5 \frac{1}{5!} (\sin^5 \theta_1 - \sin^5 \theta_2) \cdots \right\}. \qquad (7\text{-}35)$$

The lowest-order approximation to Equation 7-35 neglects terms in $\sin \theta$ higher than the first and injects the following constraint on the variables h, r, and d:

$$\frac{h}{d - r} \ll (24)^{1/2} / \pi.$$

For the first-order approximation, one obtains:

$$\frac{\partial p_h}{\partial h} \cong C' \int_{\varphi_1}^{\varphi_2} d\varphi \, (\sin \theta_1 - \sin \theta_2) \, h^{-1}$$

$$\cong C' \int_{\varphi_1}^{\varphi_2} d\varphi \left[\frac{h}{d - r} - \frac{h}{d + r} \right] h^{-1} \cong C' \cdot r(d^2 + r^2)^{-1},$$

which is seen to be independent of h since the integration was performed with respect to the angle φ, which lies in the plane of the surface (Fig. 7-5). Integrating the preceding yields for p_h:

$$p_h = ah + k_1 \qquad (7\text{-}36)$$

for $h/(d - r) \ll 24^{1/2}/\pi$. The next higher-order approximation to Equation 7-35 includes terms in $\sin^3 \theta$ and corresponds to the altitude:

$$\frac{h}{d - r} \ll 80^{1/2}/\pi.$$

For this condition, one obtains:

$$\frac{\partial p_h}{\partial h} \cong C' \int_{\varphi_1}^{\varphi_2} d\varphi \left[\frac{2r}{(d^2 - r^2)} - \frac{(2r^3 + 6rd^2)\, h^2}{(d^2 - r^2)^3} \right] \cong C_1 - C_2 h^2 .$$

Integration yields:

$$p_h = ah - bh^3 + k_2 , \qquad (7\text{-}37)$$

which is applicable for the interval, $24^{1/2}/\pi < h/(d - r) < 80^{1/2}/\pi$.

The first-order approximation to Equation 7-33 given by Equation 7-36 is compared in Fig. 7-9 with the altitude profile of incidental noise power observed above a surface location 16 miles east of Seattle.[5] The experimental data shown in Fig. 7-9 by the solid curve were obtained using a vertically polarized dipole antenna tuned to 137 MHz.[5] Plotted is the ratio of noise power at altitude h to the surface noise power expressed in decibels. The normalized curve drawn as a dashed line in Fig. 7-9 was derived by evaluating the inte-

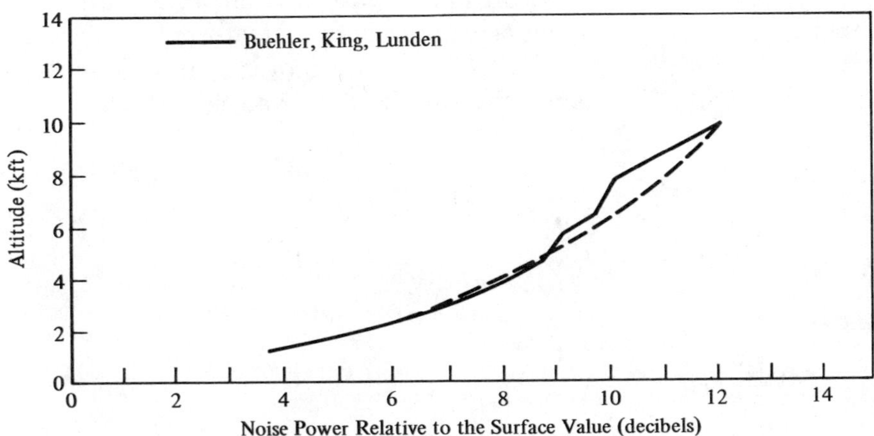

Fig. 7-9. Airborne noise power as a function of altitude at an offset metropolitan location. Noise power in decibels relative to surface value.

gration constants of Equation 7-36 at values of $h = 1500$ and 10,000 ft to obtain the equation:

$$p_h = 6.31 \times 10^{-18}\, h - 10^{-16}, \text{ in mW/kHz.} \tag{7-38}$$

Within the altitude limits shown in Fig. 7-9, the maximum deviation between the experimental and computed curves does not exceed 1 dB. The largest departure occurs at elevations for which the experimental data show notable irregularities. In the altitude interval of $1500 \leqq h \leqq 5000$ ft, the curves are nearly coincident. Equation 7-38 is representative of the experimental data with an error of 10 percent or less in the altitude interval $0 \leqq h \leqq 4$ miles. Equation 7-37, which is applicable to the extended altitude range of $0 \leqq h \leqq 7.5$ miles, demonstrates a decreasing slope for p_h at the higher altitudes, as must occur for a fixed-area surface-noise source.

Composite Incidental-Noise-Power Contours above a Metropolitan Area. Equations 7-36 and 7-37 define the contours of incidental noise power at offset locations from an urban center as a function of observer altitude and specified frequency. Directly above the urban center, Equation 7-30 provides the necessary functional representation but does not reveal the intrinsic behavior of the noise variation with altitude change because of the complexity of the expression. Simplification of Equation 7-30 is possible for two limiting cases, that is, low and very high altitudes, providing an insight into the parametric variation of composite noise above a metropolitan center.

By transforming the variable of integration of Equation 7-30 using the geometrical relationships shown in Fig. 7-7 and the identities,

$$d\theta_i = \frac{d_{i-1}}{h^2 + d_{i-1}^2}\, dh = \frac{\sin 2\theta_i}{2h}\, dh$$

$$\sin \theta_i = \frac{h}{(h^2 + d_{i-1}^2)^{1/2}},$$

one obtains

$$_cp_h = \frac{1}{2} \sum_{i=1}^{N} \left\{ [b_i(C_i' - d_{i-1}) - a_i] \cdot \int_{h_0}^{h} h^{-1} \cos^2 \left[\frac{1}{2} \pi h (h^2 + d_{i-1}^2)^{-1/2} \right] dh \right\}. \tag{7-39}$$

As the lower limit h_0 is allowed to approach zero, ${}_cp_h \rightarrow p_n$, the surface, composite incidental-noise power. The functional dependence of ${}_cp_h$ upon h may be determined for two limiting cases of the integrand, Equation 7-39, which are defined by the ratio h/d_{i-1}:

1. When h/d_{i-1} is small, the limit of the integral is

$$ {}_cp_h \rightarrow - \ln(h/h_0). \tag{7-40}$$

2. In the limit of large h/d_{i-1},

$$\cos\left[\frac{1}{2}\pi\,\frac{h}{(h^2 + d_{i-1}^2)^{1/2}}\right] \rightarrow \sin\left(\frac{1}{4}\pi\,\frac{d_{i-1}^2}{h^2}\right),$$

and the integral approaches

$$ {}_cp_h \rightarrow h^{-2}. \tag{7-41}$$

Finally, the limit of ${}_cp_h$ as $h \rightarrow \infty$ is the sky background noise power, $p_{h\ skbg}$.

Using Equation 7-3 converted to power per unit of bandwidth, and Equations 7-37, 7-40, and 7-41, composite incidental-noise contours on and above the surface of a metropolitan area may be drawn and are presented in Fig. 7-10 for a vertically polarized dipole antenna. A metropolitan-area noise distribution is shown in Fig. 7-10 as a one-dimensional construction by a line extending from $-d$ to d, which passes through the urban center. The vertical z-axis represents observer height above the surface. Noise power is plotted in the direction of the positive x-axis, i.e., normal to both the line $(-d, d)$ and to the z-axis. At the surface, composite incidental-noise power, $p_s(f, d)$, is obtained from Equation 7-3 converted to watts per bandwidth. In Fig. 7-10, $p_s(f, d)$ is shown super-imposed upon the earth ambient thermal noise, p_{ea}.

For an observer rising above the urban center, the noise power at a specific frequency is observed to decrease slowly, at first, varying with increasing altitude h as: $p_s - \ln(h/h_0)$. This functional variation evolves into an h^{-2} dependence for $h >> d$, approaching the sky background noise, $p_{h\ skbg}$, at very high altitudes. Were an observer to rise above the surface from an urban offset location, incidental-noise power would, initially, be observed to increase as the antenna aperture encompasses an increasingly greater portion of the urban surface. After increasing in a manner indicated by Equations 7-36 and 7-37, the noise power would be observed to peak and then be followed by

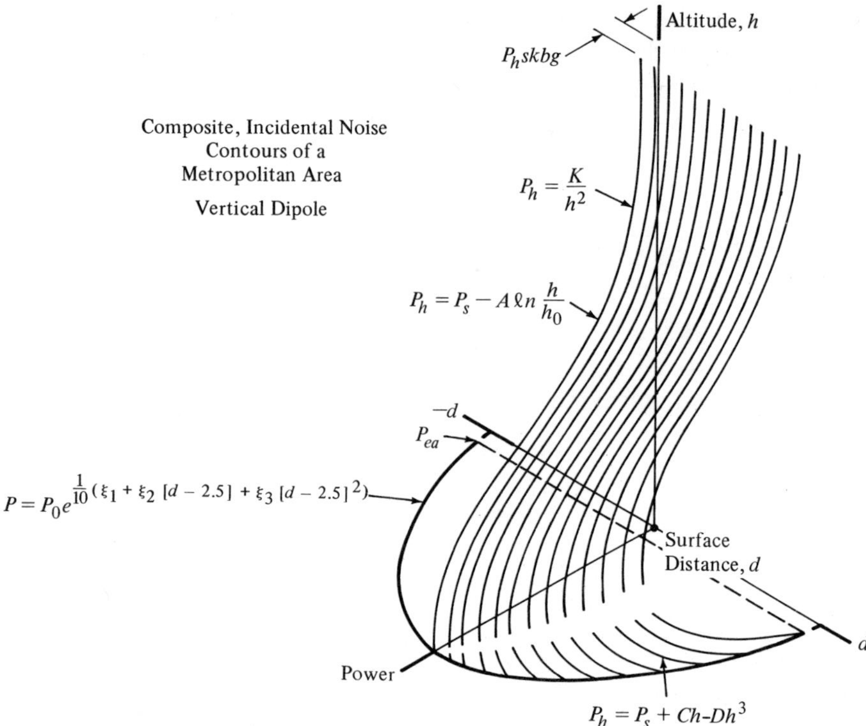

Fig. 7-10. Composite incidental-noise contours of a metropolitan-area vertical-dipole antenna.

a decrease, which at large altitudes obeys an h^{-2} dependence, ultimately attaining the sky-background level.

Horizontally Polarized Dipole

Computation of Airborne Noise Power for Urban-Centered Locations. At an altitude h, the observed noise power p_h received by a horizontally polarized dipole is representable by Equation 7-7, in which the normalized antenna pattern function $F(\gamma_1, \gamma_2)$ is given in terms of the coordinate angles of Fig. 7-5 as:

$$F(\gamma_1, \gamma_2) = F(\theta) = \frac{\cos^2 \left(\frac{1}{2}\pi \cos \theta\right)}{\sin^2 \theta}. \tag{7-42}$$

Integration of Equation 7-7 is performed over area A of the surface noise distribution, $p_s(f, d)$, which, at a fixed observation frequency, is a function of the radial surface distance, d, of each elemental area, dA. Geometrical representation of the element of surface area dA may be obtained from a projection of the spherical area element, $dA\Theta$, upon the plane X, Y as:

$$dA = (dA\Theta) \cos (\tfrac{1}{2}\pi - \alpha)$$
$$= R^2 \sin\theta \sin\alpha \, d\theta d\phi$$

or by using

$$h = R \sin\theta \cos\phi = R \sin\alpha$$
$$dA = R^2 \sin^2\theta \cos\phi \, d\theta \, d\phi. \tag{7-43}$$

Use of Equation 7-43 in Equation 7-7 provides a solution accurate for a plane earth model. To produce a finite radio horizon for any observation altitude, an earth-curvature representation may be introduced during the computation of p_h. This is accomplished by subdividing each surface interval $d_i \leqq d \leqq d_{i+1}$ appearing in Equation 7-16 into k parts and setting the surface noise power for the kth subpart, $p_{sk} = 0$ when the value of d_k lying within $d_i \leqslant d_k \leqslant d_{i+1}$ exceeds $(2ah)^{1/2}$. Here, a is the radius of the earth.

Substitution of the antenna-pattern function of Equation 7-42 and the representation for an element of surface area dA into either Equations 7-18 or 7-19 yields the necessary expression for p_h or ${}_{qp}p_h$, respectively. Selecting Equation 7-19a, which is the form most amenable to evaluation and for which a substantial body of airborne noise data are available for comparison, yields:

$${}_{qp}p_h = \frac{1}{4\pi} \sum_{i=1}^{N} \iint [A_i - B_i(R^2 - h^2)^{1/2}] \cos^2\left(\frac{1}{2}\pi \cos\theta\right) \cdot \cos\theta \, d\theta \, d\phi$$

$$= \sum_{i=1}^{N} {}_{qp}p_{hi} \tag{7-44}$$

in which ${}_{qp}p_{hi}$ is the contribution to the total observed power of the ith annular surface zone. In writing Equation 7-44, use has been made of the equality, $A_r = A_m$, to simplify the integral coefficient. By substituting for h using Equation 7-43, an expression for the observed noise power can be obtained, which may be expanded in a rapidly converging series as follows:

$$4\pi_{qp}p_h = \sum_{i=1}^{N} \iint \left[A_i - \frac{B_i h(1 - \sin^2\theta \cos^2\phi)^{1/2}}{\sin\theta \cos\phi}\right]$$

$$\cdot \cos^2\left(\frac{1}{2}\pi \cos\theta\right) \cos\phi \, d\theta \, d\phi$$

$$= \sum_{i=1}^{N} \iint \cos^2\left(\frac{1}{2}\pi \cos\theta\right) \cdot \cos\phi$$

$$\cdot \left\{ A_i - \frac{B_i h}{\cos\theta \cos\phi} \left[1 - \frac{1}{2}\sin^2\theta \cdot \cos^2\phi\right.\right.$$

$$\left.\left. - \frac{1}{8}(\sin^4\theta \cos^4\phi) \cdots \right]\right\} d\theta \, d\phi$$

Performing the integration in ϕ for $0° \leqslant \phi \leqslant \Phi$ yields:

$$_{qp}p_h = \frac{1}{\pi} \sum_{i=1}^{N} \int_{\theta_{1i}}^{\pi/2} \cos^2\left(\frac{1}{2}\pi \cos\theta\right) \left\{ A_i \sin\phi - \frac{B_i h}{\sin\theta}\right.$$

$$\cdot \left[\phi - \frac{1}{2}\sin^2\theta\left(\frac{1}{2}\phi + \frac{1}{4}\sin 2\phi\right)\right.$$

$$- \frac{1}{8}\sin^4\theta\left(\frac{3\phi}{8} + \frac{1}{4}\sin 2\phi + \frac{\sin 4\phi}{32}\right)$$

$$- \frac{\sin^6\theta}{16}\left(\frac{5\phi}{16} + \frac{15\sin 2\phi}{64} + \frac{3\sin 4\phi}{64}\right.$$

$$\left.\left.\left. + \frac{\sin 6\phi}{192}\right) \cdots \right]\right\} d\theta. \qquad (7\text{-}45)$$

The lower integration limit in Equation 7-45 is seen from Fig. 7-5 to equal:

$$\theta_{1i} = \arctan(h/d_i) \qquad (7\text{-}46)$$

corresponding to the point $\phi = 0$. However, the variables of integration, θ and ϕ, are coupled through the equation:

$$\phi = \arctan\,[(d_i^2/h^2)\sin^2\theta - \cos^2\theta]^{1/2}, \qquad (7\text{-}47)$$

which derives from the constraint, $d_i^2 = x^2 + y^2$. Consequently, before the integration indicated by Equation 7-45 can be performed, the ϕ appearing in the integrand must be replaced by Equation 7-47.

It is unnecessary to carry the expansion of Equation 7-45 beyond the sixth power in sin θ to achieve a 1 percent accuracy in the final value of ${}_{qp}p_h$. Equation 7-45 representation of the observed incidental radio-noise power consists of 25 integrals requiring separate numerical evaluation. This evaluation may be readily performed using sixth-order Gaussian quadratures applied to k equal subelements in each range increment, $d_i \leqq d_k \leqq d_{i+1}$, where $i = 1, 2, 3, \cdots, N$. In performing the evaluation of Equation 7-45 within a kth range subelement, the limits of integration can be further subdivided into m integration elements. The number of integration elements m is established by specifying the maximum acceptable difference between successive evaluations of each term of Equation 7-45. For the computations presented later, the value of m for each term was increased until the incremental change was less than one part in 10^6. All numerical evaluations converged rapidly, never requiring an m greater than 12.

The result obtained from integration of Equation 7-45 is in units of power per kilohertz of receiver bandwidth, which must be converted into electric-field intensity using Equation 7-14 before comparison is possible with existing observations. For a constant gain antenna, Equation 7-14 can be written as:

$${}_{qp}p_h = \frac{{}_{qp}E_h^2}{Z_0} \cdot \frac{G\lambda^2}{4\pi}. \qquad (7\text{-}48)$$

By solving Equation 7-48 for ${}_{qp}E_h$ and expressing the observed electric-field intensity in decibels relative to 1 μV/m/kHz of bandwidth (dBu/kHz), using $Z_0 = 120\pi$ ohms and $G = 1.64$ for a tuned dipole or monopole, one obtains:

$${}_{qp}E_h = 10\log({}_{qp}p_h) + 154.7 - 20\log\lambda, \text{ in dBu/kHz}. \qquad (7\text{-}49)$$

The units for the variables appearing on the right-hand side of Equation 7-49 are ${}_{qp}p_h$ in watts per kilohertz of detection bandwidth and λ, the observing wavelength in meters.

Appearance of the unit "1 kHz of detection bandwidth" arises from the units used for expressing the surface-noise data of Figs. 6-20, 6-21, 6-25, and 6-26. In previous treatments of experimental results, various data sets were converted to a common set of units. This process included the normalization of all data to a common detection bandwidth of 1 kHz. When the incidental radio noise observed with a quasi-peak detector is predominantly impulsive in form, the detector output voltage is known to be proportional to the detection bandwidth.[13] For any detection bandwidth, Δf (expressed in kilohertz), and observation wavelength, λ (in meters), Equation 7-49 becomes

$$ {}_{qp}E_h = 10 \log ({}_{qp}p_h) + 154.7 - 20 \log \lambda + 20 \log \Delta f. \quad (7\text{-}50) $$

The surface composite noise data presented in Figs. 6-25 and 6-26 were obtained using quasi-peak field-strength instrumentation designed to ANSI specifications of Table 3-1, and possessed detection circuits with charge-to-discharge time constant ratios, $T_c/T_d = 1/600$. A decrease in this ratio causes the factor $\mathfrak{F}$ of Equation 7-12 to increase, for a constant observed average noise power, $p_n(f, d)$.[13] A small change in $\mathfrak{F}$, called $\Delta\mathfrak{F}$, produces a computable alteration in ${}_{qp}E_h$ of Equation 7-50 when it may be assumed the observed noise manifests a Rayleigh envelope distribution. As has been shown, thermalization of impulsive incidental noise will occur when either (1) the separation of an observer from an area-noise-source distribution increases, while the radio frequency remains fixed, or (2) when the observation frequency increases, while the separation distance of the observer and area noise source is held constant. The value of $\mathfrak{F}$ as a function of T_c/T_d, receiver noise bandwidth Δf, and the average period of the noise impulses τ may be computed from Geselowitz (1961)[13] wherein $\mathfrak{F}$ is seen to be directly proportional to the quantity N. Consequently, when $\Delta\mathfrak{F} << \mathfrak{F}$ an additional term may be added to Equation 7-50 given by $\delta\mathfrak{F} = (\mathfrak{F} + \Delta\mathfrak{F})/\mathfrak{F}$, which yields:

$$ {}_{qp}E_h = 10 \log ({}_{qp}p_h) + 154.7 + 20 \log [(\Delta f \cdot \delta\mathfrak{F})/\lambda]. \quad (7\text{-}51) $$

As an example of the contribution to ${}_{qp}E_h$ arising from a change in $\mathfrak{F}$ produced by altering the ratio T_c/T_d of a quasi-peak detector used to observe a Gaussian noise sample, if T_c/T_d decreases from $\frac{1}{600}$ to $\frac{1}{5000}$, the value of $\mathfrak{F}$ increases from 8.2 to 9.8 dB. The related value of $\Delta\,\mathfrak{F}$ is 1.6 dB.

Comparison of Measured and Computed Airborne Incidental-Noise Power for Urban-Centered Locations. VHF-band, airborne, composite incidental-noise data have been measured over Tokyo with a horizontally polarized half-wave dipole antenna.[14] The altitude interval for which these data were reported was 0.3 to 3 km. The frequency to which the airborne receiver was tuned was 108 MHz. The resulting experimental data for various altitudes above Tokyo have been reproduced as solid lines in Figs. 7-11 and 7-12 where the ordinate is the quasi-peak electric-field strength in decibels above 1 μV/m for a detection bandwidth of 80 kHz. In each figure, the results of concurrent surface-noise measurements obtained with identical instrumentation are shown. The location of the surface-noise monitoring point was the roof of the Meteorological Agency building in central Tokyo. Figure 7-11 presents the incidental-noise electric-field intensity observed as a function of altitude during both the ascent and descent of a helicopter bearing the instrumentation. The maximum test altitude was 3 km; the minimum was about 0.3 km. Note that the tests were repeated one year later and that these results are presented in Fig. 7-12. Shown in Figs. 7-11 and 7-12 by dashed lines are the computed results for ${}_{qp}E_h$ obtained by evaluating ${}_{qp}P_h$ using Equation 7-45 and subsequently incorporating the results in Equation 7-51 where the following values of Δf and $\delta\,\mathfrak{F}$ and λ applicable to the experiment were employed:

$\Delta f = 80$ kHz
$\lambda = 2.78$ meters
$\delta\,\mathfrak{F} = 1.6$ dB arising from the use of a quasi-peak detector with a value of $T_c/T_d = 0.12/600 = 1/5000$.[14]

Inclusion of these parameter values in Equation 7-51 results in:

$$ {}_{qp}E_h = 10 \log ({}_{qp}p_h) + 185.5, \text{ in dBu/80 kHz} \qquad (7\text{-}52) $$

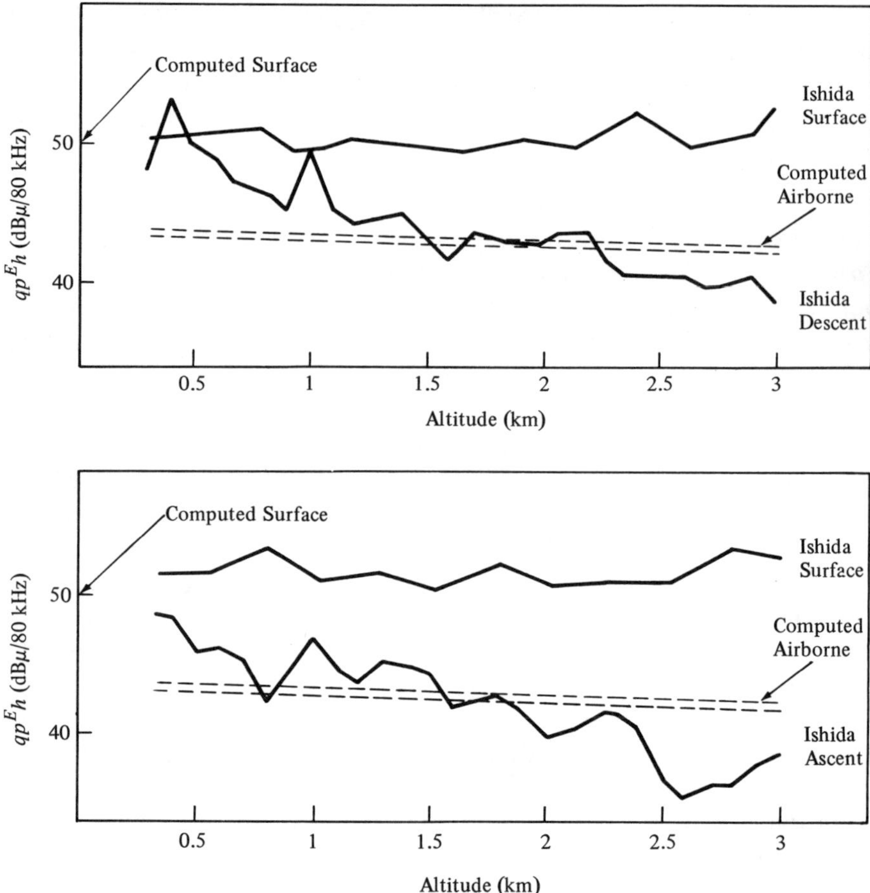

Fig. 7-11. Quasi-peak electric-field strength measured by Ishida, March 17, 1968, solid lines. Computed results shown by dashed lines. **a**, Airborne during descent. **b**, Airborne during ascent.

where $_{qp}p_h$ is in watts per kilohertz. In Figs. 7-11 and 7-12, $_{qp}E_h$ has been plotted as a function of altitude for the range 0.3 to 3 km at the observation frequency of 108 MHz.

Several comments may be made regarding these experimental and computed results:

1. The ground reference point above which the helicopter ascended and returned was located about 5 miles inland from Tokyo Bay.

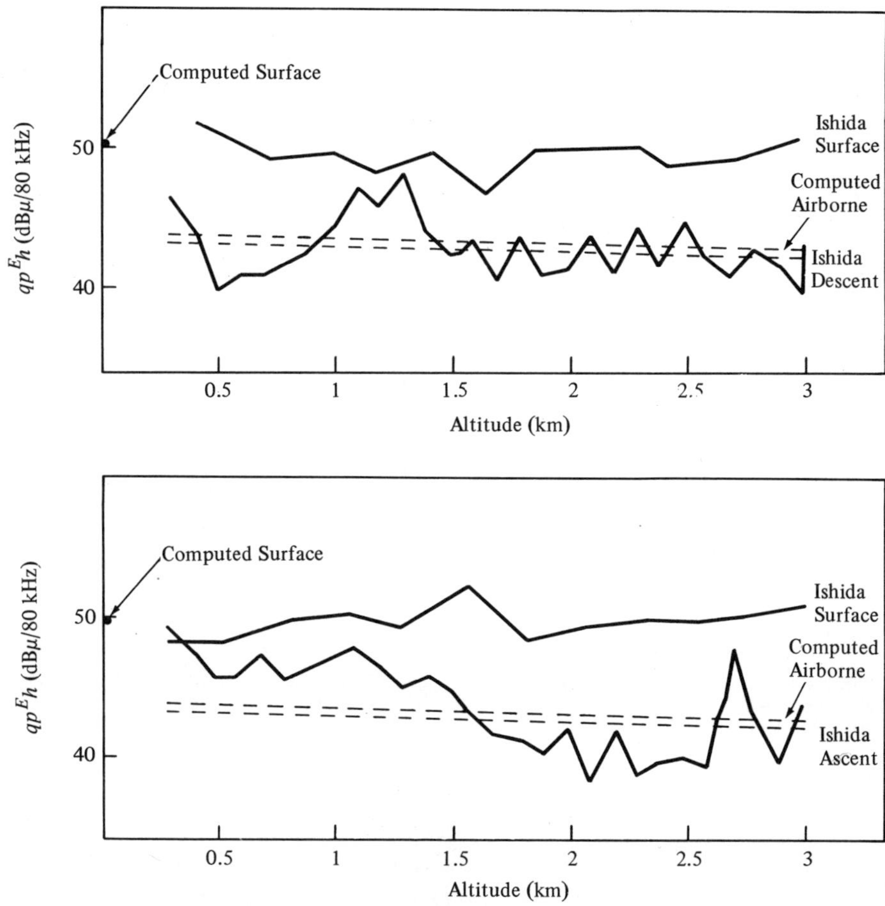

Fig. 7-12. Quasi-peak electric field strength. Measured by Ishida, March 27, 1969, solid lines. Computed results shown by dashed lines. **a**, Airborne during descent. **b**, Airborne during ascent.

However, the western bay perimeter is irregular, causing some portions to be displaced by 12 km from the ascent point. Since the expanse of the bay contained few, if any, noise sources, the absence of noise from this region must be included. The procedure used was as follows: When a circularly symmetric, planar, area-noise source of uniform noise density is sectored by a line such as (L_1, L_2) of Fig. 7-13a, lying at a distance x from the center, and when all noise sources are excluded from the segment, A_s, the resulting value of

$\tilde{p}_{hi}$ at altitude h above point O is:

$$\tilde{p}_{hi} = (1 - A_s r^{-2}/\pi)\, p_{hi}. \tag{7-53}$$

Here, p_{hi} is the value of the observed power above point O that would be present if $x = r$.

Similarly, the effect of a sectoring line at a distance x from the center of symmetry, upon each contribution, p_{hi}, arising from an annular noise source bounded by circles of radii, r_1 and r_2, Fig. 7-13b reduces the observed power to:

$$\tilde{p}_{hi} = 1 - \frac{A_{s1} - A_{s2}}{\pi(r_1^2 - r_2^2)}\, p_{hi} \tag{7-54}$$

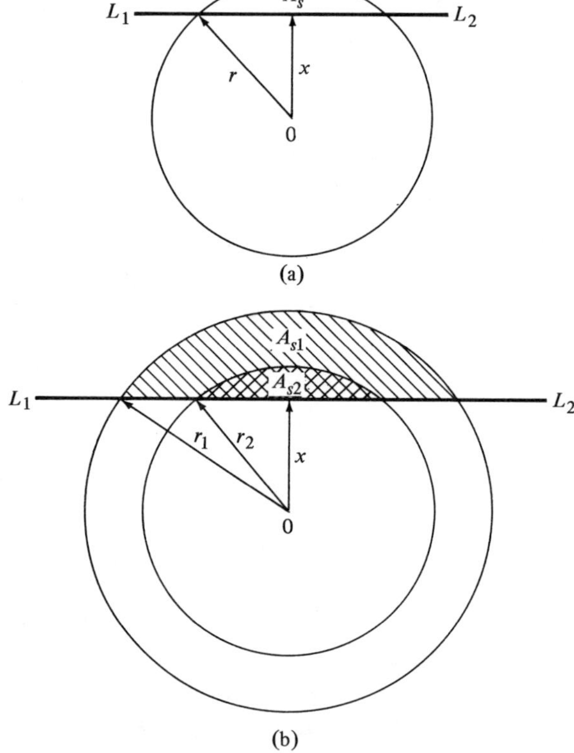

Fig. 7-13. **a**, Coordinates used for sectoring a circular surface distribution of incidental-noise sources. **b**, Coordinates used for sectoring two concentric circles yielding annular area A_{s1} and A_{s2}.

where the areas of the sectors lying beyond the sectoring line (L_1, L_2), with radii r_1 and r_2 are, respectively, A_{s1} and A_{s2}.

The sector areas A_s, A_{s1}, and A_{s2} of Equations 7-53 and 7-54 are related to the distance x and the radii as:

$$A_s = \tfrac{1}{2}\pi r^2 - x(r^2 - x^2)^{1/2} - r^2 \arcsin(x/r). \qquad (7\text{-}55)$$

For a concentric family of annular surface noise sources sectored by a line (L_1, L_2), the total observed noise power at altitude h is representable as shown in Equation 7-44 with ${}_{qp}p_{hi}$ replaced by ${}_{qp}\tilde{p}_{hi}$.

Equations 7-53 and 7-54 might be incorporated into Equation 7-15 to reduce the contributions to the computed airborne noise power arising from a set of N annular range increments such as are defined by the linear segmentation process given by Equation 7-16. However, as has been noted earlier, a much finer division of each range increment N into k circularly symmetric subelements may be used in the evaluation of Equation 7-15. To each of the k subelements of N range increments, the reduced powers $\tilde{p}_{hi}$ of Equations 7-53 and 7-54 have been applied in the computation of ${}_{qp}\tilde{p}_h$ to obtain the theoretical results shown in Figs. 7-11 and 7-12.

2. The measurements performed on March 17, 1968, contained an anomalous event. During ascent, the helicopter was wind-deflected toward the bay beginning at an altitude of 2.4 km. It passed the shoreline while at an altitude of about 2.7 km, began to correct its location thereupon, but did not reattain its ground reference ascent point until after it had begun descent. Therefore, the ascent portion of the airborne data recorded above 2.4 km and the descent data in the neighborhood of 3-km altitude of Fig. 7-11 cannot be compared with the theoretical results.

3. As the helicopter changed altitude, it circled about the ground reference point. The alternating orientation of the maximum gain direction of the horizontal dipole radiation pattern toward and away from the noise void of the bay produced serrations in the airborne noise data, the most distinguishable of which are seen in the descent data of March 27, 1969, at an altitude of from 1.5 to 2.5 km (Figure 7-12). In this portion of the record, the peak-to-peak field strength variation is 3 dB, just as expected for a case where the maximum gain and the null points of the antenna are alternately directed toward, and away from, the bay.

4. The surface-noise traces in both Figs. 7-11 and 7-12 display variations of significant magnitude during the ascent-descent cycle. The cause of these variations was judged to be local changes in automotive traffic, a major incidental noise source for this location and frequency. The average surface noise quasi-peak field strength for the cases shown was obtained by averaging the least-squares straight lines fitted to the four sets of data and found to be 50.5 dBu/80 kHz. The fine structure in both the surface and airborne incidental-noise traces shows no significant correlation. This is to be expected for an experiment employing a low-directivity airborne antenna. The broad beam of the helicopter-borne antenna accumulates radiated energy from a very much larger area than does a surface-located dipole antenna.

5. Superimposed upon Figs. 7-11 and 7-12, and shown by dashed lines, are the results computed using Equation 7-52 for altitudes less than 3.1 km and two values of the bay shore displacement: 8 and 12 km as measured from the Meteorological Agency building near the center of Tokyo. Only 0.6 dB separate the computed values of ${}_{qp}E_h$ for the pair of bay shore displacement, indicating that only a weak dependence upon boundary location exists for displacements in the range of 5 to 7.5 miles. The computed value of ${}_{qp}E_h$ at the surface was obtained using Equation 7-52 and the value of ${}_{qp}P_n$ from Fig. 6-25.

6. For altitudes above 1.5 km, the computed and measured noise-field intensity-height variations are quite similar; the only major differences seen are those arising from either flight-path anomalies, as in Fig. 7-11, for heights greater than 2.4 km or from unrepeated large-signal variations, e.g., the ascent portion of Fig. 7-12 for h = 2.7 km or the descent profile at $h = 0.45$ and $1.1 \leqslant h \leqslant 1.3$ km.

7. For altitudes above 1.1 km, the maximum deviation between computed and measured field-intensity levels is less than 5 dB except where the singular signal peak of Fig. 7-12 at $h = 2.7$ km exists.

8. Throughout the altitude interval of 0.3 to 3 km, the average difference between the experimental and theoretical results is 0.7 dB. In determining this average, the mean of the two theoretical computations for bay shore displacements of 8 and 12 km was used. Those data in Fig. 7-11 for ascent above 2.4 km and for the descent at 3 km have been omitted in preparing this average for the reasons mentioned

earlier. However, were these portions of the experimental results included, the average error would have been less.

TEMPORAL VARIATIONS OF INCIDENTAL RADIO-NOISE LEVELS ABOVE METROPOLITAN AREAS

The diurnal variations observed in surface, incidental-noise-source emissions typically arise as a consequence of the differing utilization rates existing for the dominant man-made sources, automobiles, electric power facilities, and electrically operated industrial equipment. These changing rates determine the daily variation occurring in the level of composite surface noise power, and in airborne noise data. The business-day rush-hour average noise increment of 10 to 12 dB existing in industrial-area surface data appears also in composite noise-temperature measurements made from elevated observation points. Table 7-3 summarizes the observed maximum changes in F_a occurring between business-day rush-hour values and the late-evening noise level as a function of altitude and frequency. Data for five VHF/UHF-band frequencies are presented for five major American cities. At an observation altitude of 25,000 ft, the mid-VHF-band measurements over three East Coast urban centers display a diurnal

Table 7-3. Diurnal variations of F_a in decibels of incidental noise power observed over several urban centers for radio frequencies in the VHF and UHF bands. Entries are daily maxima minus nighttime minima.

Altitude	Frequency (MHz)					
(kft)	121.5	137	243	300	1000	Location
4				11[a]		Phoenix
				8[a]	7[a]	Phoenix
				5[a]		Phoenix
				8 Average		
5		2[b]				Seattle
25	8.2[c]		1.1			Philadelphia
	10.5[c]		0.0			Baltimore
	6.3[c]		−2			New York
	8.3 Average		−0.3 Average			

[a] Morning rush hour to late evening.
[b] Morning rush hour to midday.
[c] Evening rush hour to late evening.

mean variation in F_a of approximately 11 dB, which reduces to a negligible daily change at 243 MHz.[8] A reduced variation between business-day maxima and night-time minima is reflected in the observations recorded over Phoenix at flight altitudes of 4000 ft for a frequency of 1 GHz.[6] Over Phoenix, the results obtained at 300 MHz displayed a diurnal cyclical change of 8 dB, comparable to that recorded on the East Coast at 121.5 MHz.

References

1. Ploussios, G. City noise and its effect upon airborne antenna noise temperature at UHF. *IEEE Trans. Aerospace and Electronic Systems* **AES-4** (1): 41–51 (1968).
2. Swenson, G. W., Jr., and Cochran, W. W. Radio noise from towns: Measured from an airplane. *Science* **181**: 543–545. (August 10, 1973).
3. King, C. H., and Lunden, C. D. Noise Temperature of an Airborne VHF Communications Antenna. Boeing Airplane Co. Document D6-9461, April 1964.
4. Barnard, C. R. W. VHF Noise Levels Over Large Towns. Royal Aircraft Establishment, Technical Report 67213, August 1967.
5. Buehler, W. E., King, C. H., and Lunden, C. D. VHF city noise. *IEEE Electromagnetic Compatibility Symp. Record* 113–118 (1968).
6. Anzic, G., and May, C. Results and Analyses of a Combined Aerial and Ground UHF Noise Survey in an Urban Area. NASA Tech. Memo, NASA TM X-2244, April 1971.
7. Mills, A. H. Measurement of Radio Frequency Noise in Urban, Suburban, and Rural Areas. Final Report NASA Contract NAS 3-11531, General Dynamics Corp. Convair Division, December 1970.
8. Taylor, R. E., and Hill, J. S. Aircraft measurement of radio frequency noise at 121.5 MHz, 243 MHz, and 405 MHz.. *Symp. Record, Second Electromagnetic Compatibility Symp*., July 1977, Montreux.
9. Parzen, E. *Modern Probability Theory and its Applications*. New York: John Wiley & Sons Inc., 1960.
10. Rice, S. O. Mathematical analysis of random noise. *Bell System Tech. Jour.* **23** (July 1944).
11. Lauber, W. R., and Bertrand, J. M. Preliminary urban VHF/UHF radio noise intensity measurements in Ottawa, Canada. *Second Symp. Electromagnetic Compatibility*, Montreux, Conf. Record., July 1977.
12. ESSA Laboratories. Interim Evaluation Report, Field Test Results, Area A, Warren AFB, Wyoming. ESSA Boulder, Colorado. September 10, 1968.
13. Geselowitz, D. B. Response of ideal radio noise meter to continuous sine wave, recurrent impulses, and random noise. *IRE Trans. Radio Frequency Interference* **RFI-3** 2–10 (May 1961).
14. Ishida, T. Measurement of the height dependence of the intensity of city noise. *Progress in Radio Science*, URSI **1**: 277–281 (1971).

Appendix*

CALCULATION OF THE VOLTAGE GRADIENT AT THE CONDUCTOR SURFACE

The voltage gradient E at the surface of the conductors of a high-voltage line is generally determined from the effective capacitance of the conductor C_b and the working voltage U of the line.

$$E = k\, C_b\, U.$$

For calculation of the surface voltage gradient therefore, the location of the conductors, the nature of the conductors (single or bundled conductors) and the voltage must be known. The effective capacitance is determined by the height of the conductors above ground, the distance between conductors and their shape. As the height of the conductors above ground varies due to sagging, calculations are made using a mean height which is generally given by:

$$h = H - 0.7\hat{f}$$

where

h = mean of height of conductor,
H = height of conductor above ground, measured at the pylon,
$\hat{f}$ = maximum sag of the conductor.

For a three-phase line the effective capacitance should be calculated for each conductor separately. In the case of a bundled conductor, it is necessary to determine the radius of the equivalent single conductor. The capacitance of the bundled conductor will then be equal to the capacitance of the equivalent single conductor. The radius of the equivalent single conductor ρ_0 is calculated as follows:

$$\rho_0 = \sqrt[n]{n\,\rho_T R^{(n-1)}}$$

*This appendix is taken from the International Special Committee on Radio Interference (CISPR) Publication 1, Second Edition (1972), Reference 30.

where

n = number of conductors in the bundle,
ρ_T = radius of the conductors.
R = radius of the circle on which the centres of the conductors are located [Fig. 3-36].

Comparison of the measured values and the calculated values of the effective capacitance has shown that the calculated value should be increased by about 2% in order to allow for the influence of the pylons.

The voltage gradient at the surface of a single conductor is calculated from the formula:

$$E = \frac{1.8\, U_{ph} C_b}{\rho} \text{ kV (eff)/cm}$$

where

C_b = effective capacitance per unit length of conductor in picofarads/centimetre (pF/cm),
ρ = radius of conductor in centimetres,
U_{ph} = voltage between conductor and ground (phase voltage) in kilovolts (kV).

This voltage gradient is the same at all points on the circumference of the conductor. The voltage gradient is not the same at all points on the circumference of the individual conductors in a bundled conductor but, since it is the maximum value with which we are concerned, this may be deduced from the formula:

$$E = \frac{1.8\, U_{ph} C_b}{n \rho_T} \left(1 + \frac{2(n-1) \sin \frac{\pi}{n}}{s'}\right) \text{ kV (eff)/cm}$$

where

ρ_T, U_{ph} have the same meaning as above, and
C_b = the effective capacitance per phase (equivalent single conductor)
n = number of conductors in the bundle,
$s' = \frac{s}{\rho_T}$ relative distance between conductors in the bundle,
s = distance between conductors in the bundle in centimetres [Fig. 3-36].

In all calculations the unevenness of the conductor surface, due to the protective covering, is ignored. For ordinary conductor cables, the effect is negligible.

The reference voltage gradient to be quoted when radio interference voltage or field strength measurements are made, on a three-phase line, for example, should be the r.m.s. value of the voltage gradient on the most highly stressed conductor.

C_b can be calculated from the following system of equations:

$$V_1 = K_{11} q_1 + \dots\dots\dots + K_{1n} q_n$$
$$V_n = K_{n1} q_1 + \dots\dots\dots + K_{nn} q_n$$

V_n = the potential on the equivalent conductor No. n

q_n = the charge per metre on the equivalent conductor No. n

$$K_{ij} = \frac{1}{2\pi\epsilon_0} \ln \frac{A_{ij}}{\rho_0} \qquad i = j$$

$$K_{ij} = \frac{1}{2\pi\epsilon_0} \ln \frac{A_{ij}}{a_{ij}} \qquad i \neq j$$

A_{ij} = the distance between phase i and the image of phase j

a_{ij} = the distance between phase i and phase j

ρ_0 = as defined above.

For a three-phase line, C_b is calculated from the following formula (designations as above). The most highly stressed phase is designated No. 2.

$$C_b = \left\{ \pi\epsilon_0 \log e \left[\log \frac{A_{12}}{a_{12}} \log \frac{A_{13}\rho_0}{a_{12}A_{33}} - \log \frac{A_{11}}{\rho_0} \log \frac{A_{33}^2 A_{23}}{\rho_0^2 a_{23}} + \log \frac{A_{13}}{a_{13}} \log \frac{A_{13}^2 A_{23}}{a_{13}^2 a_{23}} \right] \right\} \Big/ \left\{ \log \frac{A_{12}}{a_{12}} \left(\log \frac{A_{12}}{a_{12}} \log \frac{A_{33}}{\rho_0} - \frac{A_{13}}{a_{13}} \log \frac{A_{23}}{a_{23}} \right) + \log \frac{A_{13}}{a_{13}} \left(\log \frac{A_{13}}{a_{13}} \log \frac{A_{22}}{\rho_0} - \log \frac{A_{12}}{a_{12}} \log \frac{A_{23}}{a_{23}} \right) + \log \frac{A_{11}}{\rho_0} \left[\left(\log \frac{A_{23}}{a_{23}} \right)^2 - \log \frac{A_{22}}{\rho_0} \log \frac{A_{33}}{\rho_0} \right] \right\}$$

log means $\log_{10}$.

In the formula above, the influence of the ground wires is neglected. This is a permitted simplification when their contribution is less than 1%.

Author Index

Subject Index